A

SYSTEM OF LOGIC

RATIOCINATIVE AND INDUCTIVE

—

VOL. II.

A

SYSTEM OF LOGIC

RATIOCINATIVE AND INDUCTIVE

BEING A CONNECTED VIEW OF THE
PRINCIPLES OF EVIDENCE
AND THE
METHODS OF SCIENTIFIC INVESTIGATION

BY

JOHN STUART MILL

IN TWO VOLUMES

VOL. II.

SEVENTH EDITION

LONDON:
LONGMANS, GREEN, READER, AND DYER
MDCCCLXVIII

CONTENTS

OF

THE SECOND VOLUME.

—

BOOK III.

ON INDUCTION.—(*Continued.*)

BOOK IV.

OF OPERATIONS SUBSIDIARY TO INDUCTION.

CHAPTER I. *Of Observation and Description.*

CHAPTER II. *Of Abstraction, or the Formation of Conceptions.*

CHAPTER III. *Of Naming, as subsidiary to Induction.*

CHAPTER IV. *Of the Requisites of a Philosophical Language, and the Principles of Definition.*

BOOK VI.

ON THE LOGIC OF THE MORAL SCIENCES.

CHAPTER X. *Of the Inverse Deductive, or Historical Method.*

CHAPTER XI. *Additional Elucidations of the Science of History.*

CHAPTER XII. *Of the Logic of Practice, or Art; including
Morality and Policy.*

BOOK III.

CONTINUED.

—

k

OF INDUCTION.

" In such cases the inductive and deductive methods of inquiry may be said
to go hand in hand, the one verifying the conclusions deduced by the other ;
and the combination of experiment and theory, which may thus be brought to
bear in such cases, forms an engine of discovery infinitely more powerful than
either taken separately.   This state of any department of science is perhaps of
all others the most interesting, and that which promises the most to research."
—SIR J. HERSCHEL, *Discourse on the Study of Natural Philosophy.*

CHAPTER XIV.

OF THE LIMITS TO THE EXPLANATION OF LAWS OF NATURE; AND OF HYPOTHESES.

§ 1. THE preceding considerations have led us to recognise a distinction between two kinds of laws, or observed uniformities in nature: ultimate laws, and what may be termed derivative laws. Derivative laws are such as are deducible from, and may, in any of the modes which we have pointed out, be resolved into, other and more general ones. Ultimate laws are those which cannot. We are not sure that any of the uniformities with which we are yet acquainted are ultimate laws; but we know that there must be ultimate laws; and that every resolution of a derivative law into more general laws, brings us nearer to them.

Since we are continually discovering that uniformities, not previously known to be other than ultimate, are derivative, and resolvable into more general laws; since (in other words) we are continually discovering the explanation of some sequence which was previously known only as a fact; it becomes an interesting question whether there are any necessary limits to this philosophical operation, or whether it may proceed until all the uniform sequences in nature are resolved into some one universal law. For this seems, at first sight, to be the ultimatum towards which the progress of induction, by the Deductive Method resting on a basis of observation and experiment, is tending. Projects of this kind were universal in the infancy of philosophy; any speculations which held out a less brilliant prospect, being in those early times deemed not worth pursuing. And the idea receives so much apparent countenance from the nature of the most remarkable achievements of modern science, that speculators are even now frequently appearing, who profess either to have solved the problem, or to suggest modes in which it may one day be solved. Even where pretensions of this magnitude are not made, the character of the solutions which are given or sought of particular classes of phenomena, often involves such conceptions of what constitutes explanation, as would render the notion of explaining all phenomena whatever by means of some one cause or law, perfectly admissible.

§ 2. It is therefore useful to remark, that the ultimate Laws of Nature cannot possibly be less numerous than the distinguishable sensations or other feelings of our nature ;— those, I mean, which are distinguishable from one another in quality, and not merely in quantity or degree. For example ; since there is a phenomenon *sui generis*, called colour, which our consciousness testifies to be not a particular degree of some other phenomenon, as heat or odour or motion, but intrinsically unlike all others, it follows that there are ultimate laws of colour ; that though the facts of colour may admit of explanation, they never can be explained from laws of heat or odour alone, or of motion alone, but that however far the explanation may be carried, there will always remain in it a law of colour. I do not mean that it might not possibly be shown that some other phenomenon, some chemical or mechanical action for example, invariably precedes, and is the cause of, every phenomenon of colour. But though this, if proved, would be an important extension of our knowledge of nature, it would not explain how or why a motion, or a chemical action, can produce a sensation of colour ; and however diligent might be our scrutiny of the phenomena, whatever number of hidden links we might detect in the chain of causation terminating in the colour, the last link would still be a law of colour, not a law of motion, nor of any other phenomenon whatever. Nor does this observation apply only to colour, as compared with any other of the great classes of sensations ; it applies to every particular colour, as compared with others. White colour can in no manner be explained exclusively by the laws of the production of red colour. In any attempt to explain it, we cannot but introduce, as one element of the explanation, the proposition that some antecedent or other produces the sensation of white.

The ideal limit, therefore, of the explanation of natural phenomena (towards which as towards other ideal limits we are constantly tending, without the prospect of ever completely attaining it) would be to show that each distinguishable variety of our sensations, or other states of consciousness, has only one sort of cause; that, for example, whenever we perceive a white colour, there is some one condition or set of conditions which is always present, and the presence of which always produces in us that sensation. As long as there are several known modes of production of a phenomenon, (several different substances, for instance, which have the property of whiteness, and between which we cannot trace any other resemblance,) so long it is not impossible that one of these modes of production may be resolved into another, or that all of them may be resolved into some more general mode of production not hitherto recognised. But when the modes of production are reduced to one, we cannot, in point of simplification, go any further. This one may not, after all, be the ultimate mode; there may be other links to be discovered between the supposed cause and the effect; but we can only further resolve the known law, by introducing some other law hitherto unknown; which will not diminish the number of ultimate laws.

In what cases, accordingly, has science been most successful in explaining phenomena, by resolving their complex laws into laws of greater simplicity and generality? Hitherto chiefly in cases of the propagation of various phenomena through space: and, first and principally, the most extensive and important of all facts of that description, the fact of motion. Now this is exactly what might be expected from the principles here laid down. Not only is motion one of the most universal of all phenomena, it is also (as might be expected from that circumstance) one of those which, apparently at least, are produced in the greatest number of ways; but the phenomenon itself is always, to our sensations, the same in every respect but degree.

Differences of duration, or of velocity, are evidently differences in degree only; and differences of direction in space, which alone has any semblance of being a distinction in kind, entirely disappear (so far as our sensations are concerned) by a change in our own position; indeed the very same motion appears to us, according to our position, to take place in every variety of direction, and motions in every different direction to take place in the same. And again, motion in a straight line and in a curve are no otherwise distinct than that the one is motion continuing in the same direction, the other is motion which at each instant changes its direction. There is, therefore, according to the principles I have stated, no absurdity in supposing that all motion may be produced in one and the same way; by the same kind of cause. Accordingly, the greatest achievements in physical science have consisted in resolving one observed law of the production of motion into the laws of other known modes of production, or the laws of several such modes into one more general mode; as when the fall of bodies to the earth, and the motions of the planets, were brought under the one law of the mutual attraction of all particles of matter; when the motions said to be produced by magnetism were shown to be produced by electricity; when the motions of fluids in a lateral direction, or even contrary to the direction of gravity, were shown to be produced by gravity; and the like. There is an abundance of distinct causes of motion still unresolved into one another; gravitation, heat, electricity, chemical action, nervous action, and so forth; but whether the efforts of the present generation of savans to resolve all these different modes of production into one, are ultimately successful or not, the attempt so to resolve them is perfectly legitimate. For though these various causes produce, in other respects, sensations intrinsically different, and are not, therefore, capable of being resolved into one another, yet in so far as they all produce motion, it is quite possible that the immediate antecedent of the motion may in all these different cases be the same; nor is it impossible that these various agencies themselves may, as the new doctrines assert, all of them have for their own immediate antecedent, modes of molecular motion.

We need not extend our illustration to other cases, as for instance to the propagation of light, sound, heat, electricity, &c. through space, or any of the other phenomena which have been found susceptible of explanation by the resolution of their observed laws into more general laws. Enough has been said to display the difference between the kind of explanation and resolution of laws which is chimerical, and that of which the accomplishment is the great aim of science; and to show into what sort of elements the resolution must be effected, if at all.

§ 3. As, however, there is scarcely any one of the principles of a true method of philosophizing which does not require to be guarded against errors on both sides, I must enter a caveat against another misapprehension, of a kind directly contrary to the preceding. M. Comte, among other occasions on which he has condemned, with some asperity, any attempt to explain phenomena which are " evidently primordial," (meaning, apparently, no more than that every peculiar phenomenon must have at least one peculiar and therefore inexplicable law,) has spoken of the attempt to furnish any explanation of the colour belonging to each substance, " la couleur élémentaire propre à chaque substance," as essentially illusory. " No one," says he, " in our time attempts to explain the particular specific gravity of each substance or of each structure. Why should it be otherwise as to the specific colour, the notion of which is undoubtedly no less primordial ?"*

Now although, as he elsewhere observes, a colour must always remain a different thing from a weight or a sound, varieties of colour might nevertheless follow, or correspond to, given varieties of weight, or sound, or some other phenomenon as different as these are from colour itself. It is one question what a thing is, and another what it depends on; and though to ascertain the conditions of an elementary phenomenon is not to obtain any new insight into the nature of

---

* *Cours de Philosophie Positive,* ii. 656.

the phenomenon itself, that is no reason against attempting to discover the conditions. The interdict against endeavouring to reduce distinctions of colour to any common principle, would have held equally good against a like attempt on the subject of distinctions of sound; which nevertheless have been found to be immediately preceded and caused by distinguishable varieties in the vibrations of elastic bodies: though a sound, no doubt, is quite as different as a colour is from any motion of particles, vibratory or otherwise. We might add, that, in the case of colours, there are strong positive indications that they are not ultimate properties of the different kinds of substances, but depend on conditions capable of being superinduced upon all substances; since there is no substance which cannot, according to the kind of light thrown upon it, be made to assume almost any colour; and since almost every change in the mode of aggregation of the particles of the same substance, is attended with alterations in its colour, and in its optical properties generally.

The real defect in the attempts which have been made to account for colours by the vibrations of a fluid, is not that the attempt itself is unphilosophical, but that the existence of the fluid, and the fact of its vibratory motion, are not proved; but are assumed, on no other ground than the facility they are supposed to afford of explaining the phenomena. And this consideration leads to the important question of the proper use of scientific hypotheses; the connexion of which with the subject of the explanation of the phenomena of nature, and of the necessary limits to that explanation, needs not be pointed out.

§ 4. An hypothesis is any supposition which we make (either without actual evidence, or on evidence avowedly insufficient) in order to endeavour to deduce from it conclusions in accordance with facts which are known to be real; under the idea that if the conclusions to which the hypothesis leads are known truths, the hypothesis itself either must be, or at least is likely to be, true. If the hypothesis relates to the cause, or mode of production of a phenomenon, it will serve, if admitted, to explain such facts as are found capable of being deduced from it. And this explanation is the purpose of many, if not most, hypotheses. Since explaining, in the scientific sense, means resolving an uniformity which is not a law of causation, into the laws of causation from which it results, or a complex law of causation into simpler and more general ones from which it is capable of being deductively inferred; if there do not exist any known laws which fulfil this requirement, we may feign or imagine some which would fulfil it; and this is making an hypothesis.

An hypothesis being a mere supposition, there are no other limits to hypotheses than those of the human imagination; we may, if we please, imagine, by way of accounting for an effect, some cause of a kind utterly unknown, and acting according to a law altogether fictitious. But as hypotheses of this sort would not have any of the plausibility belonging to those which ally themselves by analogy with known laws of nature, and besides would not supply the want which arbitrary hypotheses are generally invented to satisfy, by enabling the imagination to represent to itself an obscure phenomenon in a familiar light; there is probably no hypothesis in the history of science in which both the agent itself and the law of its operation were fictitious. Either the phenomenon assigned as the cause is real, but the law according to which it acts, merely supposed; or the cause is fictitious, but is supposed to produce its effects according to laws similar to those of some known class of phenomena. An instance of the first kind is afforded by the different suppositions made respecting the law of the planetary central force, anterior to the discovery of the true law, that the force varies as the inverse square of the distance; which also suggested itself to Newton, in the first instance, as an hypothesis, and was verified by proving that it led deductively to Kepler's laws. Hypotheses of the second kind are such as the vortices of Descartes, which were ficti-tious, but were supposed to obey the known laws of rotatory motion; or the two rival hypotheses respecting the nature of light, the one ascribing the phenomena to a fluid emitted from all luminous bodies, the other (now generally received) attri-

buting them to vibratory motions among the particles of an
ether pervading all space. Of the existence of either fluid
there is no evidence, save the explanation they are calculated
to afford of some of the phenomena; but they are supposed to
produce their effects according to known laws; the ordinary
laws of continued locomotion in the one case, and in the other,
those of the propagation of undulatory movements among the
particles of an elastic fluid.

According to the foregoing remarks, hypotheses are
invented to enable the Deductive Method to be earlier applied
to phenomena. But* in order to discover the cause of any
phenomenon by the Deductive Method, the process must
consist of three parts; induction, ratiocination, and verification. Induction, (the place of which, however, may be supplied
by a prior deduction,) to ascertain the laws of the causes;
ratiocination, to compute from those laws, how the causes will
operate in the particular combination known to exist in the
case in hand; verification, by comparing this calculated effect
with the actual phenomenon. No one of these three parts of
the process can be dispensed with. In the deduction which
proves the identity of gravity with the central force of the
solar system, all the three are found. First, it is proved from
the moon's motions, that the earth attracts her with a force
varying as the inverse square of the distance. This (though
partly dependent on prior deductions) corresponds to the first,
or purely inductive, step, the ascertainment of the law of the
cause. Secondly, from this law, and from the knowledge previously obtained of the moon's mean distance from the earth,
and of the actual amount of her deflexion from the tangent, it
is ascertained with what rapidity the earth's attraction would
cause the moon to fall, if she were no further off, and no more
acted upon by extraneous forces, than terrestrial bodies are:
that is the second step, the ratiocination. Finally, this calculated velocity being compared with the observed velocity with
which all heavy bodies fall, by mere gravity, towards the
surface of the earth, (sixteen feet in the first second, forty-

---

* Vide supra, book iii. ch. xi.

eight in the second, and so forth, in the ratio of the odd numbers, 1, 3, 5, &c.,) the two quantities are found to agree. The order in which the steps are here presented, was not that of their discovery; but it is their correct logical order, as portions of the proof that the same attraction of the earth which causes the moon's motion, causes also the fall of heavy bodies to the earth: a proof which is thus complete in all its parts.

Now, the Hypothetical Method suppresses the first of the three steps, the induction to ascertain the law; and contents itself with the other two operations, ratiocination and verification; the law which is reasoned from, being assumed, instead of proved.

This process may evidently be legitimate on one supposition, namely, if the nature of the case be such that the final step, the verification, shall amount to, and fulfil the conditions of, a complete induction. We want to be assured that the law we have hypothetically assumed is a true one; and its leading deductively to true results will afford this assurance, provided the case be such that a false law cannot lead to a true result; provided no law, except the very one which we have assumed, can lead deductively to the same conclusions which that leads to. And this proviso is often realized. For example, in the very complete specimen of deduction which we just cited, the original major premise of the ratiocination, the law of the attractive force, was ascertained in this mode; by this legitimate employment of the Hypothetical Method. Newton began by an assumption, that the force which at each instant deflects a planet from its rectilineal course, and makes it describe a curve round the sun, is a force tending directly towards the sun. He then proved that if this be so, the planet will describe, as we know by Kepler's first law that it does describe, equal areas in equal times; and, lastly, he proved that if the force acted in any other direction whatever, the planet would not describe equal areas in equal times. It being thus shown that no other hypothesis would accord with the facts, the assumption was proved; the hypothesis became an inductive truth. Not only did Newton ascertain by this hypothetical process the direction of the deflecting force; he proceeded in exactly the same manner to ascertain the law of variation of the quantity of that force. He assumed that the force varied inversely as the square of the distance; showed that from this assumption the remaining two of Kepler's laws might be deduced; and finally, that any other law of variation would give results inconsistent with those laws, and inconsistent, therefore, with the real motions of the planets, of which Kepler's laws were known to be a correct expression.

I have said that in this case the verification fulfils the conditions of an induction: but an induction of what sort? On examination we find that it conforms to the canon of the Method of Difference. It affords the two instances, A B C, a b c, and B C, b c. A represents central force; A B C, the planets *plus* a central force; B C, the planets apart from a central force. The planets with a central force give a, areas proportional to the times; the planets without a central force give b c (a set of motions) without a, or with something else instead of a. This is the Method of Difference in all its strictness. It is true, the two instances which the method requires are obtained in this case, not by experiment, but by a prior deduction. But that is of no consequence. It is immaterial what is the nature of the evidence from which we derive the assurance that A B C will produce a b c, and B C only b c; it is enough that we have that assurance. In the present case, a process of reasoning furnished Newton with the very instances, which, if the nature of the case had admitted of it, he would have sought by experiment.

It is thus perfectly possible, and indeed is a very common occurrence, that what was an hypothesis at the beginning of the inquiry, becomes a proved law of nature before its close. But in order that this should happen, we must be able, either by deduction or experiment, to obtain *both* the instances which the Method of Difference requires. That we are able from the hypothesis to deduce the known facts, gives only the affirmative instance, A B C, a b c. It is equally necessary that we should be able to obtain, as Newton did, the negative instance B C, *b c* ; by showing that no antecedent, except the one assumed in the hypothesis, would in conjunction with B C produce *a*.

Now it appears to me that this assurance cannot be obtained, when the cause assumed in the hypothesis is an unknown cause, imagined solely to account for *a*. When we are only seeking to determine the precise law of a cause already ascertained, or to distinguish the particular agent which is in fact the cause, among several agents of the same kind, one or other of which it is already known to be, we may then obtain the negative instance. An inquiry, which of the bodies of the solar system causes by its attraction some particular irregularity in the orbit or periodic time of some satellite or comet, would be a case of the second description. Newton's was a case of the first. If it had not been previously known that the planets were hindered from moving in straight lines by some force tending towards the interior of their orbit, though the exact direction was doubtful ; or if it had not been known that the force increased in some proportion or other as the distance diminished, and diminished as it increased ; Newton's argument would not have proved his conclusion. These facts, however, being already certain, the range of admissible suppositions was limited to the various possible directions of a line, and the various possible numerical relations between the variations of the distance, and the variations of the attractive force : now among these it was easily shown that different suppositions could not lead to identical consequences.

Accordingly, Newton could not have performed his second great scientific operation, that of identifying terrestrial gravity with the central force of the solar system, by the same hypothetical method. When the law of the moon's attraction had been proved from the data of the moon itself, then on finding the same law to accord with the phenomena of terrestrial gravity, he was warranted in adopting it as the law of those phenomena likewise ; but it would not have been allowable for him, without any lunar data, to assume that the moon was attracted towards the earth with a force as the inverse square of the distance, merely because that ratio would enable him to account for terrestrial gravity: for it would have been impossible for him to prove that the observed law of the fall of heavy bodies to the earth could not result from any force, save one extending to the moon, and proportional to the inverse square.

It appears, then, to be a condition of a genuinely scien·tific hypothesis, that it be not destined always to remain an hypothesis, but be of such a nature as to be either proved or disproved by comparison with observed facts. This condition is fulfilled when the effect is already known to depend on the very cause supposed, and the hypothesis relates only to the precise mode of dependence; the law of the variation of the effect according to the variations in the quantity or in the relations of the cause. With these may be classed the hypotheses which do not make any supposition with regard to causation, but only with regard to the law of correspondence between facts which accompany each other in their variations, though there may be no relation of cause and effect between them. Such were the different false hypotheses which Kepler made respecting the law of the refraction of light. It was known that the direction of the line of refraction varied with every variation in the direction of the line of incidence, but it was not known how; that is, what changes of the one corresponded to the different changes of the other. In this case any law, different from the true one, must have led to false results. And, lastly, we must add to these, all hypothetical modes of merely representing, or *describing*, phenomena; such as the hypothesis of the ancient astronomers that the heavenly bodies moved in circles; the various hypotheses of excentrics, deferents, and epicycles, which were added to that original hypothesis; the nineteen false hypotheses which Kepler made and abandoned respecting the form of the planetary orbits; and even the doctrine in which he finally rested, that those orbits are ellipses, which was but an hypothesis like the rest until verified by facts.

In all these cases, verification is proof; if the supposition accords with the phenomena there needs no other evidence

of it. But in order that this may be the case, I conceive it
to be necessary, when the hypothesis relates to causation, that
the supposed cause should not only be a real phenomenon,
something actually existing in nature, but should be already
known to exercise, or at least to be capable of exercising, an
influence of some sort over the effect. In any other case, it is
no evidence of the truth of the hypothesis that we are able to
deduce the real phenomena from it.

Is it, then, never allowable, in a scientific hypothesis, to
assume a cause; but only to ascribe an assumed law to a
known cause? I do not assert this. I only say, that in the
latter case alone can the hypothesis be received as true
merely because it explains the phenomena: in the former
case it is only useful by suggesting a line of investigation
which may possibly terminate in obtaining real proof. For
this purpose, as is justly remarked by M. Comte, it is indispensable that the cause suggested by the hypothesis should
be in its own nature susceptible of being proved by other
evidence. This seems to be the philosophical import of
Newton's maxim, (so often cited with approbation by subsequent writers,) that the cause assigned for any phenomenon
must not only be such as if admitted would explain the
phenomenon, but must also be a *vera causa*. What he meant
by a *vera causa* Newton did not indeed very explicitly define;
and Dr. Whewell, who dissents from the propriety of any such
restriction upon the latitude of framing hypotheses, has had
little difficulty in showing* that his conception of it was
neither precise nor consistent with itself: accordingly his
optical theory was a signal instance of the violation of his
own rule. It is certainly not necessary that the cause assigned
should be a cause already known; else how could we ever
become acquainted with any new cause? But what is true in
the maxim is, that the cause, though not known previously,
should be capable of being known thereafter; that its existence
should be capable of being detected, and its connexion with
the effect ascribed to it should be susceptible of being proved,

---

* *Philosophy of Discovery*, pp. 185 et seqq.

by independent evidence. The hypothesis, by suggesting observations and experiments, puts us on the road to that independent evidence if it be really attainable; and till it be attained, the hypothesis ought not to count for more than a conjecture.

§ 5. This function, however, of hypotheses, is one which must be reckoned absolutely indispensable in science. When Newton said, "Hypotheses non fingo," he did not mean that he deprived himself of the facilities of investigation afforded by assuming in the first instance what he hoped ultimately to be able to prove. Without such assumptions, science could never have attained its present state: they are necessary steps in the progress to something more certain; and nearly everything which is now theory was once hypothesis. Even in purely experimental science, some inducement is necessary for trying one experiment rather than another; and though it is abstractedly possible that all the experiments which have been tried, might have been produced by the mere desire to ascertain what would happen in certain circumstances, without any previous conjecture as to the result; yet, in point of fact, those unobvious, delicate, and often cumbrous and tedious processes of experiment, which have thrown most light upon the general constitution of nature, would hardly ever have been undertaken by the persons or at the time they were, unless it had seemed to depend on them whether some general doctrine or theory which had been suggested, but not yet proved, should be admitted or not. If this be true even of merely experimental inquiry, the conversion of experimental into deductive truths could still less have been effected without large temporary assistance from hypotheses. The process of tracing regularity in any complicated, and at first sight confused set of appearances, is necessarily tentative: we begin by making any supposition, even a false one, to see what consequences will follow from it; and by observing how these differ from the real phenomena, we learn what corrections to make in our assumption. The simplest supposition which accords with the more obvious facts, is the best to begin with; because its consequences are the most easily traced. This rude hypothesis is then rudely corrected, and the operation repeated; and the comparison of the consequences deducible from the corrected hypothesis, with the oberved facts, suggests still further correction, until the deductive results are at last made to tally with the phenomena. "Some fact is as yet little understood, or some law is unknown: we frame on the subject an hypothesis as accordant as possible with the whole of the data already possessed; and the science, being thus enabled to move forward freely, always ends by leading to new consequences capable of observation, which either confirm or refute, unequivocally, the first supposition." Neither induction nor deduction would enable us to understand even the simplest phenomena, "if we did not often commence by anticipating on the results; by making a provisional supposition, at first essentially conjectural, as to some of the very notions which constitute the final object of the inquiry.* Let any one watch the manner in which he himself unravels a complicated mass of evidence; let him observe how, for instance, he elicits the true history of any occurrence from the involved statements of one or of many witnesses: he will find that he does not take all the items of evidence into his mind at once, and attempt to weave them together: he extemporises, from a few of the particulars, a first rude theory of the mode in which the facts took place, and then looks at the other statements one by one, to try whether they can be reconciled with that provisional theory, or what alterations or additions it requires to make it square with them. In this way, which has been justly compared to the Methods of Approximation of mathematicians, we arrive, by means of hypotheses, at conclusions not hypothetical.†

---

* *Philosophie Positive*, ii. 434–437.

† As an example of legitimate hypothesis according to the test here laid down, has been justly cited that of Broussais, who, proceeding on the very rational principle that every disease must originate in some definite part or other of the organism, boldly assumed that certain fevers, which not being known to be local were called constitutional, had their origin in the mucous membrane of the alimentary canal. The supposition was indeed, as is now

§ 6. It is perfectly consistent with the spirit of the method, to assume in this provisional manner not only an hypothesis

---

generally admitted, erroneous; but he was justified in making it, since by deducing the consequences of the supposition, and comparing them with the facts of those maladies, he might be certain of disproving his hypothesis if it was ill founded, and might expect that the comparison would materially aid him in framing another more conformable to the phenomena.

The doctrine now universally received, that the earth is a natural magnet, was originally an hypothesis of the celebrated Gilbert.

Another hypothesis, to the legitimacy of which no objection can lie, and which is well calculated to light the path of scientific inquiry, is that suggested by several recent writers, that the brain is a voltaic pile, and that each of its pulsations is a discharge of electricity through the system. It has been remarked that the sensation felt by the hand from the beating of a brain, bears a strong resemblance to a voltaic shock. And the hypothesis, if followed to its consequences, might afford a plausible explanation of many physiological facts, while there is nothing to discourage the hope that we may in time sufficiently understand the conditions of voltaic phenomena to render the truth of the hypothesis amenable to observation and experiment.

The attempt to localize, in different regions of the brain, the physical organs of our different mental faculties and propensities, was, on the part of its original author, a legitimate example of a scientific hypothesis; and we ought not, therefore, to blame him for the extremely slight grounds on which he often proceeded, in an operation which could only be tentative, though we may regret that materials barely sufficient for a first rude hypothesis should have been hastily worked up into the vain semblance of a science. If there be really a connexion between the scale of mental endowments and the various degrees of complication in the cerebral system, the nature of that connexion was in no other way so likely to be brought to light as by framing, in the first instance, an hypothesis similar to that of Gall. But the verification of any such hypothesis is attended, from the peculiar nature of the phenomena, with difficulties which phrenologists have not shown themselves even competent to appreciate, much less to overcome.

Mr. Darwin's remarkable speculation on the Origin of Species is another unimpeachable example of a legitimate hypothesis. What he terms "natural selection" is not only a vera causa, but one proved to be capable of producing effects of the same kind with those which the hypothesis ascribes to it: the question of possibility is entirely one of degree. It is unreasonable to accuse Mr Darwin (as has been done) of violating the rules of Induction. The rules of Induction are concerned with the conditions of Proof. Mr. Darwin has never pretended that his doctrine was proved. He was not bound by the rules of Induction, but by those of Hypothesis. And these last have seldom been more completely fulfilled. He has opened a path of inquiry full of promise, the results of which none can foresee. And is it not a wonderful feat of scientific knowledge and ingenuity to have rendered so bold a suggestion, which the first impulse of every one was to reject at once, admissible and discussable, even as a conjecture?

respecting the law of what we already know to be the cause, but an hypothesis respecting the cause itself. It is allowable, useful, and often even necessary, to begin by asking ourselves what cause *may* have produced the effect, in order that we may know in what direction to look out for evidence to determine whether it actually *did*. The vortices of Descartes would have been a perfectly legitimate hypothesis, if it had been possible, by any mode of exploration which we could entertain the hope of ever possessing, to bring the reality of the vortices, as a fact in nature, conclusively to the test of observation. The hypothesis was vicious, simply because it could not lead to any course of investigation capable of converting it from an hypothesis into a proved fact. It might chance to be *dis*proved, either by some want of correspondence with the phenomena it purported to explain, or (as actually happened) by some extraneous fact. "The free passage of comets through the spaces in which these vortices should have been, convinced men that these vortices did not exist."[*] But the hypothesis would have been false, though no such direct evidence of its falsity had been procurable. Direct evidence of its truth there could not be.

The prevailing hypothesis of a luminiferous ether, in other respects not without analogy to that of Descartes, is not in its own nature entirely cut off from the possibility of direct evidence in its favour. It is well known that the difference between the calculated and the observed times of the periodical return of Encke's comet, has led to a conjecture that a medium capable of opposing resistance to motion is diffused through space. If this surmise should be confirmed, in the course of ages, by the gradual accumulation of a similar variance in the case of the other bodies of the solar system, the luminiferous ether would have made a considerable advance towards the character of a *vera causa*, since the existence would have been ascertained of a great cosmical agent, possessing some of the attributes which the hypothesis assumes; though there would still remain many difficulties, and the identification of the

---

* Whewell's *Phil. of Discovery*, pp. 275, 276.

2—2

chance) that even an erroneous interpretation which accorded with all the visible parts of the inscription would accord also with the small remainder; as would be the case, for example, if the inscription had been designedly so contrived as to admit of a double sense. I assume that the uncovered characters afford an amount of coincidence too great to be merely casual: otherwise the illustration is not a fair one. No one supposes the agreement with the phenomena of light with the theory of undulations to be merely fortuitous. It must arise from the actual identity of some of the laws of undulations with some of those of light: and if there be that identity, it is reasonable to suppose that its consequences would not end with the phenomena which first suggested the identification, nor be even confined to such phenomena as were known at the time. But it does not follow, because some of the laws agree with those of undulations, that there are any actual undulations; no more than it followed because some (though not so many) of the same laws agreed with those of the projection of particles, that there was actual emission of particles. Even the undulatory hypothesis does not account for all the phenomena of light. The natural colours of objects, the compound nature of the solar ray, the absorption of light, and its chemical and vital action, the hypothesis leaves as mysterious as it found them; and some of these facts are, at least apparently, more reconcileable with the emission theory than with that of Young and Fresnel. Who knows but that some third hypothesis, including all these phenomena, may in time leave the undulatory theory as far behind as that has left the theory of Newton and his successors?

To the statement, that the condition of accounting for all the known phenomena is often fulfilled equally well by two conflicting hypotheses, Dr. Whewell makes answer that he knows "of no such case in the history of science, where the phenomena are at all numerous and complicated."* Such an affirmation, by a writer of Dr. Whewell's minute acquaintance with the history of science, would carry great authority, if he

---

* P. 271.

had not, a few pages before, taken pains to refute it,* by maintaining that even the exploded scientific hypotheses might
always, or almost always, have been so modified as to make
them correct representations of the phenomena. The hypothesis of vortices, he tells us, was, by successive modifications,
brought to coincide in its results with the Newtonian theory
and with the facts. The vortices did not indeed explain all
the phenomena which the Newtonian theory was ultimately
found to account for, such as the precession of the equinoxes;
but this phenomenon was not, at the time, in the contemplation
of either party, as one of the facts to be accounted for.
All the facts which they did contemplate, we may believe on
Dr. Whewell's authority to have accorded as accurately with
the Cartesian hypothesis, in its finally improved state, as with
Newton's.

But it is not, I conceive, a valid reason for accepting any
given hypothesis, that we are unable to imagine any other
which will account for the facts. There is no necessity for
supposing that the true explanation must be one which, with
only our present experience, we could imagine. Among the
natural agents with which we are acquainted, the vibrations
of an elastic fluid may be the only one whose laws bear a
close resemblance to those of light; but we cannot tell that
there does not exist an unknown cause, other than an elastic
ether diffused through space, yet producing effects identical
in some respects with those which would result from the undulations of such an ether. To assume that no such cause
can exist, appears to me an extreme case of assumption without evidence.

I do not mean to condemn those who employ themselves
in working out into detail this sort of hypotheses; it is useful
to ascertain what are the known phenomena, to the laws of
which those of the subject of inquiry bear the greatest, or
even a great analogy, since this may suggest (as in the case
of the luminiferous ether it actually did) experiments to
determine whether the analogy which goes so far does not

---

* P. 251 and the whole of Appendix G.

extend still further.   But that, in doing this, we should
imagine ourselves to be seriously inquiring whether the
hypothesis of an ether, an electric fluid, or the like, is true;
that we should fancy it possible to obtain the assurance
that the phenomena are produced in that way and no other;
seems to me, I confess, unworthy of the present improved
conceptions of the methods of physical science.   And at the
risk of being charged with want of modesty, I cannot help
expressing astonishment that a philosopher of Dr. Whewell's
abilities and attainments should have written an elaborate
treatise on the philosophy of induction, in which he recognises absolutely no mode of induction except that of trying
hypothesis after hypothesis until one is found which fits the
phenomena; which one, when found, is to be assumed as true,
with no other reservation than that if on re-examination it
should appear to assume more than is needful for explaining
the phenomena, the superfluous part of the assumption
should be cut off.   And this without the slightest distinction between the cases in which it may be known beforehand
that two different hypotheses cannot lead to the same result,
and those in which, for aught we can ever know, the range of
suppositions, all equally consistent with the phenomena, may
be infinite.*

---

* In Dr. Whewell's latest version of his theory (*Philosophy of Discovery*,
p. 331) he makes a concession respecting the medium of the transmission of
light, which, taken in conjunction with the rest of his doctrine on the subject,
is not, I confess, very intelligible to me, but which goes far towards removing,
if it does not actually remove, the whole of the difference between us.   He is
contending, against Sir William Hamilton, that all matter has weight.   Sir
William, in proof of the contrary, cited the luminiferous ether, and the calorific
and electric fluids, "which," he said, "we can neither denude of their character
of substance, nor clothe with the attribute of weight."   "To which," continues
Dr. Whewell, "my reply is, that precisely because I cannot clothe these agents
with the attribute of Weight, I *do* denude them of the character of Substance.
They are not substances, but agencies.   These Imponderable Agents, are not properly called Imponderable Fluids.   This I conceive that I have proved."   Nothing
can be more philosophical.   But if the luminiferous ether is not matter, and
fluid matter too, what is the meaning of its undulations?   Can an agency undu-
late?   Can there be alternate motion forward and backward of the particles of
an agency?   And does not the whole mathematical theory of the undulations
imply them to be material?   Is it not a series of deductions from the known

§ 7. It is necessary, before quitting the subject of hypotheses, to guard against the appearance of reflecting upon the scientific value of several branches of physical inquiry, which, though only in their infancy, I hold to be strictly inductive. There is a great difference between inventing agencies to account for classes of phenomena, and endeavouring, in conformity with known laws, to conjecture what former collocations of known agents may have given birth to individual facts still in existence. The latter is the legitimate operation of inferring from an observed effect, the existence, in time past, of a cause similar to that by which we know it to be produced in all cases in which we have actual experience of its origin. This, for example, is the scope of the inquiries of geology; and they are no more illogical or visionary than judicial inquiries, which also aim at discovering a past event by inference from those of its effects which still subsist. As we can ascertain whether a man was murdered or died a natural death, from the indications exhibited by the corpse, the presence or absence of signs of struggling on the ground or on the adjacent objects, the

properties of elastic fluids? *This* opinion of Dr. Whewell reduces the undulations to a figure of speech, and the undulatory theory to the proposition which all must admit, that the transmission of light takes place according to laws which present a very striking and remarkable agreement with those of undulations. If Dr. Whewell is prepared to stand by this doctrine, I have no difference with him on the subject.

Since this chapter was written, the hypothesis of the luminiferous ether has acquired a great accession of apparent strength, by being adopted into the new doctrine of the Conservation of Force, as affording a mechanism by which to explain the mode of production not of light only, but of heat, and probably of all the other so-called imponderable agencies. In the present immature stage of the great speculation in question, I would not undertake to define the ultimate relation of the hypothetical fluid to it; but I must remark that the essential part of the new theory, the reciprocal convertibility and interchangeability of these great cosmic agencies, is quite independent of the molecular motions which have been imagined as the immediate causes of those different manifestations and of their substitutions for one another; and the former doctrine by no means necessarily carries the latter with it. I confess that the entire theory of the vibrations of the ether, and the movements which these vibrations are supposed to communicate to the particles of solid bodies, seems to me at present the weakest part of the new system, tending rather to weigh down than to prop up those of its doctrines which rest on real scientific induction.

marks of blood, the footsteps of the supposed murderers, and so on, proceeding throughout on uniformities ascertained by a perfect induction without any mixture of hypothesis; so if we find, on and beneath the surface of our planet, masses exactly similar to deposits from water, or to results of the cooling of matter melted by fire, we may justly conclude that such has been their origin; and if the effects, though similar in kind, are on a far larger scale than any which are now produced, we may rationally, and without hypothesis, conclude either that the causes existed formerly with greater intensity, or that they have operated during an enormous length of time. Further than this no geologist of authority has, since the rise of the present enlightened school of geological speculation, attempted to go.

In many geological inquiries it doubtless happens that though the laws to which the phenomena are ascribed are known laws, and the agents known agents, those agents are not known to have been present in the particular case. In the speculation respecting the igneous origin of trap or granite, the fact does not admit of direct proof, that those substances have been actually subjected to intense heat. But the same thing might be said of all judicial inquiries which proceed on circumstantial evidence. We can conclude that a man was murdered, though it is not proved by the testimony of eye-witnesses that some person who had the intention of murdering him was present on the spot. It is enough, for most purposes, if no other known cause could have generated the effects shown to have been produced.

The celebrated speculation of Laplace concerning the origin of the earth and planets, participates essentially in the inductive character of modern geological theory. The speculation is, that the atmosphere of the sun originally extended to the present limits of the solar system; from which, by the process of cooling, it has contracted to its present dimensions; and since, by the general principles of mechanics, the rotation of the sun and of its accompanying atmosphere must increase in rapidity as its volume diminishes, the increased centrifugal force generated by the more rapid rotation, overbalancing the action of gravitation, has caused the sun to abandon successive rings of vaporous matter, which are supposed to have condensed by cooling, and to have become the planets. There is in this theory no unknown substance introduced on supposition, nor any unknown property or law ascribed to a known substance. The known laws of matter authorize us to suppose that a body which is constantly giving out so large an amount of heat as the sun is, must be progressively cooling, and that, by the process of cooling, it must contract; if, therefore, we endeavour, from the present state of that luminary, to infer its state in a time long past, we must necessarily suppose that its atmosphere extended much farther than at present, and we are entitled to suppose that it extended as far as we can trace effects such as it might naturally leave behind it on retiring; and such the planets are. These suppositions being made, it follows from known laws that successive zones of the solar atmosphere might be abandoned; that these would continue to revolve round the sun with the same velocity as when they formed part of its substance; and that they would cool down, long before the sun itself, to any given temperature, and consequently to that at which the greater part of the vaporous matter of which they consisted would become liquid or solid. The known law of gravitation would then cause them to agglomerate in masses, which would assume the shape our planets actually exhibit; would acquire, each about its own axis, a rotatory movement; and would in that state revolve, as the planets actually do, about the sun, in the same direction with the sun's rotation, but with less velocity, because in the same periodic time which the sun's rotation occupied when his atmosphere extended to that point. There is thus, in Laplace's theory, nothing, strictly speaking, hypothetical: it is an example of legitimate reasoning from a present effect to a possible past cause, according to the known laws of that cause. The theory therefore is, as I have said, of a similar character to the theories of geologists; but considerably in-

ferior to them in point of evidence. Even if it were proved
(which it is not) that the conditions necessary for determining
the breaking off of successive rings would certainly occur ;
there would still be a much greater chance of error in assuming
that the existing laws of nature are the same which existed at
the origin of the solar system, than in merely presuming (with
geologists) that those laws have lasted through a few revolutions and transformations of a single one among the bodies
of which that system is composed.

CHAPTER XV.

§ 1. In the last four chapters we have traced the general outlines of the theory of the generation of derivative laws from ultimate ones. In the present chapter our attention will be directed to a particular case of the derivation of laws from other laws, but a case so general, and so important, as not only to repay, but to require, a separate examination. This is, the case of a complex phenomenon resulting from one simple law, by the continual addition of an effect to itself.

There are some phenomena, some bodily sensations for example, which are essentially instantaneous, and whose existence can only be prolonged by the prolongation of the existence of the cause by which they are produced. But most phenomena are in their own nature permanent; having begun to exist, they would exist for ever unless some cause intervened having a tendency to alter or destroy them. Such, for example, are all the facts or phenomena which we call bodies. Water, once produced, will not of itself relapse into a state of hydrogen and oxygen; such a change requires some agent having the power of decomposing the compound. Such, again, are the positions in space, and the movements, of bodies. No object at rest alters its position without the intervention of some conditions extraneous to itself; and when once in motion, no object returns to a state of rest, or alters either its direction or its velocity, unless some new external conditions are superinduced. It, therefore, perpetually happens that a temporary cause gives rise to a permanent effect. The contact of iron with moist air for a few hours, produces a rust which may endure for centuries; or a projectile force which launches a cannon ball into space, produces a motion which would continue for ever unless some other force counteracted it.

Between the two examples which we have here given, there is a difference worth pointing out. In the former (in which the phenomenon produced is a substance, and not a motion of a substance), since the rust remains for ever and unaltered unless some new cause supervenes, we may speak of the contact of air a hundred years ago as even the proximate cause of the rust which has existed from that time until now. But when the effect is motion, which is itself a change, we must use a different language. The permanency of the effect is now only the permanency of a series of changes. The second foot, or inch, or mile of motion, is not the mere prolonged duration of the first foot, or inch, or mile, but another fact which succeeds, and which may in some respects be very unlike the former, since it carries the body through a different region of space. Now, the original projectile force which set the body moving is the remote cause of all its motion, however long continued, but the proximate cause of no motion except that which took place at the first instant. The motion at any subsequent instant is proximately caused by the motion which took place at the instant preceding. It is on that, and not on the original moving cause, that the motion at any given moment depends. For, suppose that the body passes through some resisting medium, which partially counteracts the effect of the original impulse, and retards the motion: this counteraction (it needs scarcely here be repeated) is as strict an example of obedience to the law of the impulse, as if the body had gone on moving with its original velocity; but the motion which results is different, being now a compound of the effects of two causes acting in contrary directions, instead of the single effect of one cause. Now, what cause does the body obey in its subsequent motion? The original cause of motion, or the actual motion at the preceding instant? The latter: for when the object issues from the resisting medium, it continues moving, not with its original, but with its retarded velocity. The motion having once been diminished, all that which follows is diminished. The effect changes, because the cause which it really obeys, the proximate cause, the real cause in fact, has changed. This principle is recognised by mathematicians when they enumerate among the causes by which the motion of a body is at any instant determined, the *force generated* by the previous motion; an expression which would be absurd if taken to imply that this "force" was an intermediate link between the cause and the effect, but which really means only the previous motion itself, considered as a cause of further motion. We must, therefore, if we would speak with perfect precision, consider each link in the succession of motions as the effect of the link preceding it. But if, for the convenience of discourse, we speak of the whole series as one effect, it must be as an effect produced by the original impelling force; a permanent effect produced by an instantaneous cause, and possessing the property of self-perpetuation.

Let us now suppose that the original agent or cause, instead of being instantaneous, is permanent. Whatever effect has been produced up to a given time, would (unless prevented by the intervention of some new cause) subsist permanently, even if the cause were to perish. Since, however, the cause does not perish, but continues to exist and to operate, it must go on producing more and more of the effect; and instead of an uniform effect, we have a progressive series of effects, arising from the accumulated influence of a permanent cause. Thus, the contact of iron with the atmosphere causes a portion of it to rust; and if the cause ceased, the effect already produced would be permanent, but no further effect would be added. If, however, the cause, namely, exposure to moist air, continues, more and more of the iron becomes rusted, until all which is exposed is converted into a red powder, when one of the conditions of the production of rust, namely, the presence of unoxidized iron, has ceased, and the effect cannot any longer be produced. Again, the earth causes bodies to fall towards it, that is, the existence of the earth at a given instant, causes an unsupported body to move towards it at the succeeding instant: and if the earth were annihilated, as much of the effect as is already produced would continue; the object would go on moving in the same direction, with its acquired velocity, until intercepted by some body or deflected by some other force. The earth, however, not being annihilated, goes on producing in the second instant an effect similar and of equal amount with the first, which two effects being added together, there results an accelerated velocity; and this operation being repeated at each successive instant, the mere permanence of the cause, though without increase, gives rise to a constant progressive increase of the effect, so long as all the conditions, negative and positive, of the production of that effect, continue to be realized.

It is obvious that this state of things is merely a case of the Composition of Causes. A cause which continues in action, must on a strict analysis be considered as a number of causes exactly similar, successively introduced, and producing by their combination the sum of the effects which they would severally produce if they acted singly. The progressive rusting of the iron is in strictness the sum of the effects of many particles of air acting in succession upon corresponding particles of iron. The continued action of the earth upon a falling body is equivalent to a series of forces, applied in successive instants, each tending to produce a certain constant quantity of motion; and the motion at each instant is the sum of the effects of the new force applied at the preceding instant, and the motion already acquired. In each instant, a fresh effect, of which gravity is the proximate cause, is added to the effect of which it was the remote cause: or (to express the same thing in another manner) the effect produced by the earth's influence at the instant last elapsed, is added to the sum of the effects of which the remote causes were the influences exerted by the earth at all the previous instants since the motion began. The case, therefore, comes under the principle of a concurrence of causes producing an effect equal to the sum of their separate effects. But as the causes come into play not all at once, but successively, and as the effect at each instant is the sum of the effects of those causes only which have come into action up to that instant, the result assumes the form of an ascending series; a succession of sums, each greater than that which preceded it; and we have thus a progressive effect from the continued action of a cause.

Since the continuance of the cause influences the effect only by adding to its quantity, and since the addition takes place according to a fixed law (equal quantities in equal times), the result is capable of being computed on mathematical principles. In fact, this case, being that of infinitesimal increments, is precisely the case which the differential calculus was invented to meet. The questions, what effect will result from the continual addition of a given cause to itself, and what amount of the cause, being continually added to itself, will produce a given amount of the effect, are evidently mathematical questions, and to be treated, therefore, deductively. If, as we have seen, cases of the Composition of Causes are seldom adapted for any other than deductive investigation, this is especially true in the case now examined, the continual composition of a cause with its own previous effects; since such a case is peculiarly amenable to the deductive method, while the undistinguishable manner in which the effects are blended with one another and with the causes, must make the treatment of such an instance experimentally, still more chimerical than in any other case.

§ 2. We shall next advert to a rather more intricate operation of the same principle, namely, when the cause does not merely continue in action, but undergoes, during the same time, a progressive change in those of its circumstances which contribute to determine the effect. In this case, as in the former, the total effect goes on accumulating by the continual addition of a fresh effect to that already produced, but it is no longer by the addition of equal quantities in equal times; the quantities added are unequal, and even the quality may now be different. If the change in the state of the permanent cause be progressive, the effect will go through a double series of changes, arising partly from the accumulated action of the cause, and partly from the changes in its action. The effect is still a progressive effect, produced however, not by the mere continuance of a cause, but by its continuance and its progressiveness combined.

A familiar example is afforded by the increase of the temperature as summer advances, that is, as the sun draws nearer to a vertical position, and remains a greater number of hours above the horizon. This instance exemplifies in a very interesting manner the twofold operation on the effect, arising from the continuance of the cause, and from its progressive change. When once the sun has come near enough to the zenith, and remains above the horizon long enough, to give more warmth during one diurnal rotation than the counteracting cause, the earth's radiation, can carry off, the mere continuance of the cause would progressively increase the effect, even if the sun came no nearer and the days grew no longer; but in addition to this, a change takes place in the accidents of the cause (its series of diurnal positions), tending to increase the quantity of the effect. When the summer solstice has passed, the progressive change in the cause begins to take place the reverse way; but, for some time, the accumulating effect of the mere continuance of the cause exceeds the effect of the changes in it, and the temperature continues to increase.

Again, the motion of a planet is a progressive effect, produced by causes at once permanent and progressive. The orbit of a planet is determined (omitting perturbations) by two causes: first, the action of the central body, a permanent cause, which alternately increases and diminishes as the planet draws nearer to or goes further from its perihelion, and which acts at every point in a different direction; and, secondly, the tendency of the planet to continue moving in the direction and with the velocity which it has already acquired. This force also grows greater as the planet draws nearer to its perihelion, because as it does so its velocity increases; and less, as it recedes from its perihelion: and this force as well as the other acts at each point in a different direction, because at every point the action of the central force, by deflecting the planet from its previous direction, alters the line in which it tends to continue moving. The motion at each instant is determined by the amount and direction of the motion, and the amount and direction of the sun's action, at the previous instant : and if we speak of the entire revolution of the planet as one phenomenon (which, as it is periodical and similar to itself, we often find it convenient to do,) that phenomenon is the progressive effect of two permanent and progressive causes, the central force and the acquired motion. Those causes happening to be progressive in the particular way which is called periodical, the effect necessarily is so too; because the quantities to be added together returning in a regular order, the same sums must also regularly return.

This example is worthy of consideration also in another respect. Though the causes themselves are permanent, and independent of all conditions known to us, the changes which take place in the quantities and relations of the causes are actually caused by the periodical changes in the effects. The causes, as they exist at any moment, having produced a certain motion, that motion, becoming itself a cause, reacts upon the causes, and produces a change in them. By altering the distance and direction of the central body relatively to the planet, and the direction and quantity of the force in the direction of the tangent, it alters the elements which determine the motion at the next succeeding instant. This change renders the next motion somewhat different; and this difference, by a fresh reaction upon the causes, renders the next motion again different, and so on. The original state of the causes might have been such, that this series of actions modified by reactions would not have been periodical. The sun's action, and the original impelling force, might have been in such a ratio to one another, that the reaction of the effect would have been such as to alter the causes more and more, without ever bringing them back to what they were at any former time. The planet would then have moved in a parabola, or an hyperbola, curves not returning into themselves. The quantities of the two forces were, however, originally such, that the successive reactions of the effect bring back the causes, after a certain time, to what they were before; and from that time all the variations continue to recur again and again in the same periodical order, and must so continue while the causes subsist and are not counteracted.

§ 3. In all cases of progressive effects, whether arising from the accumulation of unchanging or of changing elements, there is an uniformity of succession not merely between the cause and the effect, but between the first stages of the effect and its subsequent stages. That a body *in vacuo* falls sixteen feet in the first second, forty-eight in the second, and so on in the ratio of the odd numbers, is as much an uniform sequence as that when the supports are removed the body falls. The sequence of spring and summer is as regular and invariable as that of the approach of the sun and spring: but we do not consider spring to be the cause of summer; it is evident that both are successive effects of the heat received from the sun, and that, considered merely in itself, spring might continue for ever, without having the slightest tendency to produce summer. As we have so often remarked, not the conditional, but the unconditional invariable antecedent is termed the cause. That which would not be followed by the effect unless something else had preceded, is not the cause, however invariable the sequence may in fact be.

It is in this way that most of those uniformities of succession are generated, which are not cases of causation. When a phenomenon goes on increasing, or periodically increases and diminishes, or goes through any continued and unceasing process of variation reducible to an uniform rule or law of succession, we do not on this account presume that any two successive terms of the series are cause and effect. We presume the contrary; we expect to find that the whole series originates either from the continued action of fixed causes, or from causes which go through a corresponding process of continuous change. A tree grows from half an inch high to a hundred feet; and some trees will generally grow to that height, unless prevented by some counteracting cause. But we do not call the seedling the cause of the full-grown tree; the invariable antecedent it certainly is, and we know very imperfectly on what other antecedents the sequence is contingent, but we are convinced that it is contingent on something; because the homogeneousness of the antecedent with the consequent, the close resemblance of the seedling to the tree in all respects except magnitude, and the graduality of the growth, so exactly resembling the progressively accumulating effect produced by the long action of some one cause, leave no possibility of doubting that the seedling and the tree are two terms in a series of that description, the first term of which is yet to seek. The conclusion is further confirmed by this, that we are able to prove by strict induction the dependence of the growth of the tree, and even of the continuance of its existence, upon the continued repetition of certain processes of nutrition, the rise of the sap, the absorptions and exhalations by the leaves, &c.; and the same experiments would probably prove to us that the growth of the tree is the accumulated sum of the effects of these continued processes, were we not, for want of sufficiently microscopic eyes, unable to observe correctly and in detail what those effects are.

This supposition by no means requires that the effect should not, during its progress, undergo many modifications besides those of quantity, or that it should not sometimes appear to undergo a very marked change of character. This may be either because the unknown cause consists of several component elements or agents, whose effects, accumulating according to different laws, are compounded in different proportions at different periods in the existence of the organized being; or because, at certain points in its progress, fresh causes or agencies come in, or are evolved, which intermix their laws with those of the prime agent.

CHAPTER XVI.

OF EMPIRICAL LAWS.

§ 1. SCIENTIFIC inquirers give the name of Empirical Laws to those uniformities which observation or experiment has shown to exist, but on which they hesitate to rely in cases varying much from those which have been actually observed, for want of seeing any reason *why* such a law should exist. It is implied, therefore, in the notion of an empirical law, that it is not an ultimate law; that if true at all, its truth is capable of being, and requires to be, accounted for. It is a derivative law, the derivation of which is not yet known. To state the explanation, the *why*, of the empirical law, would be to state the laws from which it is derived; the ultimate causes on which it is contingent. And if we knew these, we should also know what are its limits; under what conditions it would cease to be fulfilled.

The periodical return of eclipses, as originally ascertained by the persevering observation of the early eastern astronomers, was an empirical law, until the general laws of the celestial motions had accounted for it. The following are empirical laws still waiting to be resolved into the simpler laws from which they are derived. The local laws of the flux and reflux of the tides in different places: the succession of certain kinds of weather to certain appearances of sky: the apparent exceptions to the almost universal truth that bodies expand by increase of temperature: the law that breeds, both animal and vegetable, are improved by crossing: that gases have a strong tendency to permeate animal membranes: that substances containing a very high proportion of nitrogen (such as hydrocyanic acid and morphia) are powerful poisons: that when different metals are fused together, the alloy is harder than the various elements: that the number of

atoms of acid required to neutralize one atom of any base,
is equal to the number of atoms of oxygen in the base:
that the solubility of substances in one another, depends*
(at least in some degree) on the similarity of their elements.

An empirical law, then, is an observed uniformity, presumed to be resolvable into simpler laws, but not yet resolved
into them. The ascertainment of the empirical laws of phenomena often precedes by a long interval the explanation of
those laws by the Deductive Method; and the verification of
a deduction usually consists in the comparison of its results
with empirical laws previously ascertained.

§ 2. From a limited number of ultimate laws of causation, there are necessarily generated a vast number of derivative uniformities, both of succession and of coexistence.
Some are laws of succession or of coexistence between
different effects of the same cause: of these we had examples in the last chapter. Some are laws of succession
between effects and their remote causes; resolvable into the
laws which connect each with the intermediate link. Thirdly,
when causes act together and compound their effects, the
laws of those causes generate the fundamental law of the
effect, namely, that it depends on the coexistence of those
causes. And, finally, the order of succession or of coexistence which obtains among effects, necessarily depends
on their causes. If they are effects of the same cause, it
depends on the laws of that cause; if on different causes, it
depends on the laws of those causes severally, and on the
circumstances which determine their coexistence. If we

---

* Thus, water, of which eight-ninths in weight are oxygen, dissolves most
bodies which contain a high proportion of oxygen, such as all the nitrates,
(which have more oxygen than any others of the common salts,) most of the
sulphates, many of the carbonates, &c. Again, bodies largely composed of
combustible elements, like hydrogen and carbon, are soluble in bodies of similar
composition; rosin, for instance, will dissolve in alcohol, tar in oil of turpentine.
This empirical generalization is far from being universally true; no doubt
because it is a remote, and therefore easily defeated, result of general laws too
deep for us at present to penetrate; but it will probably in time suggest processes of inquiry, leading to the discovery of those laws.

inquire further when and how the causes will coexist, that, again, depends on *their* causes : and we may thus trace back the phenomena higher and higher, until the different series of effects meet in a point, and the whole is shown to have depended ultimately on some common cause ; or until, instead of converging to one point, they terminate in different points, and the order of the effects is proved to have arisen from the collocation of some of the primeval causes, or natural agents. For example, the order of succession and of coexistence among the heavenly motions, which is expressed by Kepler's laws, is derived from the coexistence of two primeval causes, the sun, and the original impulse or projectile force belonging to each planet.* Kepler's laws are resolved into the laws of these causes and the fact of their coexistence.

Derivative laws, therefore, do not depend solely on the ultimate laws into which they are resolvable : they mostly depend on those ultimate laws, and an ultimate fact ; namely, the mode of coexistence of some of the component elements of the universe. The ultimate laws of causation might be the same as at present, and yet the derivative laws completely different, if the causes coexisted in different proportions, or with any difference in those of their relations by which the effects are influenced. If, for example, the sun's attraction, and the original projectile force, had existed in some other ratio to one another than they did (and we know of no reason why this should not have been the case), the derivative laws of the heavenly motions might have been quite different from what they are. The proportions which exist happen to be such as to produce regular elliptical motions ; any other proportions would have produced different ellipses, or circular, or parabolic, or hyperbolic motions, but still regular ones ; because the effects of each of the agents accumulate according to an uniform law ; and two regular series of quantities, when their corresponding terms are added, must produce a regular series of some sort, whatever the quantities themselves are.

---

* Or (according to Laplace's theory) the sun and the sun's rotation.

§ 3. Now this last-mentioned element in the resolution of a derivative law, the element which is not a law of causation, but a collocation of causes, cannot itself be reduced to any law. There is (as formerly remarked*) no uniformity, no *norma*, principle, or rule, perceivable in the distribution of the primeval natural agents through the universe. The different substances composing the earth, the powers that pervade the universe, stand in no constant relation to one another. One substance is more abundant than others, one power acts through a larger extent of space than others, without any pervading analogy that we can discover. We not only do not know of any reason why the sun's attraction and the force in the direction of the tangent coexist in the exact proportion they do, but we can trace no coincidence between it and the proportions in which any other elementary powers in the universe are intermingled. The utmost disorder is apparent in the combination of the causes; which is consistent with the most regular order in their effects; for when each agent carries on its own operations according to an uniform law, even the most capricious combination of agencies will generate a regularity of some sort; as we see in the kaleidoscope, where any casual arrangement of coloured bits of glass produces by the laws of reflection a beautiful regularity in the effect.

§ 4. In the above considerations lies the justification of the limited degree of reliance which scientific inquirers are accustomed to place in empirical laws.

A derivative law which results wholly from the operation of some one cause, will be as universally true as the laws of the cause itself; that is, it will always be true except where some one of those effects of the cause, on which the derivative law depends, is defeated by a counteracting cause. But when the derivative law results not from different effects of one cause, but from effects of several causes, we cannot be certain that it will be true under any variation in the mode of coexis-

---

* Supra, book iii. ch. v. § 7.

tence of those causes, or of the primitive natural agents on which the causes ultimately depend. The proposition that coal beds rest on certain descriptions of strata exclusively, though true on the earth so far as our observation has reached, cannot be extended to the moon or the other planets, supposing coal to exist there; because we cannot be assured that the original constitution of any other planet was such as to produce the different depositions in the same order as in our globe. The derivative law in this case depends not solely on laws, but on a collocation; and collocations cannot be reduced to any law.

Now it is the very nature of a derivative law which has not yet been resolved into its elements, in other words, an empirical law, that we do not know whether it results from the different effects of one cause, or from effects of different causes. We cannot tell whether it depends wholly on laws, or partly on laws and partly on a collocation. If it depends on a collocation, it will be true in all the cases in which that particular collocation exists. But, since we are entirely ignorant, in case of its depending on a collocation, what the collocation is, we are not safe in extending the law beyond the limits of time and place in which we have actual experience of its truth. Since within those limits the law has always been found true, we have evidence that the collocations, whatever they are, on which it depends, do really exist within those limits. But, knowing of no rule or principle to which the collocations themselves conform, we cannot conclude that because a collocation is proved to exist within certain limits of place or time, it will exist beyond those limits. Empirical laws, therefore, can only be received as true within the limits of time and place in which they have been found true by observation: and not merely the limits of time and place, but of time, place, and circumstance: for since it is the very meaning of an empirical law that we do not know the ultimate laws of causation on which it is dependent, we cannot foresee, without actual trial, in what manner or to what extent the introduction of any new circumstance may affect it.

§ 5. But how are we to know that an uniformity, ascertained by experience, is only an empirical law? Since, by the supposition, we have not been able to resolve it into any other laws, how do we know that it is not an ultimate law of causation?

I answer, that no generalization amounts to more than an empirical law when the only proof on which it rests is that of the Method of Agreement. For it has been seen that by that method alone we never can arrive at causes. The utmost that the Method of Agreement can do is, to ascertain the whole of the circumstances common to all cases in which a phenomenon is produced: and this aggregate includes not only the cause of the phenomenon, but all phenomena with which it is connected by any derivative uniformity, whether as being collateral effects of the same cause, or effects of any other cause which, in all the instances we have been able to observe, coexisted with it. The method affords no means of determining which of these uniformities are laws of causation, and which are merely derivative laws, resulting from those laws of causation and from the collocation of the causes. None of them, therefore, can be received in any other character than that of derivative laws, the derivation of which has not been traced; in other words, empirical laws: in which light, all results obtained by the Method of Agreement (and therefore almost all truths obtained by simple observation without experiment) must be considered, until either confirmed by the Method of Difference, or explained deductively, in other words accounted for *à priori.*

These empirical laws may be of greater or less authority, according as there is reason to presume that they are resolvable into laws only, or into laws and collocations together. The sequences which we observe in the production and subsequent life of an animal or a vegetable, resting on the Method of Agreement only, are mere empirical laws; but though the antecedents in those sequences may not be the causes of the consequents, both the one and the other are doubtless, in the main, successive stages of a progressive effect originating in a

common cause, and therefore independent of collocations. The
uniformities, on the other hand, in the order of superposition
of strata on the earth, are empirical laws of a much weaker
kind, since they not only are not laws of causation, but there
is no reason to believe that they depend on any common
cause: all appearances are in favour of their depending on the
particular collocation of natural agents which at some time or
other existed on our globe, and from which no inference can
be drawn as to the collocation which exists or has existed in
any other portion of the universe.

§ 6. Our definition of an empirical law including not
only those uniformities which are not known to be laws of
causation, but also those which are, provided there be reason
to presume that they are not ultimate laws; this is the proper
place to consider by what signs we may judge that even if an
observed uniformity be a law of causation, it is not an ultimate
but a derivative law.

The first sign is, if between the antecedent a and the consequent b there be evidence of some intermediate link; some
phenomenon of which we can surmise the existence, though
from the imperfection of our senses or of our instruments we
are unable to ascertain its precise nature and laws. If there
be such a phenomenon (which may be denoted by the letter
x), it follows that even if a be the cause of b, it is but the
remote cause, and that the law, a causes b, is resolvable into
at least two laws, a causes x, and x causes b. This is a very
frequent case, since the operations of nature mostly take place
on so minute a scale, that many of the successive steps are
either imperceptible, or very indistinctly perceived.

Take, for example, the laws of the chemical composition
of substances; as that hydrogen and oxygen being combined,
water is produced. All we see of the process is, that the
two gases being mixed in certain proportions, and heat or
electricity being applied, an explosion takes place, the gases
disappear, and water remains. There is no doubt about the
law, or about its being a law of causation. But between the
antecedent (the gases in a state of mechanical mixture,

heated or electrified), and the consequent (the production of water), there must be an intermediate process which we do not see. For if we take any portion whatever of the water, and subject it to analysis, we find that it always contains hydrogen and oxygen; nay, the very same proportions of them, namely, two thirds, in volume, of hydrogen, and one third oxygen. This is true of a single drop; it is true of the minutest portion which our instruments are capable of appreciating. Since, then, the smallest perceptible portion of the water contains both those substances, portions of hydrogen and oxygen smaller than the smallest perceptible must have come together in every such minute portion of space; must have come closer together than when the gases were in a state of mechanical mixture, since (to mention no other reasons) the water occupies far less space than the gases. Now, as we cannot see this contact or close approach of the minute particles, we cannot observe with what circumstances it is attended, or according to what laws it produces its effects. The production of water, that is, of the sensible phenomena which characterize the compound, may be a very remote effect of those laws. There may be innumerable intervening links; and we are sure that there must be some. Having full proof that corpuscular action of some kind takes place previous to any of the great transformations in the sensible properties of substances, we can have no doubt that the laws of chemical action, as at present known, are not ultimate but derivative laws; however ignorant we may be, and even though we should for ever remain ignorant, of the nature of the laws of corpuscular action from which they are derived.

In like manner, all the processes of vegetative life, whether in the vegetable properly so called or in the animal body, are corpuscular processes. Nutrition is the addition of particles to one another, sometimes merely replacing other particles separated and excreted, sometimes occasioning an increase of bulk or weight, so gradual, that only after a long continuance does it become perceptible. Various organs, by means of peculiar vessels, secrete from the blood, fluids, the component

particles of which must have been in the blood, but which
differ from it most widely both in mechanical properties and
in chemical composition. Here, then, are abundance of unknown links to be filled up; and there can be no doubt that
the laws of the phenomena of vegetative or organic life are
derivative laws, dependent on properties of the corpuscles, and
of those elementary tissues which are comparatively simple
combinations of corpuscles.

The first sign, then, from which a law of causation, though
hitherto unresolved, may be inferred to be a derivative law,
is any indication of the existence of an intermediate link or
links between the antecedent and the consequent. The
second is, when the antecedent is an extremely complex
phenomenon, and its effects therefore, probably, in part at
least, compounded of the effects of its different elements;
since we know that the case in which the effect of the whole
is not made up of the effects of its parts, is exceptional,
the Composition of Causes being by far the more ordinary
case.

We will illustrate this by two examples, in one of which
the antecedent is the sum of many homogeneous, in the other
of heterogeneous, parts. The weight of a body is made up
of the weights of its minute particles: a truth which astronomers express in its most general terms, when they say that
bodies, at equal distances, gravitate to one another in proportion to their quantity of matter. All true propositions,
therefore, which can be made concerning gravity, are derivative laws; the ultimate law into which they are all resolvable
being, that every particle of matter attracts every other. As
our second example, we may take any of the sequences observed in meteorology: for instance, a diminution of the pressure of the atmosphere (indicated by a fall of the barometer)
is followed by rain. The antecedent is here a complex
phenomenon, made up of heterogeneous elements; the column
of the atmosphere over any particular place consisting of two
parts, a column of air, and a column of aqueous vapour mixed
with it; and the change in the two together manifested by a
fall of the barometer, and followed by rain, must be either a

change in one of these, or in the other, or in both. We might, then, even in the absence of any other evidence, form a reasonable presumption, from the invariable presence of both these elements in the antecedent, that the sequence is probably not an ultimate law, but a result of the laws of the two different agents; a presumption only to be destroyed when we had made ourselves so well acquainted with the laws of both, as to be able to affirm that those laws could not by themselves produce the observed result.

There are but few known cases of succession from very complex antecedents, which have not either been actually accounted for from simpler laws, or inferred with great probability (from the ascertained existence of intermediate links of causation not yet understood) to be capable of being so accounted for. It is, therefore, highly probable that all sequences from complex antecedents are thus resolvable, and that ultimate laws are in all cases comparatively simple. If there were not the other reasons already mentioned for believing that the laws of organized nature are resolvable into simpler laws, it would be almost a sufficient reason that the antecedents in most of the sequences are so very complex.

§ 7. In the preceding discussion we have recognised two kinds of empirical laws : those known to be laws of causation, but presumed to be resolvable into simpler laws; and those not known to be laws of causation at all. Both these kinds of laws agree in the demand which they make for being explained by deduction, and agree in being the appropriate means of verifying such deduction, since they represent the experience with which the result of the deduction must be compared. They agree, further, in this, that until explained, and connected with the ultimate laws from which they result, they have not attained the highest degree of certainty of which laws are susceptible. It has been shown on a former occasion that laws of causation which are derivative, and compounded of simpler laws, are not only, as the nature of the case implies, less general, but even less certain, than the

simpler laws from which they result; not in the same degree
to be relied on as universally true. The inferiority of evidence,
however, which attaches to this class of laws, is trifling, compared with that which is inherent in uniformities not known
to be laws of causation at all. So long as these are unresolved,
we cannot tell on how many collocations, as well as laws, their
truth may be dependent; we can never, therefore, extend
them with any confidence to cases in which we have not
assured ourselves, by trial, that the necessary collocation of
causes, whatever it may be, exists. It is to this class of laws
alone that the property, which philosophers usually consider
as characteristic of empirical laws, belongs in all its strictness;
the property of being unfit to be relied on beyond the limits
of time, place, and circumstance, in which the observations
have been made. These are empirical laws in a more emphatic sense; and when I employ that term (except where the
context manifestly indicates the reverse) I shall generally mean
to designate those uniformities only, whether of succession
or of coexistence, which are not known to be laws of
causation.

CHAPTER XVII.

OF CHANCE AND ITS ELIMINATION.

§ 1. CONSIDERING then as empirical laws only those observed uniformities respecting which the question whether they are laws of causation must remain undecided until they can be explained deductively, or until some means are found of applying the Method of Difference to the case, it has been shown in the preceding chapter, that until an uniformity can, in one or the other of these modes, be taken out of the class of empirical laws, and brought either into that of laws of causation or of the demonstrated results of laws of causation, it cannot with any assurance be pronounced true beyond the local and other limits within which it has been found so by actual observation. It remains to consider how we are to assure ourselves of its truth even within those limits; after what quantity of experience a generalization which rests solely on the Method of Agreement, can be considered sufficiently established, even as an empirical law. In a former chapter, when treating of the Methods of Direct Induction, we expressly reserved this question,[*] and the time is now come for endeavouring to solve it.

We found that the Method of Agreement has the defect of not proving causation, and can therefore only be employed for the ascertainment of empirical laws. But we also found that besides this deficiency, it labours under a characteristic imperfection, tending to render uncertain even such conclusions as it is in itself adapted to prove. This imperfection arises from Plurality of Causes. Although two or more cases in which the phenomenon *a* has been met with, may have no common antecedent except A, this does not prove that there

---

[*] Supra, book iii. ch. x. § 2.

is any connexion between *a* and A, since *a* may have many causes, and may have been produced, in these different instances, not by anything which the instances had in common, but by some of those elements in them which were different. We nevertheless observed, that in proportion to the multiplication of instances pointing to A as the antecedent, the characteristic uncertainty of the method diminishes, and the existence of a law of connexion between A and *a* more nearly approaches to certainty. It is now to be determined, after what amount of experience this certainty may be deemed to be practically attained, and the connexion between A and *a* may be received as an empirical law.

This question may be otherwise stated in more familiar terms :—After how many and what sort of instances may it be concluded, that an observed coincidence between two phenomena is not the effect of chance?

It is of the utmost importance for understanding the logic of induction, that we should form a distinct conception of what is meant by chance, and how the phenomena which common language ascribes to that abstraction are really produced.

§ 2. Chance is usually spoken of in direct antithesis to law; whatever (it is supposed) cannot be ascribed to any law, is attributed to chance. It is, however, certain, that whatever happens is the result of some law; is an effect of causes, and could have been predicted from a knowledge of the existence of those causes, and from their laws. If I turn up a particular card, that is a consequence of its place in the pack. Its place in the pack was a consequence of the manner in which the cards were shuffled, or of the order in which they were played in the last game; which, again, were effects of prior causes. At every stage, if we had possessed an accurate knowledge of the causes in existence, it would have been abstractedly possible to foretell the effect.

An event occurring by chance, may be better described as a coincidence from which we have no ground to infer an uniformity: the occurrence of a phenomenon in certain circumstances, without our having reason on that account to infer that it will happen again in those circumstances. This, however, when looked closely into, implies that the enumeration of the circumstances is not complete. Whatever the fact be, since it has occurred once, we may be sure that if *all* the same circumstances were repeated, it would occur again; and not only if all, but there is some particular portion of those circumstances, on which the phenomenon is invariably consequent. With most of them, however, it is not connected in any permanent manner: its conjunction with those is said to be the effect of chance, to be merely casual. Facts casually conjoined are separately the effects of causes, and therefore of laws; but of different causes, and causes not connected by any law.

It is incorrect, then, to say that any phenomenon is produced by chance; but we may say that two or more phenomena are conjoined by chance, that they coexist or succeed one another only by chance: meaning that they are in no way related through causation; that they are neither cause and effect, nor effects of the same cause, nor effects of causes between which there subsists any law of coexistence, nor even effects of the same collocation of primeval causes.

If the same casual coincidence never occurred a second time, we should have an easy test for distinguishing such from the coincidences which are the results of a law. As long as the phenomena had been found together only once, so long, unless we knew some more general laws from which the coincidence might have resulted, we could not distinguish it from a casual one; but if it occurred twice, we should know that the phenomena so conjoined must be in some way connected through their causes.

There is, however, no such test. A coincidence may occur again and again, and yet be only casual. Nay, it would be inconsistent with what we know of the order of nature, to doubt that every casual coincidence will sooner or later be repeated, as long as the phenomena between which it occurred do not cease to exist, or to be reproduced. The recurrence, therefore, of the same coincidence more than once, or even its frequent recurrence, does not prove that it is an instance of any law; does not prove that it is not casual, or, in common language, the effect of chance.

And yet, when a coincidence cannot be deduced from known laws, nor proved by experiment to be itself a case of causation, the frequency of its occurrence is the only evidence from which we can infer that it is the result of a law. Not, however, its absolute frequency. The question is not whether the coincidence occurs often or seldom, in the ordinary sense of those terms; but whether it occurs more often than chance will account for; more often than might rationally be expected if the coincidence were casual. We have to decide, therefore, what degree of frequency in a coincidence, chance will account for. And to this there can be no general answer. We can only state the principle by which the answer must be determined: the answer itself will be different in every different case.

Suppose that one of the phenomena, A, exists always, and the other phenomenon, B, only occasionally: it follows that every instance of B will be an instance of its coincidence with A, and yet the coincidence will be merely casual, not the result of any connexion between them. The fixed stars have been constantly in existence since the beginning of human experience, and all phenomena that have come under human observation have, in every single instance, coexisted with them; yet this coincidence, though equally invariable with that which exists between any of those phenomena and its own cause, does not prove that the stars are its cause, nor that they are in anywise connected with it. As strong a case of coincidence, therefore, as can possibly exist, and a much stronger one in point of mere frequency than most of those which prove laws, does not here prove a law: why? because, since the stars exist always, they *must* coexist with every other phenomenon, whether connected with them by causation or not. The uniformity, great though it be, is no greater than would occur on the supposition that no such connexion exists.

On the other hand, suppose that we were inquiring whether

there be any connexion between rain and any particular wind.
Rain, we know, occasionally occurs with every wind; therefore the connexion, if it exists, cannot be an actual law; but
still, rain may be connected with some particular wind through
causation; that is, though they cannot be always effects of the
same cause (for if so they would regularly coexist), there may
be some causes common to the two, so that in so far as either
is produced by those common causes, they will, from the laws
of the causes, be found to coexist. How, then, shall we ascertain this? The obvious answer is, by observing whether rain
occurs with one wind more frequently than with any other.
That, however, is not enough; for perhaps that one wind
blows more frequently than any other; so that its blowing
more frequently in rainy weather is no more than would
happen, although it had no connexion with the causes of rain,
provided it were not connected with causes adverse to rain. In
England, westerly winds blow during about twice as great a
portion of the year as easterly. If, therefore, it rains only
twice as often with a westerly, as with an easterly wind, we
have no reason to infer that any law of nature is concerned in
the coincidence. If it rains more than twice as often, we may
be sure that some law is concerned; either there is some cause
in nature which, in this climate, tends to produce both rain
and a westerly wind, or a westerly wind has itself some tendency to produce rain. But if it rains less than twice as often,
we may draw a directly opposite inference: the one, instead of
being a cause, or connected with causes, of the other, must be
connected with causes adverse to it, or with the absence of
some cause which produces it; and though it may still rain
much oftener with a westerly wind than with an easterly, so
far would this be from proving any connexion between the
phenomena, that the connexion proved would be between rain
and an easterly wind, to which, in mere frequency of coincidence, it is less allied.

Here, then, are two examples: in one, the greatest possible frequency of coincidence, with no instance whatever to
the contrary, does not prove that there is any law; in the
other, a much less frequency of coincidence, even when non-

coincidence is still more frequent, does prove that there is a law. In both cases the principle is the same. In both we consider the positive frequency of the phenomena themselves, and how great frequency of coincidence that must of itself bring about, without supposing any connexion between them, provided there be no repugnance; provided neither be connected with any cause tending to frustrate the other. If we find a greater frequency of coincidence than this, we conclude that there is some connexion; if a less frequency, that there is some repugnance. In the former case, we conclude that one of the phenomena can under some circumstances cause the other, or that there exists something capable of causing them both; in the latter, that one of them, or some cause which produces one of them, is capable of counteracting the production of the other. We have thus to deduct from the observed frequency of coincidence, as much as may be the effect of chance, that is, of the mere frequency of the phenomena themselves; and if anything remains, what does remain is the residual fact which proves the existence of a law.

The frequency of the phenomena can only be ascertained within definite limits of space and time; depending as it does on the quantity and distribution of the primeval natural agents, of which we can know nothing beyond the boundaries of human observation, since no law, no regularity, can be traced in it, enabling us to infer the unknown from the known. But for the present purpose this is no disadvantage, the question being confined within the same limits as the data. The coincidences occurred in certain places and times, and within those we can estimate the frequency with which such coincidences would be produced by chance. If, then, we find from observation that A exists in one case out of every two, and B in one case out of every three; then if there be neither connexion nor repugnance between them, or between any of their causes, the instances in which A and B will both exist, that is to say will coexist, will be one case in every six. For A exists in three cases out of six: and B, existing in one case out of every three without

regard to the presence or absence of A, will exist in one
case out of those three. There will therefore be, of the
whole number of cases, two in which A exists without B;
one case of B without A; two in which neither B nor A
exists, and one case out of six in which they both exist.
If then, in point of fact, they are found to coexist oftener
than in one case out of six; and, consequently, A does not
exist without B so often as twice in three times, nor B without A so often as once in every twice; there is some cause
in existence which tends to produce a conjunction between A
and B.

Generalizing the result, we may say, that if A occurs in
a larger proportion of the cases where B is, than of the cases
where B is not; then will B also occur in a larger proportion
of the cases where A is, than of the cases where A is not; and
there is some connexion, through causation, between A and B.
If we could ascend to the causes of the two phenomena, we
should find, at some stage, either proximate or remote, some
cause or causes common to both; and if we could ascertain
what these are, we could frame a generalization which would
be true without restriction of place or time: but until we can
do so, the fact of a connexion between the two phenomena
remains an empirical law.

§ 8. Having considered in what manner it may be determined whether any given conjunction of phenomena is casual,
or the result of some law; to complete the theory of chance,
it is necessary that we should now consider those effects which
are partly the result of chance and partly of law, or, in other
words, in which the effects of casual conjunctions of causes
are habitually blended in one result with the effects of a
constant cause.

This is a case of Composition of Causes; and the peculiarity of it is, that instead of two or more causes intermixing
their effects in a regular manner with those of one another,
we have now one constant cause, producing an effect which
is successively modified by a series of variable causes. Thus,
as summer advances, the approach of the sun to a vertical

position tends to produce a constant increase of temperature ;
but with this effect of a constant cause, there are blended the
effects of many variable causes, winds, clouds, evaporation,
electric agencies and the like, so that the temperature of any
given day depends in part on these fleeting causes, and only in
part on the constant cause.   If the effect of the constant cause
is always accompanied and disguised by effects of variable
causes, it is impossible to ascertain the law of the constant
cause in the ordinary manner, by separating it from all other
causes and observing it apart.   Hence arises the necessity of
an additional rule of experimental inquiry.

When the action of a cause A is liable to be interfered
with, not steadily by the same cause or causes, but by different causes at different times, and when these are so frequent,
or so indeterminate, that we cannot possibly exclude all of
them from any experiment, though we may vary them ;
our resource is, to endeavour to ascertain what is the effect
of all the variable causes taken together.   In order to do
this, we make as many trials as possible, preserving A invariable.   The results of these different trials will naturally
be different, since the indeterminate modifying causes are
different in each : if, then, we do not find these results to be
progressive, but, on the contrary, to oscillate about a certain
point, one experiment giving a result a little greater, another
a little less, one a result tending a little more in one direction,
another a little more in the contrary direction ; while the
average or middle point does not vary, but different sets of
experiments (taken in as great a variety of circumstances as
possible) yield the same mean, provided only they be sufficiently numerous ; then that mean or average result, is the
part, in each experiment, which is due to the cause A, and
is the effect which would have been obtained if A could have
acted alone : the variable remainder is the effect of chance,
that is, of causes the coexistence of which with the cause A
was merely casual.   The test of the sufficiency of the induction in this case is, when any increase of the number of trials
from which the average is struck, does not materially alter the
average.

This kind of elimination, in which we do not eliminate any one assignable cause, but the multitude of floating unassignable ones, may be termed the Elimination of Chance. We afford an example of it when we repeat an experiment, in order, by taking the mean of different results, to get rid of the effects of the unavoidable errors of each individual experiment. When there is no permanent cause such as would produce a tendency to error peculiarly in one direction, we are warranted by experience in assuming that the errors on one side will, in a certain number of experiments, about balance the errors on the contrary side. We therefore repeat the experiment, until any change which is produced in the average of the whole by further repetition, falls within limits of error consistent with the degree of accuracy required by the purpose we have in view.*

§ 4. In the supposition hitherto made, the effect of the constant cause A has been assumed to form so great and conspicuous a part of the general result, that its existence never could be a matter of uncertainty, and the object of the eliminating process was only to ascertain *how much* is attributable to that cause; what is its exact law. Cases, however, occur in which the effect of a constant cause is so small, compared with that of some of the changeable causes with which

---

* In the preceding discussion, the *mean* is spoken of as if it were exactly the same thing with the *average*. But the mean for purposes of inductive inquiry, is not the average, or arithmetical mean, though in a familiar illustration of the theory the difference may be disregarded. If the deviations on one side of the average are much more numerous than those on the other (these last being fewer but greater), the effect due to the invariable cause, as distinct from the variable ones, will not coincide with the average, but will be either below or above the average, whichever be the side on which the greatest number of the instances are found. This follows from a truth, ascertained both inductively and deductively, that small deviations from the true central point are greatly more frequent than large ones. The mathematical law is, "that the most probable determination of one or more invariable elements from observation is that in which *the sum of the squares* of the individual aberrations," or deviations, "*shall be the least possible.*" See this principle stated, and its grounds popularly explained, by Sir John Herschel, in his review of Quetelet on Probabilities, *Essays*, pp. 395 *et seq.*

it is liable to be casually conjoined, that of itself it escapes
notice, and the very existence of any effect arising from a
constant cause is first learnt, by the process which in general
serves only for ascertaining the quantity of that effect. This
case of induction may be characterized as follows. A given
effect is known to be chiefly, and not known not to be wholly,
determined by changeable causes. If it be wholly so produced, then if the aggregate be taken of a sufficient number of
instances, the effects of these different causes will cancel one
another. If, therefore, we do not find this to be the case, but,
on the contrary, after such a number of trials has been made
that no further increase alters the average result, we find that
average to be, not zero, but some other quantity, about which,
though small in comparison with the total effect, the effect
nevertheless oscillates, and which is the middle point in its
oscillation; we may conclude this to be the effect of some
constant cause: which cause, by some of the methods already
treated of, we may hope to detect. This may be called *the
discovery of a residual phenomenon by eliminating the effects
of chance.*

It is in this manner, for example, that loaded dice may
be discovered. Of course no dice are so clumsily loaded that
they must always throw certain numbers; otherwise the fraud
would be instantly detected. The loading, a constant cause,
mingles with the changeable causes which determine what
cast will be thrown in each individual instance. If the dice
were not loaded, and the throw were left to depend entirely
on the changeable causes, these in a sufficient number of instances would balance one another, and there would be no
preponderant number of throws of any one kind. If, therefore, after such a number of trials that no further increase of
their number has any material effect upon the average, we find
a preponderance in favour of a particular throw; we may conclude with assurance that there is some constant cause acting
in favour of that throw, or in other words, that the dice are
not fair; and the exact amount of the unfairness. In a
similar manner, what is called the diurnal variation of the
barometer, which is very small compared with the variations

arising from the irregular changes in the state of the atmosphere, was discovered by comparing the average height of the barometer at different hours of the day. When this comparison was made, it was found that there was a small difference, which on the average was constant, however the absolute quantities might vary, and which difference, therefore, must be the effect of a constant cause. This cause was afterwards ascertained, deductively, to be the rarefaction of the air, occasioned by the increase of temperature as the day advances.

§ 5. After these general remarks on the nature of chance, we are prepared to consider in what manner assurance may be obtained that a conjunction between two phenomena, which has been observed a certain number of times, is not casual, but a result of causation, and to be received therefore as one of the uniformities of nature, though (until accounted for *à priori*) only as an empirical law.

We will suppose the strongest case, namely, that the phenomenon B has never been observed except in conjunction with A. Even then, the probability that they are connected is not measured by the total number of instances in which they have been found together, but by the excess of that number above the number due to the absolute frequency of A. If, for example, A exists always, and therefore coexists with everything, no number of instances of its coexistence with B would prove a connexion; as in our example of the fixed stars. If A be a fact of such common occurrence that it may be presumed to be present in half of all the cases that occur, and therefore in half the cases in which B occurs, it is only the proportional excess above half, that is to be reckoned as evidence towards proving a connexion between A and B.

In addition to the question, What is the number of coincidences which, on an average of a great multitude of trials, may be expected to arise from chance alone? there is also another question, namely, Of what extent of deviation from that average is the occurrence credible, from chance alone, in some number of instances smaller than that required for striking a fair average? It is not only to be considered what is the general result of the chances in the long run, but also what are the extreme limits of variation from the general result, which may occasionally be expected as the result of some smaller number of instances.

The consideration of the latter question, and any consideration of the former beyond that already given to it, belong to what mathematicians term the doctrine of chances, or, in a phrase of greater pretension, the Theory of Probabilities.

CHAPTER XVIII.

OF THE CALCULATION OF CHANCES.

§ 1. "PROBABILITY," says Laplace,* " has reference partly to our ignorance, partly to our knowledge. We know that among three or more events, one, and only one, must happen; but there is nothing leading us to believe that any one of them will happen rather than the others. In this state of indecision, it is impossible for us to pronounce with certainty on their occurrence. It is, however, probable that any one of these events, selected at pleasure, will not take place; because we perceive several cases, all equally possible, which exclude its occurrence, and only one which favours it.

"The theory of chances consists in reducing all events of the same kind to a certain number of cases equally possible, that is, such that we are *equally undecided* as to their existence; and in determining the number of these cases which are favourable to the event of which the probability is sought. The ratio of that number to the number of all the possible cases, is the measure of the probability; which is thus a fraction, having for its numerator the number of cases favourable to the event, and for its denominator the number of all the cases which are possible."

To a calculation of chances, then, according to Laplace, two things are necessary: we must know that of several events some one will certainly happen, and no more than one; and we must not know, nor have any reason to expect, that it will be one of these events rather than another. It has been contended that these are not the only requisites, and that Laplace has overlooked, in the general theoretical statement, a necessary part of the foundation of the doctrine of chances.

* *Essai Philosophique sur les Probabilités*, fifth Paris Edition, p. 7.

To be able (it has been said) to pronounce two events equally probable, it is not enough that we should know that one or the other must happen, and should have no grounds for conjecturing which. Experience must have shown that the two events are of equally frequent occurrence. Why, in tossing up a halfpenny, do we reckon it equally probable that we shall throw cross or pile? Because we know that in any great number of throws, cross and pile are thrown about equally often; and that the more throws we make, the more nearly the equality is perfect. We may know this if we please by actual experiment; or by the daily experience which life affords of events of the same general character; or deductively, from the effect of mechanical laws on a symmetrical body acted upon by forces varying indefinitely in quantity and direction. We may know it, in short, either by specific experience, or on the evidence of our general knowledge of nature. But, in one way or the other, we must know it, to justify us in calling the two events equally probable; and if we knew it not, we should proceed as much at haphazard in staking equal sums on the result, as in laying odds.

This view of the subject was taken in the first edition of the present work: but I have since become convinced, that the theory of chances, as conceived by Laplace and by mathematicians generally, has not the fundamental fallacy which I had ascribed to it.

We must remember that the probability of an event is not a quality of the event itself, but a mere name for the degree of ground which we, or some one else, have for expecting it. The probability of an event to one person is a different thing from the probability of the same event to another, or to the same person after he has acquired additional evidence. The probability to me, that an individual of whom I know nothing but his name, will die within the year, is totally altered by my being told, the next minute, that he is in the last stage of a consumption. Yet this makes no difference in the event itself, nor in any of the causes on which it depends. Every event is in itself certain, not probable: if we knew all, we should either know positively that it will happen, or positively that it will not. But its probability to us means the degree of expectation of its occurrence, which we are warranted in entertaining by our present evidence.

Bearing this in mind, I think it must be admitted, that even when we have no knowledge whatever to guide our expectations, except the knowledge that what happens must be some one of a certain number of possibilities, we may still reasonably judge, that one supposition is more probable *to us* than another supposition; and if we have any interest at stake, we shall best provide for it by acting conformably to that judgment.

§ 2. Suppose that we are required to take a ball from a box, of which we only know that it contains balls both black and white, and none of any other colour. We know that the ball we select will be either a black or a white ball; but we have no ground for expecting black rather than white, or white rather than black. In that case, if we are obliged to make a choice, and to stake something on one or the other supposition, it will, as a question of prudence, be perfectly indifferent which; and we shall act precisely as we should have acted if we had known beforehand that the box contained an equal number of black and white balls. But though our conduct would be the same, it would not be founded on any surmise that the balls were in fact thus equally divided; for we might, on the contrary, know, by authentic information, that the box contained ninety-nine balls of one colour, and only one of the other; still, if we are not told which colour has only one, and which has ninety-nine, the drawing of a white and of a black ball will be equally probable to us; we shall have no reason for staking anything on the one event rather than on the other; the option between the two will be a matter of indifference; in other words it will be an even chance.

But let it now be supposed that instead of two there are three colours—white, black, and red; and that we are entirely ignorant of the proportion in which they are mingled. We should then have no reason for expecting one more than

another, and if obliged to bet, should venture our stake on
red, white, or black, with equal indifference. But should we
be indifferent whether we betted for or against some one
colour, as, for instance, white? Surely not. From the very
fact that black and red are each of them separately equally
probable to us with white, the two together must be twice as
probable. We should in this case expect not-white rather
than white, and so much rather, that we would lay two to one
upon it. It is true, there might for aught we knew be more
white balls than black and red together; and if so, our bet
would, if we knew more, be seen to be a disadvantageous one.
But so also, for aught we knew, might there be more red balls
than black and white, or more black balls than white and red,
and in such case the effect of additional knowledge would be
to prove to us that our bet was more advantageous than we
had supposed it to be. There is in the existing state of our
knowledge a rational probability of two to one against white;
a probability fit to be made a basis of conduct. No reasonable
person would lay an even wager in favour of white, against
black and red; though against black alone, or red alone, he
might do so without imprudence.

The common theory, therefore, of the calculation of chances,
appears to be tenable. Even when we know nothing except
the number of the possible and mutually excluding contingencies, and are entirely ignorant of their comparative frequency, we may have grounds, and grounds numerically
appreciable, for acting on one supposition rather than on
another; and this is the meaning of Probability.

§ 3. The principle, however, on which the reasoning
proceeds, is sufficiently evident. It is the obvious one, that
when the cases which exist are shared among several kinds,
it is impossible that *each* of those kinds should be a majority
of the whole: on the contrary, there must be a majority
against each kind, except one at most; and if any kind has
more than its share in proportion to the total number, the
others collectively must have less. Granting this axiom, and
assuming that we have no ground for selecting any one kind

as more likely than the rest to surpass the average proportion, it follows that we cannot rationally presume this of any; which we should do, if we were to bet in favour of it, receiving less odds than in the ratio of the number of the other kinds. Even, therefore, in this extreme case of the calculation of probabilities, which does not rest on special experience at all, the logical ground of the process is our knowledge, such knowledge as we then have, of the laws governing the frequency of occurrence of the different cases; but in this case the knowledge is limited to that which, being universal and axiomatic, does not require reference to specific experience, or to any considerations arising out of the special nature of the problem under discussion.

Except, however, in such cases as games of chance, where the very purpose in view requires ignorance instead of knowledge, I can conceive no case in which we ought to be satisfied with such an estimate of chances as this; an estimate founded on the absolute minimum of knowledge respecting the subject. It is plain that, in the case of the coloured balls, a very slight ground of surmise that the white balls were really more numerous than either of the other colours, would suffice to vitiate the whole of the calculations made in our previous state of indifference. It would place us in that position of more advanced knowledge, in which the probabilities, to us, would be different from what they were before; and in estimating these new probabilities we should have to proceed on a totally different set of data, furnished no longer by mere counting of possible suppositions, but by specific knowledge of facts. Such data it should always be our endeavour to obtain; and in all inquiries, unless on subjects equally beyond the range of our means of knowledge and our practical uses, they may be obtained, if not good, at least better than none at all.*

---

* It even appears to me that the calculation of chances, where there are no data grounded either on special experience or on special inference, must, in an immense majority of cases, break down, from sheer impossibility of assigning any principle by which to be guided in setting out the list of possibilities. In the case of the coloured balls we have no difficulty in making

It is obvious, too, that even when the probabilities are
derived from observation and experiment, a very slight improvement in the data, by better observations, or by taking
into fuller consideration the special circumstances of the case,
is of more use than the most elaborate application of the
calculus to probabilities founded on the data in their previous
state of inferiority.   The neglect of this obvious reflection has
given rise to misapplications of the calculus of probabilities
which have made it the real opprobrium of mathematics.   It
is sufficient to refer to the applications made of it to the credibility of witnesses, and to the correctness of the verdicts of
juries.   In regard to the first, common sense would dictate
that it is impossible to strike a general average of the veracity,
and other qualifications for true testimony, of mankind, or of
any class of them ; and even if it were possible, the employment of it for such a purpose implies a misapprehension of
the use of averages : which serve indeed to protect those whose
interest is at stake, against mistaking the general result of
large masses of instances, but are of extremely small value as
grounds of expectation in any one individual instance, unless
the case be one of those in which the great majority of individual instances do not differ much from the average.   In the
case of a witness, persons of common sense would draw their
conclusions from the degree of consistency of his statements,
his conduct under cross-examination, and the relation of the
case itself to his interests, his partialities, and his mental

---

the enumeration, because we ourselves determine what the possibilities shall
be. But suppose a case more analogous to those which occur in nature : instead
of three colours, let there be in the box all possible colours : we being supposed
ignorant of the comparative frequency with which different colours occur in
nature, or in the productions of art.   How is the list of cases to be made out ?
Is every distinct shade to count as a colour ? If so, is the test to be a common
eye, or an educated eye, a painter's for instance ? On the answer to these
questions would depend whether the chances against some particular colour
would be estimated at ten, twenty, or perhaps five hundred to one.   While
if we knew from experience that the particular colour occurs on an average a
certain number of times in every hundred or thousand, we should not require
to know anything either of the frequency or of the number of the other possibilities.

capacity, instead of applying so rude a standard (even if it were capable of being verified) as the ratio between the number of true and the number of erroneous statements which he may be supposed to make in the course of his life.

Again, on the subject of juries, or other tribunals, some mathematicians have set out from the proposition that the judgment of any one judge, or juryman, is, at least in some small degree, more likely to be right than wrong, and have concluded that the chance of a number of persons concurring in a wrong verdict is diminished, the more the number is increased; so that if the judges are only made sufficiently numerous, the correctness of the judgment may be reduced almost to certainty. I say nothing of the disregard shown to the effect produced on the moral position of the judges by multiplying their numbers; the virtual destruction of their individual responsibility, and weakening of the application of their minds to the subject. I remark only the fallacy of reasoning from a wide average, to cases necessarily differing greatly from any average. It may be true that taking all causes one with another, the opinion of any one of the judges would be oftener right than wrong; but the argument forgets that in all but the more simple cases, in all cases in which it is really of much consequence what the tribunal is, the proposition might probably be reversed; besides which, the cause of error, whether arising from the intricacy of the case or from some common prejudice or mental infirmity, if it acted upon one judge, would be extremely likely to affect all the others in the same manner, or at least a majority, and thus render a wrong instead of a right decision more probable, the more the number was increased.

These are but samples of the errors frequently committed by men who, having made themselves familiar with the difficult formulæ which algebra affords for the estimation of chances under suppositions of a complex character, like better to employ those formulæ in computing what are the probabilities to a person half informed about a case, than to look out for means of being better informed. Before applying the doctrine of chances to any scientific purpose, the foundation must be laid for an evaluation of the chances, by possessing ourselves of the utmost attainable amount of positive knowledge. The knowledge required is that of the comparative frequency with which the different events in fact occur. For the purposes, therefore, of the present work, it is allowable to suppose, that conclusions respecting the probability of a fact of a particular kind, rest on our knowledge of the proportion between the cases in which facts of that kind occur, and those in which they do not occur: this knowledge being either derived from specific experiment, or deduced from our knowledge of the causes in operation which tend to produce, compared with those which tend to prevent, the fact in question.

Such calculation of chances is grounded on an induction; and to render the calculation legitimate, the induction must be a valid one. It is not less an induction, though it does not prove that the event occurs in all cases of a given description, but only that out of a given number of such cases, it occurs in about so many. The fraction which mathematicians use to designate the probability of an event, is the ratio of these two numbers; the ascertained proportion between the number of cases in which the event occurs, and the sum of all the cases, those in which it occurs and in which it does not occur taken together. In playing at cross and pile, the description of cases concerned are throws, and the probability of cross is one-half, because if we throw often enough, cross is thrown about once in every two throws. In the cast of a die, the probability of ace is one-sixth; not simply because there are six possible throws, of which ace is one, and because we do not know any reason why one should turn up rather than another; though I have admitted the validity of this ground in default of a better; but because we do actually know, either by reasoning or by experience, that in a hundred, or a million of throws, ace is thrown about one-sixth of that number, or once in six times.

§ 4. I say, "either by reasoning or by experience;" meaning specific experience. But in estimating probabilities, it is not a matter of indifference from which of these two sources we derive our assurance. The probability of events as calculated from their mere frequency in past experience, affords a less secure basis for practical guidance, than their probability as deduced from an equally accurate knowledge of the frequency of occurrence of their causes.

The generalization, that an event occurs in ten out of every hundred cases of a given description, is as real an induction as if the generalization were that it occurs in all cases. But when we arrive at the conclusion by merely counting instances in actual experience, and comparing the number of cases in which A has been present with the number in which it has been absent, the evidence is only that of the method of Agreement, and the conclusion amounts only to an empirical law. We can make a step beyond this when we can ascend to the causes on which the occurrence of A or its non-occurrence will depend, and form an estimate of the comparative frequency of the causes favourable and of those unfavourable to the occurrence. These are data of a higher order, by which the empirical law derived from a mere numerical comparison of affirmative and negative instances will be either corrected or confirmed, and in either case we shall obtain a more correct measure of probability than is given by that numerical comparison. It has been well remarked that in the kind of examples by which the doctrine of chances is usually illustrated, that of balls in a box, the estimate of probabilities is supported by reasons of causation, stronger than specific experience. "What is the reason that in a box where there are nine black balls and one white, we expect to draw a black ball nine times as much (in other words, nine times as often, frequency being the gauge of intensity in expectation) as a white? Obviously because the local conditions are nine times as favourable, because the hand may alight in nine places and get a black ball, while it can only alight in one place and find a white ball; just for the same reason that we do not expect to succeed in finding a friend in a crowd, the conditions in order that we and he should come together being many and difficult. This of course would not hold to the same extent were the white balls of smaller size than the black, neither would the probability remain the same: the larger ball would be much more likely to meet the hand."*

It is, in fact, evident, that when once causation is admitted as an universal law, our expectation of events can only be rationally grounded on that law. To a person who recognises that every event depends on causes, a thing's having happened once is a reason for expecting it to happen again, only because proving that there exists, or is liable to exist, a cause adequate to produce it.† The frequency of the particular event, apart from all surmise respecting its cause, can give rise to no other induction than that *per enumerationem simplicem;* and the precarious inferences derived from this, are superseded, and disappear from the field, as soon as the principle of causation makes its appearance there.

Notwithstanding, however, the abstract superiority of an estimate of probability grounded on causes, it is a fact that in almost all cases in which chances admit of estimation suffi-

---

* *Prospective Review* for February 1850.

† "If this be not so, why do we feel so much more probability added by the first instance, than by any single subsequent instance? Why, except that the first instance gives us its possibility (a cause *adequate* to it), while every other only gives us the frequency of its conditions? If no reference to a cause be supposed, possibility would have no meaning; yet it is clear, that, antecedent to its happening, we might have supposed the event impossible, *i.e.*, have believed that there was no physical energy really existing in the world equal to producing it. . . . . After the first time of happening, which is, then, more important to the whole probability than any other single instance (because proving the possibility), the *number* of times becomes important as an index to the intensity or extent of the cause, and its independence of any particular time. If we took the case of a tremendous leap, for instance, and wished to form an estimate of the probability of its succeeding a certain number of times; the first instance, by showing its possibility (before doubtful) is of the most importance; but every succeeding leap shows the power to be more perfectly under control, greater and more invariable, and so increases the probability; and no one would think of reasoning in this case straight from one instance to the next, without referring to the physical energy which each leap indicated. Is it not then clear that we do not ever" (let us rather say, that we do not in an advanced state of our knowledge) "conclude directly from the happening of an event to the probability of its happening again; but that we refer to the cause, regarding the past cases as an index to the cause, and the cause as our guide to the future?"—*Ibid.*

ciently precise to render their numerical appreciation of any practical value, the numerical data are not drawn from knowledge of the causes, but from experience of the events themselves. The probabilities of life at different ages, or in different climates; the probabilities of recovery from a particular disease; the chances of the birth of male or female offspring; the chances of the destruction of houses or other property by fire; the chances of the loss of a ship in a particular voyage; are deduced from bills of mortality, returns from hospitals, registers of births, of shipwrecks, &c., that is, from the observed frequency not of the causes, but of the effects. The reason is, that in all these classes of facts, the causes are either not amenable to direct observation at all, or not with the requisite precision, and we have no means of judging of their frequency except from the empirical law afforded by the frequency of the effects. The inference does not the less depend on causation alone. We reason from an effect to a similar effect by passing through the cause. If the actuary of an insurance office infers from his tables that among a hundred persons now living, of a particular age, five on the average will attain the age of seventy, his inference is legitimate, not for the simple reason that this is the proportion who have lived till seventy in times past, but because the fact of their having so lived shows that this is the proportion existing, at that place and time, between the causes which prolong life to the age of seventy, and those tending to bring it to an earlier close.*

---

* The writer last quoted says that the valuation of chances by comparing the number of cases in which the event occurs with the number in which it does not occur, "would generally be wholly erroneous," and "is not the true theory of probability." It is at least that which forms the foundation of insurance, and of all those calculations of chances in the business of life which experience so abundantly verifies. The reason which the reviewer gives for rejecting the theory, is that it "would regard an event as certain which had hitherto never failed; which is exceedingly far from the truth, even for a very large number of constant successes." This is not a defect in a particular theory, but in any theory of chances. No principle of evaluation can provide for such a case as that which the reviewer supposes. If an event has never once failed, in a number of trials sufficient to eliminate chance, it really has all the certainty which can be given by an empirical law: it *is* certain during the

§ 5. From the preceding principles it is easy to deduce
the demonstration of that theorem of the doctrine of probabilities, which is the foundation of its application to inquiries
for ascertaining the occurrence of a given event, or the reality
of an individual fact. The signs or evidences by which a fact
is usually proved, are some of its consequences: and the inquiry hinges upon determining what cause is most likely to
have produced a given effect. The theorem applicable to
such investigations is the Sixth Principle in Laplace's *Essai
Philosophique sur les Probabilités*, which is described by him
as the " fundamental principle of that branch of the Analysis
of Chances, which consists in ascending from events to their
causes."*

Given an effect to be accounted for, and there being several
causes which might have produced it, but of the presence of
which in the particular case nothing is known; the probability that the effect was produced by any one of these causes
*is as the antecedent probability of the cause, multiplied by the
probability that the cause, if it existed, would have produced
the given effect.*

Let M be the effect, and A, B, two causes, by either of
which it might have been produced. To find the probability
that it was produced by the one and not by the other, ascertain which of the two is most likely to have existed, and
which of them, if it did exist, was most likely to produce the
effect M: the probability sought is a compound of these two
probabilities.

CASE I. Let the causes be both alike in the second
respect; either A or B, when it exists, being supposed
equally likely (or equally certain) to produce M; but let A
be in itself twice as likely as B to exist, that is, twice as

continuance of the same collocation of causes which existed during the observations. If it ever fails, it is in consequence of some change in that collocation.
Now, no theory of chances will enable us to infer the future probability of an
event from the past, if the causes in operation, capable of influencing the
event, have intermediately undergone a change.

* Pp. 18, 19. The theorem is not stated by Laplace in the exact terms in
which I have stated it; but the identity of import of the two modes of expression is easily demonstrable.

frequent a phenomenon. Then it is twice as likely to have existed in this case, and to have been the cause which produced M.

For, since A exists in nature twice as often as B; in any 300 cases in which one or other existed, A has existed 200 times and B 100. But either A or B must have existed wherever M is produced: therefore in 300 times that M is produced, A was the producing cause 200 times, B only 100, that is, in the ratio of 2 to 1. Thus, then, if the causes are alike in their capacity of producing the effect, the probability as to which actually produced it, is in the ratio of their antecedent probabilities.

CASE II. Reversing the last hypothesis, let us suppose that the causes are equally frequent, equally likely to have existed, but not equally likely, if they did exist, to produce M: that in three times in which A occurs, it produces that effect twice, while B, in three times, produces it only once. Since the two causes are equally frequent in their occurrence; in every six times that either one or the other exists, A exists three times and B three times. A, of its three times, produces M in two; B, of its three times, produces M in one. Thus, in the whole six times, M is only produced thrice; but of that thrice it is produced twice by A, once only by B. Consequently, when the antecedent probabilities of the causes are equal, the chances that the effect was produced by them are in the ratio of the probabilities that if they did exist they would produce the effect.

CASE III. The third case, that in which the causes are unlike in both respects, is solved by what has preceded. For, when a quantity depends on two other quantities, in such a manner that while either of them remains constant it is proportional to the other, it must necessarily be proportional to the product of the two quantities, the product being the only function of the two which obeys that law of variation. Therefore, the probability that M was produced by either cause, is as the antecedent probability of the cause, multiplied by the probability that if it existed it would produce M. Which was to be demonstrated.

Or we may prove the third case as we proved the first and second. Let A be twice as frequent as B ; and let them also be unequally likely, when they exist, to produce M: let A produce it twice in four times, B thrice in four times. The antecedent probability of A is to that of B as 2 to 1 ; the probabilities of their producing M are as 2 to 3 ; the product of these ratios is the ratio of 4 to 3 : and this will be the ratio of the probabilities that A or B was the producing cause in the given instance. For, since A is twice as frequent as B, out of twelve cases in which one or other exists, A exists in 8 and B in 4. But of its eight cases, A, by the supposition, produces M in only 4, while B of its four cases produces M in 3. M, therefore, is only produced at all in seven of the twelve cases ; but in four of these it is produced by A, in three by B : hence, the probabilities of its being produced by A and by B are as 4 to 3, and are expressed by the fractions $\frac{4}{7}$ and $\frac{3}{7}$. Which was to be demonstrated.

§ 6. It remains to examine the bearing of the doctrine of chances on the peculiar problem which occupied us in the preceding chapter, namely, how to distinguish coincidences which are casual from those which are the result of law; from those in which the facts which accompany or follow one another are somehow connected through causation.

The doctrine of chances affords means by which, if we knew the *average* number of coincidences to be looked for between two phenomena connected only casually, we could determine how often any given deviation from that average will occur by chance. If the probability of any casual coincidence, considered in itself, be $\frac{1}{m}$, the probability that the same coincidence will be repeated n times in succession is $\frac{1}{m^n}$. For example, in one throw of a die the probability of ace being $\frac{1}{6}$; the probability of throwing ace twice in succession will be 1 divided by the square of 6, or $\frac{1}{36}$. For ace is

thrown at the first throw once in six, or six in thirty-six
times, and of those six, the die being cast again, ace will be
thrown but once; being altogether once in thirty-six times.
The chance of the same cast three times successively is, by
a similar reasoning, $\frac{1}{6^3}$ or $\frac{1}{216}$: that is, the event will happen,
on a large average, only once in two hundred and sixteen
throws.

We have thus a rule by which to estimate the probability
that any given series of coincidences arises from chance;
provided we can measure correctly the probability of a single
coincidence. If we can obtain an equally precise expression
for the probability that the same series of coincidences arises
from causation, we should only have to compare the numbers.
This however, can rarely be done. Let us see what degree of
approximation can practically be made to the necessary
precision.

The question falls within Laplace's sixth principle, just
demonstrated. The given fact, that is to say, the series of
coincidences, may have originated either in a casual conjunction of causes, or in a law of nature. The probabilities,
therefore, that the fact originated in these two modes, are as
their antecedent probabilities, multiplied by the probabilities
that if they existed they would produce the effect. But the
particular combination of chances, if it occurred, or the law of
nature if real, would certainly produce the series of coincidences. The probabilities, therefore, that the coincidences
are produced by the two causes in question, are as the antecedent probabilities of the causes. One of these, the antecedent probability of the combination of mere chances which
would produce the given result, is an appreciable quantity.
The antecedent probability of the other supposition may be
susceptible of a more or less exact estimation, according to
the nature of the case.

In some cases, the coincidence, supposing it to be the
result of causation at all, must be the result of a known
cause: as the succession of aces, if not accidental, must arise
from the loading of the die. In such cases we may be able to

form a conjecture as to the antecedent probability of such a circumstance, from the characters of the parties concerned, or other such evidence; but it would be impossible to estimate that probability with anything like numerical precision. The counter-probability, however, that of the accidental origin of the coincidence, dwindling so rapidly as it does at each new trial; the stage is soon reached at which the chance of unfairness in the die, however small in itself, must be greater than that of a casual coincidence : and on this ground, a practical decision can generally be come to without much hesitation, if there be the power of repeating the experiment.

When, however, the coincidence is one which cannot be accounted for by any known cause, and the connexion between the two phenomena, if produced by causation, must be the result of some law of nature hitherto unknown ; which is the case we had in view in the last chapter; then, though the probability of a casual coincidence may be capable of appreciation, that of the counter-supposition, the existence of an undiscovered law of nature, is clearly unsusceptible of even an approximate valuation. In order to have the data which such a case would require, it would be necessary to know what proportion of all the individual sequences or coexistences occurring in nature are the result of law, and what proportion are mere casual coincidences. It being evident that we cannot form any plausible conjecture as to this proportion, much less appreciate it numerically, we cannot attempt any precise estimation of the comparative probabilities. But of this we are sure, that the detection of an unknown law of nature—of some previously unrecognised constancy of conjunction among phenomena—is no uncommon event. If, therefore, the number of instances in which a coincidence is observed, over and above that which would arise on the average from the mere concurrence of chances, be such that so great an amount of coincidences from accident alone would be an extremely uncommon event; we have reason to conclude that the coincidence is the effect of causation, and may be received (subject to correction from further experience) as an empirical law. Further than this, in point of precision, we cannot go; nor, in most cases, is greater precision required, for the solution of any practical doubt.*

---

* For a fuller treatment of the many interesting questions raised by the theory of probabilities, I may now refer to a recent work by Mr. Venn, Fellow of Caius College, Cambridge, "The Logic of Chance;" one of the most thoughtful and philosophical treatises on any subject connected with Logic and Evidence, which have been produced in this or any other country for many years. Some criticisms contained in it have been very useful to me in revising the corresponding chapters of the present work. In several of Mr. Venn's opinions, however, I do not agree. What these are will be obvious to any reader of Mr. Venn's work who is also a reader of this.

CHAPTER XIX.

OF THE EXTENSION OF DERIVATIVE LAWS TO ADJACENT CASES.

§ 1. WE have had frequent occasion to notice the inferior generality of derivative laws, compared with the ultimate laws from which they are derived. This inferiority, which affects not only the extent of the propositions themselves, but their degree of certainty within that extent, is most conspicuous in the uniformities of coexistence and sequence obtaining between effects which depend ultimately on different primeval causes. Such uniformities will only obtain where there exists the same collocation of those primeval causes. If the collocation varies, though the laws themselves remain the same, a totally different set of derivative uniformities may, and generally will, be the result.

Even where the derivative uniformity is between different effects of the same cause, it will by no means obtain as universally as the law of the cause itself. If *a* and *b* accompany or succeed one another as effects of the cause A, it by no means follows that A is the only cause which can produce them, or that if there be another cause, as B, capable of producing *a*, it must produce *b* likewise. The conjunction therefore of *a* and *b* perhaps does not hold universally, but only in the instances in which *a* arises from A. When it is produced by a cause other than A, *a* and *b* may be dissevered. Day (for example) is always in our experience followed by night; but day is not the cause of night; both are successive effects of a common cause, the periodical passage of the spectator into and out of the earth's shadow, consequent on the earth's rotation, and on the illuminating property of the sun. If, therefore, day is ever produced by a different cause or set of causes from this, day will not, or at least may not, be followed by night. On the sun's own surface, for instance, this may be the case.

Finally, even when the derivative uniformity is itself a law of causation (resulting from the combination of several causes), it is not altogether independent of collocations. If a cause supervenes, capable of wholly or partially counteracting the effect of any one of the conjoined causes, the effect will no longer conform to the derivative law. While, therefore, each ultimate law is only liable to frustration from one set of counteracting causes, the derivative law is liable to it from several. Now, the possibility of the occurrence of counteracting causes which do not arise from any of the conditions involved in the law itself, depends on the original collocations.

It is true that (as we formerly remarked) laws of causation, whether ultimate or derivative, are, in most cases, fulfilled even when counteracted; the cause produces its effect, though that effect is destroyed by something else. That the effect may be frustrated, is, therefore, no objection to the universality of laws of causation. But it is fatal to the universality of the sequences or coexistences of effects, which compose the greater part of the derivative laws flowing from laws of causation. When, from the law of a certain combination of causes, there results a certain order in the effects; as from the combination of a single sun with the rotation of an opaque body round its axis, there results, on the whole surface of that opaque body, an alternation of day and night; then if we suppose one of the combined causes counteracted, the rotation stopped, the sun extinguished, or a second sun superadded, the truth of that particular law of causation is in no way affected; it is still true that one sun shining on an opaque revolving body will alternately produce day and night; but since the sun no longer does shine on such a body, the derivative uniformity, the succession of day and night on the given planet, is no longer true. Those derivative uniformities, therefore, which are not laws of causation, are (except in the rare case of their depending on one cause alone, not on a combination of causes,) always more or less contingent on collocations; and are hence subject to the characteristic infirmity of empirical laws, that of being admissible only where the collocations are known by experience to be such as are requisite for the truth of the law, that is, only within the conditions of time and place confirmed by actual observation.

§ 2. This principle, when stated in general terms, seems clear and indisputable; yet many of the ordinary judgments of mankind, the propriety of which is not questioned, have at least the semblance of being inconsistent with it. On what grounds, it may be asked, do we expect that the sun will rise to-morrow? To-morrow is beyond the limits of time comprehended in our observations. They have extended over some thousands of years past, but they do not include the future. Yet we infer with confidence that the sun will rise to-morrow; and nobody doubts that we are entitled to do so. Let us consider what is the warrant for this confidence.

In the example in question, we know the causes on which the derivative uniformity depends. They are, the sun giving out light, the earth in a state of rotation and intercepting light. The induction which shows these to be the real causes, and not merely prior effects of a common cause, being complete; the only circumstances which could defeat the derivative law are. such as would destroy or counteract one or other of the combined causes. While the causes exist, and are not counteracted, the effect will continue. If they exist and are not counteracted to-morrow, the sun will rise to-morrow.

Since the causes, namely the sun and the earth, the one in the state of giving out light, the other in a state of rotation, will exist until something destroys them; all depends on the probabilities of their destruction, or of their counteraction. We know by observation (omitting the inferential proofs of an existence for thousands of ages anterior), that these phenomena have continued for (say) five thousand years. Within that time there has existed no cause sufficient to diminish them appreciably; nor which has counteracted their effect in any appreciable degree. The chance, therefore, that the sun may not rise to-morrow, amounts to the chance that some cause, which has not manifested itself in the smallest degree during five thousand years, will exist to-morrow in such intensity as to destroy the sun or the earth, the sun's light or the earth's rotation, or to produce an immense disturbance in the effect resulting from those causes.

Now, if such a cause will exist to-morrow, or at any future time, some cause, proximate or remote, of that cause must exist now, and must have existed during the whole of the five thousand years. If, therefore, the sun do not rise to-morrow, it will be because some cause has existed, the effects of which though during five thousand years they have not amounted to a perceptible quantity, will in one day become overwhelming. Since this cause has not been recognised during such an interval of time, by observers stationed on our earth, it must, if it exist, be either some agent whose effects develop themselves gradually and very slowly, or one which existed in regions beyond our observation, and is now on the point of arriving in our part of the universe. Now all causes which we have experience of, act according to laws incompatible with the supposition that their effects, after accumulating so slowly as to be imperceptible for five thousand years, should start into immensity in a single day. No mathematical law of proportion between an effect and the quantity or relations of its cause, could produce such contradictory results. The sudden development of an effect of which there was no previous trace, always arises from the coming together of several distinct causes, not previously conjoined; but if such sudden conjunction is destined to take place, the causes, or *their* causes, must have existed during the entire five thousand years; and their not having once come together during that period, shows how rare that particular combination is. We have, therefore, the warrant of a rigid induction for considering it probable, in a degree undistinguishable from certainty, that the known conditions requisite for the sun's rising will exist to-morrow.

§ 3. But this extension of derivative laws, not causative, beyond the limits of observation, can only be to *adjacent* cases. If instead of to-morrow we had said this day twenty thousand years, the inductions would have been anything but conclusive. That a cause which, in opposition to very powerful causes, produced no perceptible effect during five thousand years, should produce a very considerable one by the end of twenty thousand, has nothing in it which is not in conformity with our experience of causes. We know many agents, the effect of which in a short period does not amount to a perceptible quantity, but by accumulating for a much longer period becomes considerable. Besides, looking at the immense multitude of the heavenly bodies, their vast distances, and the rapidity of the motion of such of them as are known to move, it is a supposition not at all contradictory to experience that some body may be in motion towards us, or we towards it, within the limits of whose influence we have not come during five thousand years, but which in twenty thousand more may be producing effects upon us of the most extraordinary kind. Or the fact which is capable of preventing sunrise may be, not the cumulative effect of one cause, but some new combination of causes; and the chances favourable to that combination, though they have not produced it once in five thousand years, may produce it once in twenty thousand. So that the inductions which authorize us to expect future events, grow weaker and weaker the further we look into the future, and at length become inappreciable.

We have considered the probabilities of the sun's rising to-morrow, as derived from the real laws, that is, from the laws of the causes on which that uniformity is dependent. Let us now consider how the matter would have stood if the uniformity had been known only as an empirical law; if we had not been aware that the sun's light, and the earth's rotation (or the sun's motion), were the causes on which the periodical occurrence of daylight depends. We could have extended this empirical law to cases adjacent in time, though not to so great a distance of time as we can now. Having evidence that the effects had remained unaltered and been punctually conjoined for five thousand years, we could infer that the unknown causes on which the conjunction is dependent had existed undiminished and uncounteracted during the same period. The same conclusions, therefore, would follow as in the preceding case ; except that we should only know that during five thousand years nothing had occurred to defeat perceptibly this particular effect; while, when we know the causes, we have the additional assurance, that during that interval no such change has been noticeable in the causes themselves, as by any degree of multiplication or length of continuance could defeat the effect.

To this must be added, that when we know the causes, we may be able to judge whether there exists any known cause capable of counteracting them ; while as long as they are unknown, we cannot be sure but that if we did know them, we could predict their destruction from causes actually in existence. A bedridden savage, who had never seen the cataract of Niagara, but who lived within hearing of it, might imagine that the sound he heard would endure for ever ; but if he knew it to be the effect of a rush of waters over a barrier of rock which is progressively wearing away, he would know that within a number of ages which may be calculated, it will be heard no more. In proportion, therefore, to our ignorance of the causes on which the empirical law depends, we can be less assured that it will continue to hold good ; and the farther we look into futurity, the less improbable is it that some one of the causes, whose coexistence gives rise to the derivative uniformity, may be destroyed or counteracted. With every prolongation of time, the chances multiply of such an event, that is to say, its non-occurrence hitherto becomes a less guarantee of its not occurring within the given time. If, then, it is only to cases which in point of time are adjacent (or nearly adjacent) to those which we have actually observed, that *any* derivative law, not of causation, can be extended with an assurance equivalent to certainty, much more is this true of a merely empirical law. Happily, for the purposes of life it is to such cases alone that we can almost ever have occasion to extend them.

In respect of place, it might seem that a merely empirical law could not be extended even to adjacent cases; that we could have no assurance of its being true in any place where it has not been specially observed. The past duration of a cause is a guarantee for its future existence, unless something occurs to destroy it; but the existence of a cause in one or any number of places, is no guarantee for its existence in any other place, since there is no uniformity in the collocations of primeval causes. When, therefore, an empirical law is extended beyond the local limits within which it has been found true by observation, the cases to which it is thus extended must be such as are presumably within the influence of the same individual agents. If we discover a new planet within the known bounds of the solar system (or even beyond those bounds, but indicating its connexion with the system by revolving round the sun), we may conclude, with great probability, that it revolves on its axis. For all the known planets do so; and this uniformity points to some common cause, antecedent to the first records of astronomical observation: and though the nature of this cause can only be matter of conjecture, yet if it be, as is not unlikely, and as Laplace's theory supposes, not merely the same kind of cause, but the same individual cause (such as an impulse given to all the bodies at once), that cause, acting at the extreme points of the space occupied by the sun and planets, is likely, unless defeated by some counteracting cause, to have acted at every intermediate point, and probably somewhat beyond; and therefore acted, in all probability, upon the supposed newly-discovered planet.

When, therefore, effects which are always found conjoined, can be traced with any probability to an identical (and not merely a similar) origin, we may with the same probability extend the empirical law of their conjunction to all places within the extreme local boundaries within which the fact has been observed; subject to the possibility of counteracting causes in some portion of the field. Still more confidently may we do so when the law is not merely empirical; when the phenomena which we find conjoined are effects of ascertained causes, from the laws of which the conjunction of their effects is deducible. In that case, we may both extend the derivative uniformity over a larger space, and with less abatement for the chance of counteracting causes. The first, because instead of the local boundaries of our observation of the fact itself, we may include the extreme boundaries of the ascertained influence of its causes. Thus the succession of day and night, we know, holds true of all the bodies of the solar system except the sun itself; but we know this only because we are acquainted with the causes: if we were not, we could not extend the proposition beyond the orbits of the earth and moon, at both extremities of which we have the evidence of observation for its truth. With respect to the probability of counteracting causes, it has been seen that this calls for a greater abatement of confidence, in proportion to our ignorance of the causes on which the phenomena depend. On both accounts, therefore, a derivative law which we know how to resolve, is susceptible of a greater extension to cases adjacent in place, than a merely empirical law.

CHAPTER XX.

OF ANALOGY.

§ 1. THE word Analogy, as the name of a mode of reasoning, is generally taken for some kind of argument supposed to be of an inductive nature, but not amounting to a complete induction. There is no word, however, which is used more loosely, or in a greater variety of senses, than Analogy. It sometimes stands for arguments which may be examples of the most rigorous Induction. Archbishop Whately, for instance, following Ferguson and other writers, defines Analogy conformably to its primitive acceptation, that which was given to it by mathematicians, Resemblance of Relations. In this sense, when a country which has sent out colonies is termed the mother country, the expression is analogical, signifying that the colonies of a country stand in the same *relation* to her in which children stand to their parents. And if any inference be drawn from this resemblance of relations, as, for instance, that obedience or affection is due from colonies to the mother country, this is called reasoning by analogy. Or if it be argued that a nation is most beneficially governed by an assembly elected by the people, from the admitted fact that other associations for a common purpose, such as joint-stock companies, are best managed by a committee chosen by the parties interested ; this, too, is an argument from analogy in the preceding sense, because its foundation is, not that a nation is like a joint stock company, or Parliament like a board of directors, but that Parliament stands in the same *relation* to the nation in which a board of directors stands to a joint stock company. Now, in an argument of this nature, there is no inherent inferiority of conclusiveness. Like other arguments from resemblance, it may amount to nothing, or it may be a perfect and conclusive induction. The circumstance

in which the two cases resemble, may be capable of being
shown to be the *material* circumstance; to be that on which
all the consequences, necessary to be taken into account in
the particular discussion, depend. In the example last given,
the resemblance is one of relation; the *fundamentum relationis*
being the management by a few persons, of affairs in which
a much greater number are interested along with them.
Now, some may contend that this circumstance which is
common to the two cases, and the various consequences
which follow from it, have the chief share in determining all
the effects which make up what we term good or bad
administration. If they can establish this, their argument
has the force of a rigorous induction; if they cannot, they
are said to have failed in proving the analogy between the
two cases; a mode of speech which implies that when the
analogy can be proved, the argument founded on it cannot be
resisted.

§ 2. It is on the whole more usual, however, to extend
the name of analogical evidence to arguments from any sort
of resemblance, provided they do not amount to a complete
induction: without peculiarly distinguishing resemblance of
relations. Analogical reasoning, in this sense, may be reduced
to the following formula :—Two things resemble each other in
one or more respects; a certain proposition is true of the one;
therefore it is true of the other. But we have nothing here
by which to discriminate analogy from induction, since this
type will serve for all reasoning from experience. In the
strictest induction, equally with the faintest analogy, we
conclude because A resembles B in one or more properties,
that it does so in a certain other property. The difference is,
that in the case of a complete induction it has been previously
shown, by due comparison of instances, that there is an invariable conjunction between the former property or properties
and the latter property; but in what is called analogical
reasoning, no such conjunction has been made out. There
have been no opportunities of putting in practice the Method
of Difference, or even the Method of Agreement; but we

conclude (and that is all which the argument of analogy
amounts to) that a fact *m*, known to be true of A, is more
likely to be true of B if B agrees with A in some of its properties (even though no connexion is known to exist between
m and those properties), than if no resemblance at all could be
traced between B and any other thing known to possess the
attribute *m*.

To this argument it is of course requisite, that the properties common to A with B shall be merely not known to be
connected with *m*; they must not be properties known to be
unconnected with it. If, either by processes of elimination,
or by deduction from previous knowledge of the laws of the
properties in question, it can be concluded that they have
nothing to do with *m*, the argument of analogy is put out of
court. The supposition must be that *m* is an effect really dependent on some property of A, but we know not on which.
We cannot point out any of the properties of A, which is the
cause of *m*, or united with it by any law. After rejecting all
which we know to have nothing to do with it, there remain
several between which we are unable to decide: of which
remaining properties, B possesses one or more. This accordingly, we consider as affording grounds, of more or less
strength, for concluding by analogy that B possesses the
attribute *m*.

There can be no doubt that every such resemblance which
can be pointed out between B and A, affords some degree of
probability, beyond what would otherwise exist, in favour of
the conclusion drawn from it. If B resembled A in all its
ultimate properties, its possessing the attribute *m* would be a
certainty, not a probability: and every resemblance which can
be shown to exist between them, places it by so much the
nearer to that point. If the resemblance be in an ultimate
property, there will be resemblance in all the derivative properties dependent on that ultimate property, and of these *m*
may be one. If the resemblance be in a derivative property,
there is reason to expect resemblance in the ultimate property
on which it depends, and in the other derivative properties
dependent on the same ultimate property. Every resemblance

which can be shown to exist, affords ground for expecting an indefinite number of other resemblances: the particular resemblance sought will, therefore, be oftener found among things thus known to resemble, than among things between which we know of no resemblance.*

For example, I might infer that there are probably inhabitants in the moon, because there are inhabitants on the earth, in the sea, and in the air: and this is the evidence of analogy. The circumstance of having inhabitants is here assumed not to be an ultimate property, but (as is reasonable to suppose) a consequence of other properties; and depending, therefore, in the case of the earth, on some of its properties as a portion of the universe, but on which of those properties we know not. Now the moon resembles the earth in being a solid, opaque, nearly spherical substance, appearing to contain, or to have contained, active volcanoes; receiving heat and light from the sun, in about the same quantity as our earth; revolving on its axis; composed of materials which gravitate, and obeying all the various laws resulting from that property. And I think no one will deny that if this were all that was known of the moon, the existence of inhabitants in that luminary would derive from these various resemblances to the earth, a greater degree of probability than it would otherwise have: though the amount of the augmentation it would be useless to attempt to estimate.

If, however, every resemblance proved between B and A, in any point not known to be immaterial with respect to m,

---

* There was no greater foundation for this than for Newton's celebrated conjecture that the diamond was combustible. He grounded his guess on the very high refracting power of the diamond, comparatively to its density; a peculiarity which had been observed to exist in combustible substances; and on similar grounds he conjectured that water, though not combustible, contained a combustible ingredient. Experiment having subsequently shown that in both instances he guessed right, the prophecy is considered to have done great honour to his scientific sagacity; but it is to this day uncertain whether the guess was, in truth, what there are so many examples of in the history of science, a farsighted anticipation of a law afterwards to be discovered. The progress of science has not hitherto shown ground for believing that there is any real connexion between combustibility and a high refracting power.

forms some additional reason for presuming that B has the attribute *m*; it is clear, *è contra*, that every dissimilarity which can be proved between them, furnishes a counter-probability of the same nature on the other side. It is not indeed unusual that different ultimate properties should, in some particular instances, produce the same derivative property; but on the whole it is certain that things which differ in their ultimate properties, will differ at least as much in the aggregate of their derivative properties, and that the differences which are unknown will on the average of cases bear some proportion to those which are known. There will, therefore, be a competition between the known points of agreement and the known points of difference in A and B; and according as the one or the other may be deemed to preponderate, the probability derived from analogy will be for or against B's having the property *m*. The moon, for instance, agrees with the earth in the circumstances already mentioned; but differs in being smaller, in having its surface more unequal, and apparently volcanic throughout, in having, at least on the side next the earth, no atmosphere sufficient to refract light, no clouds, and (it is therefore concluded) no water. These differences, considered merely as such, might perhaps balance the resemblances, so that analogy would afford no presumption either way. But considering that some of the circumstances which are wanting on the moon are among those which, on the earth, are found to be indispensable conditions of animal life, we may conclude that if that phenomenon does exist in the moon, (or at all events on the nearer side,) it must be as an effect of causes totally different from those on which it depends here; as a consequence, therefore, of the moon's differences from the earth, not of the points of agreement. Viewed in this light, all the resemblances which exist become presumptions against, not in favour of, the moon's being inhabited. Since life cannot exist there in the manner in which it exists here, the greater the resemblance of the lunar world to the terrestrial in other respects, the less reason we have to believe that it can contain life.

There are, however, other bodies in our system, between

which and the earth there is a much closer resemblance; which possess an atmosphere, clouds, consequently water (or some fluid analogous to it), and even give strong indications of snow in their polar regions; while the cold, or heat, though differing greatly on the average from ours, is, in some parts at least of those planets, possibly not more extreme than in some regions of our own which are habitable. To balance these agreements, the ascertained differences are chiefly in the average light and heat, velocity of rotation, density of material, intensity of gravity, and similar circumstances of a secondary kind. With regard to these planets, therefore, the argument of analogy gives a decided preponderance in favour of their resembling the earth in any of its derivative properties, such as that of having inhabitants; though, when we consider how immeasurably multitudinous are those of their properties which we are entirely ignorant of, compared with the few which we know, we can attach but trifling weight to any considerations of resemblance in which the known elements bear so inconsiderable a proportion to the unknown.

Besides the competition between analogy and diversity, there may be a competition of conflicting analogies. The new case may be similar in some of its circumstances to cases in which the fact m exists, but in others to cases in which it is known not to exist. Amber has some properties in common with vegetable, others with mineral products. A painting of unknown origin, may resemble, in certain of its characters, known works of a particular master, but in others it may as strikingly resemble those of some other painter. A vase may bear some analogy to works of Grecian, and some to those of Etruscan, or Egyptian art. We are of course supposing that it does not possess any quality which has been ascertained, by a sufficient induction, to be a conclusive mark either of the one or of the other.

§ 3. Since the value of an analogical argument inferring one resemblance from other resemblances without any antecedent evidence of a connexion between them, depends on the extent of ascertained resemblance, compared first with the

amount of ascertained difference, and next with the extent of
the unexplored region of unascertained properties; it follows
that where the resemblance is very great, the ascertained
difference very small, and our knowledge of the subject-matter
tolerably extensive, the argument from analogy may approach
in strength very near to a valid induction. If, after much
observation of B, we find that it agrees with A in nine out of
ten of its known properties, we may conclude with a probability of nine to one, that it will possess any given derivative
property of A. If we discover, for example, an unknown
animal or plant, resembling closely some known one in the
greater number of the properties we observe in it, but
differing in some few, we may reasonably expect to find
in the unobserved remainder of its properties, a general
agreement with those of the former; but also a difference
corresponding proportionately to the amount of observed
diversity.

It thus appears that the conclusions derived from analogy
are only of any considerable value, when the case to which we
reason is an adjacent case; adjacent, not as before, in place or
time, but in circumstances. In the case of effects of which
the causes are imperfectly or not at all known, when consequently the observed order of their occurrence amounts only to
an empirical law, it often happens that the conditions which
have coexisted whenever the effect was observed, have been
very numerous. Now if a new case presents itself, in which
all these conditions do not exist, but the far greater part of
them do, some one or a few only being wanting, the inference
that the effect will occur, notwithstanding this deficiency of
complete resemblance to the cases in which it has been
observed, may, though of the nature of analogy, possess a high
degree of probability. It is hardly necessary to add that, however considerable this probability may be, no competent
inquirer into nature will rest satisfied with it when a complete
induction is attainable; but will consider the analogy as a
mere guide-post, pointing out the direction in which more
rigorous investigations should be prosecuted.

It is in this last respect that considerations of analogy have the highest scientific value. The cases in which analogical evidence affords in itself any very high degree of probability, are, as we have observed, only those in which the resemblance is very close and extensive; but there is no analogy, however faint, which may not be of the utmost value in suggesting experiments or observations that may lead to more positive conclusions. When the agents and their effects are out of the reach of further observation and experiment, as in the speculations already alluded to respecting the moon and planets, such slight probabilities are no more than an interesting theme for the pleasant exercise of imagination; but any suspicion, however slight, that sets an ingenious person at work to contrive an experiment, or affords a reason for trying one experiment rather than another, may be of the greatest benefit to science.

On this ground, though I cannot accept as positive doctrines any of those scientific hypotheses which are unsusceptible of being ultimately brought to the test of actual induction, such, for instance, as the two theories of light, the emission theory of the last century, and the undulatory theory which predominates in the present, I am yet unable to agree with those who consider such hypotheses to be worthy of entire disregard. As is well said by Hartley (and concurred in by a thinker in general so diametrically opposed to Hartley's opinions as Dugald Stewart), "any hypothesis which has so much plausibility as to explain a considerable number of facts, helps us to digest these facts in proper order, to bring new ones to light, and make *experimenta crucis* for the sake of future inquirers."* If an hypothesis both explains known facts, and has led to the prediction of others previously unknown, and since verified by experience, the laws of the phenomenon which is the subject of inquiry must bear at least a great similarity to those of the class of phenomena to which the hypothesis assimilates it; and since the analogy which extends so far may probably extend farther, nothing is more

* Hartley's *Observations on Man*, vol. i. p. 16. The passage is not in Priestley's curtailed edition.

no other than A ; but, that if it be no other than A it must be A, is not proved, by these instances at least, but taken for granted. There is no need to spend time in proving that the same thing is true of the other Inductive Methods. The universality of the law of causation is assumed in them all.

But is this assumption warranted ? Doubtless (it may be said) *most* phenomena are connected as effects with some antecedent or cause, that is, are never produced unless some assignable fact has preceded them ; but the very circumstance that complicated processes of induction are sometimes necessary, shows that cases exist in which this regular order of succession is not apparent to our unaided apprehension. If, then, the processes which bring these cases within the same category with the rest, require that we should assume the universality of the very law which they do not at first sight appear to exemplify, is not this a *petitio principii ?* Can we prove a proposition, by an argument which takes it for granted ? And if not so proved, on what evidence does it rest ?

For this difficulty, which I have purposely stated in the strongest terms it will admit of, the school of metaphysicians who have long predominated in this country find a ready salvo. They affirm, that the universality of causation is a truth which we cannot help believing ; that the belief in it is an instinct, one of the laws of our believing faculty. As the proof of this, they say, and they have nothing else to say, that everybody does believe it ; and they number it among the propositions, rather numerous in their catalogue, which may be logically argued against, and perhaps cannot be logically proved, but which are of higher authority than logic, and so essentially inherent in the human mind, that even he who denies them in speculation, shows by his habitual practice that his arguments make no impression upon himself.

Into the merits of this question, considered as one of psychology, it would be foreign to my purpose to enter here : but I must protest against adducing, as evidence of the truth of a fact in external nature, the disposition, however strong or however general, of the human mind to believe it. Belief is not proof, and does not dispense with the necessity of proof. I am aware, that to ask for evidence of a proposition which we are supposed to believe instinctively, is to expose oneself to the charge of rejecting the authority of the human faculties; which of course no one can consistently do, since the human faculties are all which any one has to judge by: and inasmuch as the meaning of the word evidence is supposed to be, something which when laid before the mind, induces it to believe; to demand evidence when the belief is ensured by the mind's own laws, is supposed to be appealing to the intellect against the intellect. But this, I apprehend, is a misunderstanding of the nature of evidence. By evidence is not meant anything and everything which produces belief. There are many things which generate belief besides evidence. A mere strong association of ideas often causes a belief so intense as to be unshakeable by experience or argument. Evidence is not that which the mind does or must yield to, but that which it ought to yield to, namely, that, by yielding to which, its belief is kept conformable to fact. There is no appeal from the human faculties generally, but there is an appeal from one human faculty to another; from the judging faculty, to those which take cognizance of fact, the faculties of sense and consciousness. The legitimacy of this appeal is admitted whenever it is allowed that our judgments ought to be conformable to fact. To say that belief suffices for its own justification is making opinion the test of opinion; it is denying the existence of any outward standard, the conformity of an opinion to which constitutes its truth. We call one mode of forming opinions right and another wrong, because the one does, and the other does not, tend to make the opinion agree with the fact—to make people believe what really is, and expect what really will be. Now a mere disposition to believe, even if supposed instinctive, is no guarantee for the truth of the thing believed. If, indeed, the belief ever amounted to an irresistible necessity, there would then be no *use* in appealing from it, because there would be no possibility of altering it. But even then the

truth of the belief would not follow; it would only follow
that mankind were under a permanent necessity of believing
what might possibly not be true; in other words, that a case
might occur in which our senses or consciousness, if they
could be appealed to, might testify one thing, and our reason
believe another. But in fact there is no such permanent
necessity. There is no proposition of which it can be asserted
that every human mind must eternally and irrevocably believe
it. Many of the propositions of which this is most confidently
stated, great numbers of human beings have disbelieved. The
things which it has been supposed that nobody could possibly
help believing, are innumerable; but no two generations
would make out the same catalogue of them. One age or
nation believes implicitly what to another seems incredible
and inconceivable; one individual has not a vestige of a
belief which another deems to be absolutely inherent in
humanity. There is not one of these supposed instinctive
beliefs which is really inevitable. It is in the power of every
one to cultivate habits of thought which make him independent of them. The habit of philosophical analysis, (of
which it is the surest effect to enable the mind to command,
instead of being commanded by, the laws of the merely passive
part of its own nature,) by showing to us that things are not
necessarily connected in fact because their ideas are connected
in our minds, is able to loosen innumerable associations which
reign despotically over the undisciplined or early-prejudiced
mind. And this habit is not without power even over those
associations which the school of which I have been speaking
regard as connate and instinctive. I am convinced that any
one accustomed to abstraction and analysis, who will fairly
exert his faculties for the purpose, will, when his imagination
has once learnt to entertain the notion, find no difficulty in
conceiving that in some one for instance of the many firma-
ments into which sidereal astronomy now divides the universe,
events may succeed one another at random, without any fixed
law; nor can anything in our experience, or in our mental
nature, constitute a sufficient, or indeed any, reason for believing that this is nowhere the case.

Were we to suppose (what it is perfectly possible to imagine) that the present order of the universe were brought to an end, and that a chaos succeeded in which there was no fixed succession of events, and the past gave no assurance of the future; if a human being were miraculously kept alive to witness this change, he surely would soon cease to believe in any uniformity, the uniformity itself no longer existing. If this be admitted, the belief in uniformity either is not an instinct, or it is an instinct conquerable, like all other instincts, by acquired knowledge.

But there is no need to speculate on what might be, when we have positive and certain knowledge of what has been. It is not true as a matter of fact, that mankind have always believed that all the successions of events were uniform and according to fixed laws. The Greek philosophers, not even excepting Aristotle, recognised Chance and Spontaneity (τύχη and τὸ αὐτόματον) as among the agents in nature; in other words, they believed that to that extent there was no guarantee that the past had been similar to itself, or that the future would resemble the past. Even now a full half of the philosophical world, including the very same metaphysicians who contend most for the instinctive character of the belief in uniformity, consider one important class of phenomena, volitions, to be an exception to the uniformity, and not governed by a fixed law.*

---

* I am happy to be able to quote the following excellent passage from Mr. Baden Powell's *Essay on the Inductive Philosophy*, in confirmation, both in regard to history and to doctrine, of the statement made in the text. Speaking of the "conviction of the universal and permanent uniformity of nature," Mr. Powell says (pp. 98—100),

"We may remark that this idea, in its proper extent, is by no means one of popular acceptance or natural growth. Just so far as the daily experience of every one goes, so far indeed he comes to embrace a certain persuasion of this kind, but merely to this limited extent, that what is going on around him at present, in his own narrow sphere of observation, will go on in like manner in future. The peasant believes that the sun which rose to-day will rise again to-morrow; that the seed put into the ground will be followed in due time by the harvest this year as it was last year, and the like; but has no notion of such inferences in subjects beyond his immediate observation. And it should be observed that each class of persons, in admitting this belief within the

7—2

§ 2. As was observed in a former place,* the belief we entertain in the universality, throughout nature, of the law of cause and effect, is itself an instance of induction; and by no means one of the earliest which any of us, or which mankind in general, can have made. We arrive at this universal law, by generalization from many laws of inferior generality. We should never have had the notion of causation (in the philosophical meaning of the term) as a condition of all phenomena, unless many cases of causation, or in other words, many partial uniformities of sequence, had previously become familiar. The more obvious of the particular uniformities suggest, and give evidence of, the general uniformity, and the general uniformity, once established, enables us to prove the remainder of the particular uniformities of which it is made up. As, however, all rigorous processes of induction presuppose the general uniformity, our knowledge of the particular uniformities from which it was first inferred was not, of course, derived from rigorous induction, but from the loose and uncertain mode of induction *per enumerationem simplicem*: and the law of universal causation, being collected

---

limited range of his own experience, though he doubt or deny it in everything beyond, is, in fact, bearing unconscious testimony to its universal truth. Nor, again, is it only among the *most* ignorant that this limitation is put upon the truth. There is a very general propensity to believe that everything beyond common experience, or especially ascertained laws of nature, is left to the dominion of chance or fate or arbitrary intervention; and even to object to any attempted explanation by physical causes, if conjecturally thrown out for an apparently unaccountable phenomenon.

"The precise doctrine of the *generalisation* of this idea of the uniformity of nature, so far from being obvious, natural, or intuitive, is utterly beyond the attainment of the many. In all the extent of its universality it is characteristic of the philosopher. It is clearly the result of philosophic cultivation and training, and by no means the spontaneous offspring of any primary principle naturally inherent in the mind, as some seem to believe. It is no mere vague persuasion taken up without examination, as a common prepossession to which we are always accustomed; on the contrary, all common prejudices and associations are against it. It is pre-eminently *an acquired idea*. It is not attained without deep study and reflection. The best informed philosopher is the man who most firmly believes it, even in opposition to received notions; its acceptance depends on the extent and profoundness of his inductive studies."

* Supra, book iii. ch. iii. § 1.

from results so obtained, cannot itself rest on any better foundation.

It would seem, therefore, that induction *per enumerationem simplicem* not only is not necessarily an illicit logical process, but is in reality the only kind of induction possible; since the more elaborate process depends for its validity on a law, itself obtained in that inartificial mode. Is there not then an inconsistency in contrasting the looseness of one method with the rigidity of another, when that other is indebted to the looser method for its own foundation?

The inconsistency, however, is only apparent. Assuredly, if induction by simple enumeration were an invalid process, no process grounded on it could be valid; just as no reliance could be placed on telescopes, if we could not trust our eyes. But though a valid process, it is a fallible one, and fallible in very different degrees: if therefore we can substitute for the more fallible forms of the process, an operation grounded on the same process in a less fallible form, we shall have effected a very material improvement. And this is what scientific induction does.

A mode of concluding from experience must be pronounced untrustworthy, when subsequent experience refuses to confirm it. According to this criterion, induction by simple enumeration—in other words, generalization of an observed fact from the mere absence of any known instance to the contrary—affords in general a precarious and unsafe ground of assurance; for such generalizations are incessantly discovered, on further experience, to be false. Still, however, it affords some assurance, sufficient, in many cases, for the ordinary guidance of conduct. It would be absurd to say, that the generalizations arrived at by mankind in the outset of their experience, such as these, Food nourishes, Fire burns, Water drowns, were unworthy of reliance.* There is a scale of trustworthiness in

---

* It deserves remark, that these early generalizations did not, like scientific inductions, presuppose causation. What they did presuppose, was *uniformity* in physical facts. But the observers were as ready to presume uniformity in the coexistences of facts as in the sequences. On the other hand, they never thought of assuming that this uniformity was a principle pervading all nature:

the results of the original unscientific Induction; and on this
diversity (as observed in the fourth chapter of the present
book) depend the rules for the improvement of the process.
The improvement consists in correcting one of these inartificial generalizations by means of another. As has been already
pointed out, this is all that art can do. To test a generalization, by showing that it either follows from, or conflicts with,
some stronger induction, some generalization resting on a
broader foundation of experience, is the beginning and end of
the logic of Induction.

§ 3. Now the precariousness of the method of simple
enumeration is in an inverse ratio to the largeness of the
generalization. The process is delusive and insufficient, exactly
in proportion as the subject-matter of the observation is
special and limited in extent. As the sphere widens, this
unscientific method becomes less and less liable to mislead;
and the most universal class of truths, the law of causation for
instance, and the principles of number and of geometry, are
duly and satisfactorily proved by that method alone, nor are
they susceptible of any other proof.

With respect to the whole class of generalizations of which
we have recently treated, the uniformities which depend on
causation, the truth of the remark just made follows by
obvious inference from the principles laid down in the preceding chapters. When a fact has been observed a certain
number of times to be true, and is not in any instance known

---

their generalizations did not imply that there was uniformity in everything,
but only that as much uniformity as existed within their observation, existed
also beyond it. The induction, Fire burns, does not require for its validity
that all nature should observe uniform laws, but only that there should be
uniformity in one particular class of natural phenomena; the effects of fire on
the senses and on combustible substances. And uniformity to this extent
was not assumed, anterior to the experience, but proved by the experience.
The same observed instances which proved the narrower truth, proved as much
of the wider one as corresponded to it. It is from losing sight of this fact, and
considering the law of causation in its full extent as necessarily presupposed in
the very earliest generalizations, that persons have been led into the belief that
the law of causation is known à priori, and is not itself a conclusion from
experience.

to be false; if we at once affirm that fact as an universal truth or law of nature, without testing it by any of the four methods of induction, nor deducing it from other known laws, we shall in general err grossly: but we are perfectly justified in affirming it as an empirical law, true within certain limits of time, place, and circumstance, provided the number of coincidences be greater than can with any probability be ascribed to chance. The reason for not extending it beyond those limits is, that the fact of its holding true within them may be a consequence of collocations, which cannot be concluded to exist in one place because they exist in another; or may be dependent on the accidental absence of counteracting agencies, which any variation of time, or the smallest change of circumstances, may possibly bring into play. If we suppose, then, the subject-matter of any generalization to be so widely diffused that there is no time, no place, and no combination of circumstances, but must afford an example either of its truth or of its falsity, and if it be never found otherwise than true, its truth cannot depend on any collocations, unless such as exist at all times and places; nor can it be frustrated by any counteracting agencies, unless by such as never actually occur. It is, therefore, an empirical law coextensive with all human experience; at which point the distinction between empirical laws and laws of nature vanishes, and the proposition takes its place among the most firmly established as well as largest truths accessible to science.

Now, the most extensive in its subject-matter of all generalizations which experience warrants, respecting the sequences and coexistences of phenomena, is the law of causation. It stands at the head of all observed uniformities, in point of universality, and therefore (if the preceding observations are correct) in point of certainty. And if we consider, not what mankind would have been justified in believing in the infancy of their knowledge, but what may rationally be believed in its present more advanced state, we shall find ourselves warranted in considering this fundamental law, though itself obtained by induction from particular laws of causation, as not less certain, but on the contrary, more so, than any of those from which it

was drawn. It adds to them as much proof as it receives from
them. For there is probably no one even of the best established laws of causation which is not sometimes counteracted,
and to which, therefore, apparent exceptions do not present
themselves, which would have necessarily and justly shaken
the confidence of mankind in the universality of those laws, if
inductive processes founded on the universal law had not
enabled us to refer those exceptions to the agency of counteracting causes, and thereby reconcile them with the law with
which they apparently conflict. Errors, moreover, may have
slipped into the statement of any one of the special laws,
through inattention to some material circumstance : and instead of the true proposition, another may have been enunciated, false as an universal law, though leading, in all cases
hitherto observed, to the same result. To the law of causation,
on the contrary, we not only do not know of any exception,
but the exceptions which limit or apparently invalidate the
special laws, are so far from contradicting the universal one,
that they confirm it; since in all cases which are sufficiently
open to our observation, we are able to trace the difference of
result, either to the absence of a cause which had been present
in ordinary cases, or to the presence of one which had been
absent.

The law of cause and effect, being thus certain, is capable
of imparting its certainty to all other inductive propositions
which can be deduced from it; and the narrower inductions
may be regarded as receiving their ultimate sanction from
that law, since there is no one of them which is not rendered
more certain than it was before, when we are able to connect
it with that larger induction, and to show that it cannot be
denied, consistently with the law that everything which
begins to exist has a cause. And hence we are justified in
the seeming inconsistency, of holding induction by simple
enumeration to be good for proving this general truth, the
foundation of scientific induction, and yet refusing to rely on
it for any of the narrower inductions. I fully admit that if the
law of causation were unknown, generalization in the more
obvious cases of uniformity in phenomena would nevertheless

be possible, and though in all cases more or less precarious, and in some extremely so, would suffice to constitute a certain measure of probability: but what the amount of this probability might be, we are dispensed from estimating, since it never could amount to the degree of assurance which the proposition acquires, when, by the application to it of the Four Methods, the supposition of its falsity is shown to be inconsistent with the Law of Causation. We are therefore logically entitled, and, by the necessities of scientific Induction, required, to disregard the probabilities derived from the early rude method of generalizing, and to consider no minor generalization as proved except so far as the law of causation confirms it, nor probable except so far as it may reasonably be expected to be so confirmed.

§ 4. The assertion, that our inductive processes assume the law of causation, while the law of causation is itself a case of induction, is a paradox, only on the old theory of reasoning, which supposes the universal truth, or major premise, in a ratiocination, to be the real proof of the particular truths which are ostensibly inferred from it. According to the doctrine maintained in the present treatise,* the major premise is not the proof of the conclusion, but is itself proved, along with the conclusion from the same evidence. "All men are mortal" is not the proof that Lord Palmerston is mortal; but our past experience of mortality authorizes us to infer *both* the general truth and the particular fact, and the one with exactly the same degree of assurance as the other. The mortality of Lord Palmerston is not an inference from the mortality of all men, but from the experience which proves the mortality of all men; and is a correct inference from experience, if that general truth is so too. This relation between our general beliefs and their particular applications holds equally true in the more comprehensive case which we are now discussing. Any new fact of causation inferred by induction, is rightly inferred, if no other objection can be made to the inference than can be

---

* Book ii. chap. iii.

made to the general truth that every event has a cause. The
utmost certainty which can be given to a conclusion arrived at
in the way of inference, stops at this point. When we have
ascertained that the particular conclusion must stand or fall
with the general uniformity of the laws of nature—that it is
liable to no doubt except the doubt whether every event has a
cause—we have done all that can be done for it. The strongest
assurance we can obtain of any theory respecting the cause of
a given phenomenon, is that the phenomenon has either that
cause or none.

The latter supposition might have been an admissible one in
a very early period of our study of nature. But we have been
able to perceive that in the stage which mankind have now
reached, the generalization which gives the Law of Universal
Causation has grown into a stronger and better induction, one
deserving of greater reliance, than any of the subordinate
generalizations. We may even, I think, go a step further
than this, and regard the certainty of that great induction
as not merely comparative, but, for all practical purposes,
absolute.

The considerations which, as I apprehend, give, at the
present day, to the proof of the law of uniformity of succession
as true of all phenomena without exception, this character of
completeness and conclusiveness, are the following:—First,
that we now know it directly to be true of far the greatest
number of phenomena; that there are none of which we know
it not to be true, the utmost that can be said being, that of
some we cannot positively from direct evidence affirm its
truth; while phenomenon after phenomenon, as they become
better known to us, are constantly passing from the latter
class into the former; and in all cases in which that transition
has not yet taken place, the absence of direct proof is accounted for by the rarity or the obscurity of the phenomena,
our deficient means of observing them, or the logical difficulties arising from the complication of the circumstances in
which they occur; insomuch that, notwithstanding as rigid a
dependence on given conditions as exists in the case of any
other phenomenon, it was not likely that we should be better

acquainted with those conditions than we are. Besides this first class of considerations, there is a second, which still further corroborates the conclusion. Although there are phenomena the production and changes of which elude all our attempts to reduce them universally to any ascertained law; yet in every such case, the phenomenon, or the objects concerned in it, are found in some instances to obey the known laws of nature. The wind, for example, is the type of uncertainty and caprice, yet we find it in some cases obeying with as much constancy as any phenomenon in nature the law of the tendency of fluids to distribute themselves so as to equalize the pressure on every side of each of their particles; as in the case of the trade winds, and the monsoons. Lightning might once have been supposed to obey no laws; but since it has been ascertained to be identical with electricity, we know that the very same phenomenon in some of its manifestations is implicitly obedient to the action of fixed causes. I do not believe that there is now one object or event in all our experience of nature, within the bounds of the solar system at least, which has not either been ascertained by direct observation to follow laws of its own, or been proved to be closely similar to objects and events which, in more familiar manifestations, or on a more limited scale, follow strict laws: our inability to trace the same laws on a larger scale and in the more recondite instances, being accounted for by the number and complication of the modifying causes, or by their inaccessibility to observation.

The progress of experience, therefore, has dissipated the doubt which must have rested on the universality of the law of causation while there were phenomena which seemed to be *sui generis*, not subject to the same laws with any other class of phenomena, and not as yet ascertained to have peculiar laws of their own. This great generalization, however, might reasonably have been, as it in fact was, acted on as a probability of the highest order, before there were sufficient grounds for receiving it as a certainty. For, whatever has been found true in innumerable instances, and never found to be false after due examination in any, we are safe in acting on as universal provisionally, until an undoubted exception appears; provided the nature of the case be such that a real exception could scarcely have escaped our notice. When every phenomenon that we ever knew sufficiently well to be able to answer the question, had a cause on which it was invariably consequent, it was more rational to suppose that our inability to assign the causes of other phenomena arose from our ignorance, than that there were phenomena which were uncaused, and which happened to be exactly those which we had hitherto had no sufficient opportunity of studying.

It must, at the same time, be remarked, that the reasons for this reliance do not hold in circumstances unknown to us, and beyond the possible range of our experience. In distant parts of the stellar regions, where the phenomena may be entirely unlike those with which we are acquainted, it would be folly to affirm confidently that this general law prevails, any more than those special ones which we have found to hold universally on our own planet. The uniformity in the succession of events, otherwise called the law of causation, must be received not as a law of the universe, but of that portion of it only which is within the range of our means of sure observation, with a reasonable degree of extension to adjacent cases. To extend it further is to make a supposition without evidence, and to which, in the absence of any ground from experience for estimating its degree of probability, it would be idle to attempt to assign any.*

---

* One of the most rising thinkers of the new generation in France, M. Taine (who has given, in the Revue des Deux Mondes, the most masterly analysis, at least in one point of view, ever made of the present work), though he rejects, on this and similar points of psychology, the intuition theory in its ordinary form nevertheless assigns to the law of causation, and to some other of the most universal laws, that certainty beyond the bounds of human experience, which I have not been able to accord to them. He does this on the faith of our faculty of abstraction, in which he seems to recognise an independent source of evidence, not indeed disclosing truths not contained in our experience, but affording an assurance which experience cannot give, of the universality of those which it

does contain. By abstraction M. Taine seems to think that we are able, not merely to analyse that part of nature which we see, and exhibit apart the elements which pervade it, but to distinguish such of them as are elements of the system of nature considered as a whole, not incidents belonging to our limited terrestrial experience. I am not sure that I fully enter into M. Taine's meaning ; but I confess I do not see how any mere abstract conception, elicited by our minds from our experience, can be evidence of an objective fact in universal Nature, beyond what the experience itself bears witness of ; or how, in the process of interpreting in general language the testimony of experience, the limitations of the testimony itself can be cast off.

CHAPTER XXII.

OF UNIFORMITIES OF COEXISTENCE NOT DEPENDENT ON CAUSATION.

§ 1. THE order of the occurrence of phenomena in time, is either successive or simultaneous; the uniformities, therefore, which obtain in their occurrence, are either uniformities of succession or of coexistence. Uniformities of succession are all comprehended under the law of causation and its consequences. Every phenomenon has a cause, which it invariably follows; and from this are derived other invariable sequences among the successive stages of the same effect, as well as between the effects resulting from causes which invariably succeed one another.

In the same manner with these derivative uniformities of succession, a great variety of uniformities of coexistence also take their rise. Coordinate effects of the same cause naturally coexist with one another. High water at any point on the earth's surface, and high water at the point diametrically opposite to it, are effects uniformly simultaneous, resulting from the direction in which the combined attractions of the sun and moon act upon the waters of the ocean. An eclipse of the sun to us, and an eclipse of the earth to a spectator situated in the moon, are in like manner phenomena invariably coexistent; and their coexistence can equally be deduced from the laws of their production.

It is an obvious question, therefore, whether all the uniformities of coexistence among phenomena may not be accounted for in this manner. And it cannot be doubted that between phenomena which are themselves effects, the coexistences must necessarily depend on the causes of those phenomena. If they are effects immediately or remotely of the same cause, they cannot coexist except by virtue of some laws or properties of

that cause : if they are effects of different causes, they cannot
coexist unless it be because their causes coexist ; and the uniformity of coexistence, if such there be, between the effects,
proves that those particular causes, within the limits of our
observation, have uniformly been coëxistent.

§ 2. But these same considerations compel us to recognise
that there must be one class of coexistences which cannot
depend on causation ; the coëxistences between the ultimate
properties of things—those properties which are the causes
of all phenomena, but are not themselves caused by any
phenomenon, and a cause for which could only be sought by
ascending to the origin of all things. Yet among these
ultimate properties there are not only coexistences, but uniformities of coexistence. General propositions may be, and
are, formed, which assert that whenever certain properties are
found, certain others are found along with them. We perceive an object ; say, for instance, water. We recognise it to
be water, of course by certain of its properties. Having
recognised it, we are able to affirm of it innumerable other
properties ; which we could not do unless it were a general
truth, a law or uniformity in nature, that the set of properties
by which we identify the substance as water, always have those
other properties conjoined with them.

In a former place,* it has been explained in some detail
what is meant by the Kinds of objects ; those classes which
differ from one another not by a limited and definite, but by
an indefinite and unknown, number of distinctions. To this
we have now to add, that every proposition by which anything
is asserted of a Kind, affirms an uniformity of coexistence.
Since we know nothing of Kinds but their properties, the
Kind, to us, is the set of properties by which it is identified,
and which must of course be sufficient to distinguish it from
every other kind.† In affirming anything, therefore, of a Kind,

---

* Book i. chap. vii.

† In some cases, a Kind is sufficiently identified by some one remarkable
property : but most commonly several are required ; each property considered

we are affirming something to be uniformly coexistent with the properties by which the kind is recognised; and that is the sole meaning of the assertion.

Among the uniformities of coexistence which exist in nature, may hence be numbered all the properties of Kinds. The whole of these, however, are not independent of causation, but only a portion of them. Some are ultimate properties, others derivative; of some, no cause can be assigned, but others are manifestly dependent on causes. Thus, pure atmospheric air is a Kind, and one of its most unequivocal properties is its gaseous form: this property, however, has for its cause the presence of a certain quantity of latent heat; and if that heat could be taken away (as has been done from so many gases in Faraday's experiments), the gaseous form would doubtless disappear, together with numerous other properties which depend on, or are caused by, that property.

In regard to all substances which are chemical compounds, and which therefore may be regarded as products of the juxtaposition of substances different in Kind from themselves, there is considerable reason to presume that the specific properties of the compound are consequent, as effects, on some of the properties of the elements, though little progress has yet been made in tracing any invariable relation between the latter and the former. Still more strongly will a similar presumption exist, when the object itself, as in the case of organized beings, is no primeval agent, but an effect, which depends on a cause or causes for its very existence. The Kinds therefore which are called in chemistry simple substances, or elementary natural agents, are the only ones, any of whose properties can with

---

singly, being a joint property of that and of other Kinds. The colour and brightness of the diamond are common to it with the paste from which false diamonds are made; its octohedral form is common to it with alum, and magnetic iron ore; but the colour and brightness and the form together, identify its Kind; that is, are a mark to us that it is combustible; that when burnt it produces carbonic acid; that it cannot be cut with any known substance; together with many other ascertained properties, and the fact that \ere exist an indefinite number still unascertained.

certainty be considered ultimate; and of these the ultimate properties are probably much more numerous that we at present recognise, since every successful instance of the resolution of the properties of their compounds into simpler laws, generally leads to the recognition of properties in the elements distinct from any previously known. The resolution of the laws of the heavenly motions, established the previously unknown ultimate property of a mutual attraction between all bodies: the resolution, so far as it has yet proceeded, of the laws of crystallization, of chemical composition, electricity, magnetism, &c., points to various polarities, ultimately inherent in the particles of which bodies are composed; the comparative atomic weights of different kinds of bodies were ascertained by resolving, into more general laws, the uniformities observed in the proportions in which substances combine with one another; and so forth. Thus although every resolution of a complex uniformity into simpler and more elementary laws has an apparent tendency to diminish the number of the ultimate properties, and really does remove many properties from the list; yet, (since the result of this simplifying process is to trace up an ever greater variety of different effects to the same agents,) the further we advance in this direction, the greater number of distinct properties we are forced to recognise in one and the same object: the coexistences of which properties must accordingly be ranked among the ultimate generalities of nature.

§ 8. There are, therefore, only two kinds of propositions which assert uniformity of coexistence between properties. Either the properties depend on causes, or they do not. If they do, the proposition which affirms them to be coexistent is a derivative law of coexistence between effects, and until resolved into the laws of causation on which it depends, is an empirical law, and to be tried by the principles of induction to which such laws are amenable. If, on the other hand, the properties do not depend on causes, but are ultimate properties; then if it be true that they invariably coexist, they must all be ultimate properties of one and the same Kind; and it is of these only that the coexistences can be classed as a peculiar sort of laws of nature.

When we affirm that all crows are black, or that all negroes have woolly hair, we assert an uniformity of coexistence. We assert that the property of blackness, or of having woolly hair, invariably coexists with the properties which, in common language, or in the scientific classification that we adopt, are taken to constitute the class crow, or the class negro. Now, supposing blackness to be an ultimate property of black objects, or woolly hair an ultimate property of the animals which possess it; supposing that these properties are not results of causation, are not connected with antecedent phenomena by any law; then if all crows are black, and all negroes have woolly hair, these must be ultimate properties of the Kind *crow*, or *negro*, or of some Kind which includes them. If, on the contrary, blackness or woolly hair be an effect depending on causes, these general propositions are manifestly empirical laws; and all that has already been said respecting that class of generalizations may be applied without modification to these.

Now, we have seen that in the case of all compounds—of all things, in short, except the elementary substances and primary powers of nature—the presumption is, that the properties do really depend upon causes; and it is impossible in any case whatever to be certain that they do not. We therefore should not be safe in claiming for any generalization respecting the coexistence of properties, a degree of certainty to which, if the properties should happen to be the result of causes, it would have no claim. A generalization respecting coexistence, or in other words respecting the properties of Kinds, may be an ultimate truth, but it may, also, be merely a derivative one; and since, if so, it is one of those derivative laws which are neither laws of causation, nor have been resolved into the laws of causation on which they depend, it can possess no higher degree of evidence than belongs to an empirical law.

§ 4. This conclusion will be confirmed by the considera-

tion of one great deficiency, which precludes the application
to the ultimate uniformities of coexistence, of a system of
rigorous scientific induction, such as the uniformities in the
succession of phenomena have been found to admit of. The
basis of such a system is wanting: there is no general axiom,
standing in the same relation to the uniformities of coexistence
as the law of causation does to those of succession. The
Methods of Induction applicable to the ascertainment of causes
and effects, are grounded on the principle that everything
which has a beginning must have some cause or other; that
among the circumstances which actually existed at the time of
its commencement, there is certainly some one combination,
on which the effect in question is unconditionally consequent,
and on the repetition of which it would certainly again recur.
But in an inquiry whether some kind (as crow) universally
possesses a certain property (as blackness), there is no room
for any assumption analogous to this. We have no previous
certainty that the property must have something which constantly coexists with it; must have an invariable coexistent, in
the same manner as an event must have an invariable antecedent. When we feel pain, we must be in some circumstances
under which if exactly repeated we should always feel pain.
But when we are conscious of blackness, it does not follow that
there is something else present of which blackness is a constant
accompaniment. There is, therefore, no room for elimination;
no Method of Agreement or Difference, or of Concomitant
Variations (which is but a modification either of the Method
of Agreement or of the Method of Difference). We cannot
conclude that the blackness we see in crows must be an invariable property of crows, merely because there is nothing else
present of which it can be an invariable property. We therefore inquire into the truth of a proposition like "All crows are
black," under the same disadvantage as if, in our inquiries into
causation, we were compelled to let in, as one of the possibilities, that the effect may in that particular instance have
arisen without any cause at all.

To overlook this grand distinction was, as it seems to me,
the capital error in Bacon's view of inductive philosophy.

The principle of elimination, that great logical instrument
which he had the immense merit of first bringing into general
use, he deemed applicable in the same sense, and in as unqualified a manner, to the investigation of the coexistences,
as to that of the successions of phenomena. He seems to have
thought that as every event has a cause, or invariable antecedent, so every property of an object has an invariable coexistent, which he called its Form: and the examples he
chiefly selected for the application and illustration of his
method, were inquiries into such Forms; attempts to determine in what else all those objects resembled, which agreed
in some one general property, as hardness or softness, dryness or moistness, heat or coldness. Such inquiries could
lead to no result. The objects seldom have any such circumstances in common. They usually agree in the one point
inquired into, and in nothing else. A great proportion of the
properties which, so far as we can conjecture, are the likeliest
to be really ultimate, would seem to be inherently properties
of many different Kinds of things, not allied in any other
respect. And as for the properties which, being effects of
causes, we are able to give some account of, they have generally
nothing to do with the ultimate resemblances or diversities in
the objects themselves, but depend on some outward circumstances, under the influence of which any objects whatever are
capable of manifesting those properties; as is emphatically
the case with those favourite subjects of Bacon's scientific
inquiries, hotness and coldness; as well as with hardness and
softness, solidity and fluidity, and many other conspicuous
qualities.

In the absence, then, of any universal law of coexistence,
similar to the universal law of causation which regulates
sequence, we are thrown back upon the unscientific induction
of the ancients, *per enumerationem simplicem, ubi non reperitur
instantia contradictoria*. The reason we have for believing that
all crows are black, is simply that we have seen and heard of
many black crows, and never one of any other colour. It
remains to be considered how far this evidence can reach, and
how we are to measure its strength in any given case.

§ 5. It sometimes happens that a mere change in the mode of verbally enunciating a question, though nothing is really added to the meaning expressed, is of itself a considerable step towards its solution. This, I think, happens in the present instance. The degree of certainty of any generalization which rests on no other evidence than the agreement, so far as it goes, of all past observation, is but another phrase for the degree of improbability that an exception, if any existed, could have hitherto remained unobserved. The reason for believing that all crows are black, is measured by the improbability that crows of any other colour should have existed to the present time without our being aware of it. Let us state the question in this last mode, and consider what is implied in the supposition that there may be crows which are not black, and under what conditions we can be justified in regarding this as incredible.

If there really exist crows which are not black, one of two things must be the fact. Either the circumstance of blackness, in all crows hitherto observed, must be, as it were, an accident, not connected with any distinction of Kind; or if it be a property of Kind, the crows which are not black must be a new Kind, a Kind hitherto overlooked, though coming under the same general description by which crows have hitherto been characterized. The first supposition would be proved true if we were to discover casually a white crow among black ones, or if it were found that black crows sometimes turn white. The second would be shown to be the fact if in Australia or Central Africa a species or a race of white or grey crows were found to exist.

§ 6. The former of these suppositions necessarily implies that the colour is an effect of causation. If blackness, in the crows in which it has been observed, be not a property of Kind, but can be present or absent without any difference generally in the properties of the object; then it is not an ultimate fact in the individuals themselves, but is certainly dependent on a cause. There are, no doubt, many properties which vary from individual to individual of the same Kind,

even the same *infima species*, or lowest Kind. Some flowers may
be either white or red, without differing in any other respect.
But these properties are not ultimate; they depend on causes.
So far as the properties of a thing belong to its own nature,
and do not arise from some cause extrinsic to it, they are
always the same in the same Kind. Take, for instance, all
simple substances and elementary powers; the only things
of which we are certain that some at least of their properties
are really ultimate. Colour is generally esteemed the most
variable of all properties: yet we do not find that sulphur is
sometimes yellow and sometimes white, or that it varies in
colour at all, except so far as colour is the effect of some
extrinsic cause, as of the sort of light thrown upon it, the
mechanical arrangement of the particles, (as after fusion) &c.
We do not find that iron is sometimes fluid and sometimes
solid at the same temperature; gold sometimes malleable and
sometimes brittle; that hydrogen will sometimes combine
with oxygen and sometimes not; or the like. If from simple
substances we pass to any of their definite compounds, as
water, lime, or sulphuric acid, there is the same constancy in
their properties. When properties vary from individual to
individual, it is either in the case of miscellaneous aggrega-
tions, such as atmospheric air or rock, composed of heterogeneous substances, and not constituting or belonging to
any real Kind,* or it is in the case of organic beings. In
them, indeed, there is variability in a high degree. Animals
of the same species and race, human beings of the same
age, sex, and country, will be most different, for example, in
face and figure. But organized beings (from the extreme
complication of the laws by which they are regulated) being
more eminently modifiable, that is, liable to be influenced by
a greater number and variety of causes, than any other
phenomena whatever; having also themselves had a beginning, and therefore a cause; there is reason to believe that

---

* This doctrine of course assumes that the allotropic forms of what is
chemically the same substance are so many different Kinds; and such, in the
sense in which the word Kind is used in this treatise, they really are.

none of their properties are ultimate, but all of them derivative, and produced by causation. And the presumption is confirmed, by the fact that the properties which vary from one individual to another, also generally vary more or less at different times in the same individual; which variation, like any other event, supposes a cause, and implies, consequently, that the properties are not independent of causation.

If, therefore, blackness be merely accidental in crows, and capable of varying while the Kind remains the same, its presence or absence is doubtless no ultimate fact, but the effect of some unknown cause: and in that case the universality of the experience that all crows are black is sufficient proof of a common cause, and establishes the generalization as an empirical law. Since there are innumerable instances in the affirmative, and hitherto none at all in the negative, the causes on which the property depends must exist everywhere in the limits of the observations which have been made; and the proposition may be received as universal within those limits, and with the allowable degree of extension to adjacent cases.

§ 7. If, in the second place, the property, in the instances in which it has been observed, is not an effect of causation, it is a property of Kind; and in that case the generalization can only be set aside by the discovery of a new Kind of crow. That, however, a peculiar Kind, not hitherto discovered, should exist in nature, is a supposition so often realized, that it cannot be considered at all improbable. We have nothing to authorize us in attempting to limit the Kinds of things which exist in nature. The only unlikelihood would be that a new Kind should be discovered in localities which there was previously reason to believe had been thoroughly explored; and even this improbability depends on the degree of conspicuousness of the difference between the newly-discovered Kind and all others, since new Kinds of minerals, plants, and even animals, previously overlooked or confounded with known species, are still continually detected in the most frequented situations. On this second ground, therefore, as well as

on the first, the observed uniformity of coexistence can only
hold good as an empirical law, within the limits not only
of actual observation, but of an observation as accurate as the
nature of the case required. And hence it is that (as remarked
in an early chapter of the present Book) we so often give up
generalizations of this class at the first summons. If any
credible witness stated that he had seen a white crow, under
circumstances which made it not incredible that it should have
escaped notice previously, we should give full credence to the
statement.

It appears, then, that the uniformities which obtain in the
coexistence of phenomena,—those which we have reason to
consider as ultimate, no less than those which arise from the
laws of causes yet undetected—are entitled to reception only
as empirical laws; are not to be presumed true except within
the limits of time, place, and circumstance, in which the observations were made, or except in cases strictly adjacent.

§ 8. We have seen in the last chapter that there is a
point of generality at which empirical laws become as certain
as laws of nature, or rather, at which there is no longer any
distinction between empirical laws and laws of nature. As
empirical laws approach this point, in other words, as they
rise in their degree of generality, they become more certain;
their universality may be more strongly relied on. For, in
the first place, if they are results of causation (which, even
in the class of uniformities treated of in the present chapter,
we never can be certain that they are not) the more general
they are, the greater is proved to be the space over which the
necessary collocations prevail, and within which no causes
exist capable of counteracting the unknown causes on which
the empirical law depends. To say that anything is an
invariable property of some very limited class of objects, is
to say that it invariably accompanies some very numerous
and complex group of distinguishing properties; which, if
causation be at all concerned in the matter, argues a combination of many causes, and therefore a great liability to
counteraction; while the comparatively narrow range of the

observations renders it impossible to predict to what extent unknown counteracting causes may be distributed throughout nature. But when a generalization has been found to hold good of a very large proportion of all things whatever, it is already proved that nearly all the causes which exist in nature have no power over it; that very few changes in the combination of causes can effect it; since the greater number of possible combinations must have already existed in some one or other of the instances in which it has been found true. If, therefore, any empirical law is a result of causation, the more general it is, the more it may be depended on. And even if it be no result of causation, but an ultimate coexistence, the more general it is, the greater amount of experience it is derived from, and the greater therefore is the probability that if exceptions had existed, some would already have presented themselves.

For these reasons, it requires much more evidence to establish an exception to one of the more general empirical laws than to the more special ones. We should not have any difficulty in believing that there might be a new Kind of crow; or a new kind of bird resembling a crow in the properties hitherto considered distinctive of that Kind. But it would require stronger proof to convince us of the existence of a Kind of crow having properties at variance with any generally recognised universal property of birds; and a still higher degree if the properties conflict with any recognised universal property of animals. And this is conformable to the mode of judgment recommended by the common sense and general practice of mankind, who are more incredulous as to any novelties in nature, according to the degree of generality of the experience which these novelties seem to contradict.

§ 9. Still, however, even these greater generalizations, which embrace comprehensive Kinds, containing under them a great number and variety of *infimæ species*, are only empirical laws, resting on induction by simple enumeration merely, and not on any process of elimination, a process wholly inapplicable to this sort of case. Such generalizations, therefore, ought to be grounded on an examination of all the *infimæ species* comprehended in them, and not of a portion only. We cannot conclude (where causation is not concerned), because a proposition is true of a number of things resembling one another only in being animals, that it is therefore true of all animals. If, indeed, anything be true of species which differ more from one another than either differs from a third, (especially if that third species occupies in most of its known properties a position between the two former,) there is some probability that the same thing will also be true of that intermediate species; for it is often, though by no means universally, found, that there is a sort of parallelism in the properties of different Kinds, and that their degree of unlikeness in one respect bears some proportion to their unlikeness in others. We see this parallelism in the properties of the different metals; in those of sulphur, phosphorus, and carbon; of chlorine, iodine, and bromine; in the natural orders of plants and animals, &c. But there are innumerable anomalies and exceptions to this sort of conformity; if indeed the conformity itself be anything but an anomaly and an exception in nature.

Universal propositions, therefore, respecting the properties of superior Kinds, unless grounded on proved or presumed connexion by causation, ought not to be hazarded except after separately examining every known sub-kind included in the larger Kind. And even then such generalizations must be held in readiness to be given up on the occurrence of some new anomaly, which, when the uniformity is not derived from causation, can never, even in the case of the most general of these empirical laws, be considered very improbable. Thus all the universal propositions which it has been attempted to lay down respecting simple substances, or concerning any of the classes which have been formed among simple substances, (and the attempt has been often made,) have, with the progress of experience, either faded into inanity, or been proved to be erroneous; and each Kind of simple substance remains with its own collection of properties apart from the rest, saving a certain parallelism with a few other Kinds, the most similar to itself. In organized beings, indeed, there are abundance of propositions ascertained to be universally true of superior genera, to many of which the discovery hereafter of any exceptions must be regarded as extremely improbable. But these, as already observed, are, we have every reason to believe, properties dependent on causation.

Uniformities of coexistence, then, not only when they are consequences of laws of succession, but also when they are ultimate truths, must be ranked, for the purpose of logic, among empirical laws; and are amenable in every respect to the same rules with those unresolved uniformities which are known to be dependent on causation.

CHAPTER XXIII.

OF APPROXIMATE GENERALIZATIONS, AND PROBABLE EVIDENCE.

§ 1. IN our inquiries into the nature of the inductive process, we must not confine our notice to such generalizations from experience as profess to be universally true. There is a class of inductive truths avowedly not universal; in which it is not pretended that the predicate is always true of the subject; but the value of which, as generalizations, is nevertheless extremely great. An important portion of the field of inductive knowledge does not consist of universal truths, but of approximations to such truths; and when a conclusion is said to rest on probable evidence, the premises it is drawn from are usually generalizations of this sort.

As every certain inference respecting a particular case, implies that there is ground for a general proposition, of the form, Every A is B; so does every probable inference suppose that there is ground for a proposition of the form, Most A are B: and the degree of probability of the inference in an average case, will depend on the proportion between the number of instances existing in nature which accord with the generalization, and the number of those which conflict with it.

§ 2. Propositions in the form, Most A are B, are of a very different degree of importance in science, and in the practice of life. To the scientific inquirer they are valuable chiefly as materials for, and steps towards, universal truths. The discovery of these is the proper end of science: its work is not done if it stops at the proposition that a majority of A are B, without circumscribing that majority by some common character, fitted to distinguish them from the minority. Independently of the inferior precision of such imperfect generalizations, and the inferior assurance with which they can be applied to individual cases, it is plain that, compared with exact generalizations, they are almost useless as means of discovering ulterior truths by way of deduction. We may, it is true, by combining the proposition Most A are B, with an universal proposition, Every B is C, arrive at the conclusion that Most A are C. But when a second proposition of the approximate kind is introduced,—or even when there is but one, if that one be the major premise,—nothing can in general be positively concluded. When the major is Most B are D, then, even if the minor be Every A is B, we cannot infer that most A are D, or with any certainty that even some A are D. Though the majority of the class B have the attribute signified by D, the whole of the sub-class A may belong to the minority.*

Though so little use can be made, in science, of approximate generalizations, except as a stage on the road to something better, for practical guidance they are often all we have to rely on. Even when science has really determined the universal laws of any phenomenon, not only are those laws generally too much encumbered with conditions to be adapted for every-day use, but the cases which present themselves in life are too complicated, and our decisions require to be taken too rapidly, to admit of waiting till the existence of a phenomenon can be proved by what have been scientifically ascertained to be universal marks of it. To be indecisive and reluctant to act, because we have not evidence of a perfectly conclusive character to act on, is a defect sometimes incident to scientific minds, but which, wherever it exists, renders them unfit for practical emergencies. If we would succeed in action, we must judge by indications which, though they do not

* Mr. De Morgan, in his *Formal Logic*, makes the just remark, that from two such premises as Most A are B, and Most A are C, we may infer with certainty that some B are C. But this is the utmost limit of the conclusions which can be drawn from two approximate generalizations, when the precise degree of their approximation to universality is unknown or undefined.

generally mislead us, sometimes do ; and must make up, as far
as possible, for the incomplete conclusiveness of any one indication, by obtaining others to corroborate it. The principles
of induction applicable to approximate generalization are
therefore a not less important subject of inquiry, than the
rules for the investigation of universal truths; and might
reasonably be expected to detain us almost as long, were it not
that these principles are mere corollaries from those which
have been already treated of.

§ 3. There are two sorts of cases in which we are forced
to guide ourselves by generalizations of the imperfect form,
Most A are B. The first is, when we have no others; when
we have not been able to carry our investigation of the laws of
the phenomena any farther ; as in the following propositions :
Most dark-eyed persons have dark hair ; Most springs contain
mineral substances ; Most stratified formations contain fossils.
The importance of this class of generalizations is not very
great ; for, though it frequently happens that we see no reason
why that which is true of most individuals of a class is not true
of the remainder, nor are able to bring the former under any
general description which can distinguish them from the latter,
yet if we are willing to be satisfied with propositions of a less
degree of generality, and to break down the class A into sub-
classes, we may generally obtain a collection of propositions
exactly true. We do not know why most wood is lighter than
water, nor can we point out any general property which discri-
minates wood that is lighter than water from that which is
heavier. But we know exactly what species are the one and
what the other. And.if we meet with a specimen not conformable to any known species (the only case in which our
previous knowledge affords no other guidance than the approximate generalization), we can generally make a specific
experiment, which is a surer resource.

It often happens, however, that the proposition, Most A
are B, is not the ultimatum of our scientific progress, though
the knowledge we possess beyond it cannot conveniently be
brought to bear upon the particular instance. In such a case

we know well enough what circumstances distinguish the portion of A which has the attribute B from the portion which has it not, but have no means, or have not time, to examine whether those characteristic circumstances exist or not in the individual case. This is the situation we are generally in when the inquiry is of the kind called moral, that is, of the kind which has in view to predict human actions. To enable us to affirm anything universally concerning the actions of classes of human beings, the classification must be grounded on the circumstances of their mental culture and habits, which in an individual case are seldom exactly known; and classes grounded on these distinctions would never precisely accord with those into which mankind are divided for social purposes. All propositions which can be framed respecting the actions of human beings as ordinarily classified, or as classified according to any kind of outward indications, are merely approximate. We can only say, Most persons of a particular age, profession, country, or rank in society, have such and such qualities; or, Most persons when placed in certain circumstances act in such and such a way. Not that we do not often know well enough on what causes the qualities depend, or what sort of persons they are who act in that particular way; but we have seldom the means of knowing whether any individual person has been under the influence of those causes, or is a person of that particular sort. We could replace the approximate generalizations by propositions universally true; but these would hardly ever be capable of being applied to practice. We should be sure of our majors, but we should not be able to get minors to fit: we are forced, therefore, to draw our conclusions from coarser and more fallible indications.

§ 4. Proceeding now to consider, what is to be regarded as sufficient evidence of an approximate generalization; we can have no difficulty in at once recognising that when admissible at all, it is admissible only as an empirical law. Propositions of the form, Every A is B, are not necessarily laws of causation, or ultimate uniformities of coexistence; propositions like Most A are B *cannot* be so. Propositions hitherto found true in every observed instance, may yet be no necessary consequence of laws of causation, or of ultimate uniformities, and unless they are so, may, for aught we know, be false beyond the limits of actual observation : still more evidently must this be the case with propositions which are only true in a mere majority of the observed instances.

There is some difference, however, in the degree of certainty of the proposition, Most A are B, according as that approximate generalization composes the whole of our knowledge of the subject, or not. Suppose, first, that the former is the case. We know only that most A are B, not why they are so, nor in what respect those which are, differ from those which are not. How then did we learn that most A are B? Precisely in the manner in which we should have learnt, had such happened to be the fact, that all A are B. We collected a number of instances sufficient to eliminate chance, and having done so, compared the number of instances in the affirmative with the number in the negative. The result, like other unresolved derivative laws, can be relied on solely within the limits not only of place and time, but also of circumstance, under which its truth has been actually observed; for as we are supposed to be ignorant of the causes which make the proposition true, we cannot tell in what manner any new circumstance might perhaps affect it. The proposition, Most judges are inaccessible to bribes, would be found true of Englishmen, Frenchmen, Germans, North Americans, and so forth; but if on this evidence alone we extended the assertion to Orientals, we should step beyond the limits, not only of place but of circumstance, within which the fact had been observed, and should let in possibilities of the absence of the determining causes, or the presence of counteracting ones, which might be fatal to the approximate generalization.

In the case where the approximate proposition is not the ultimatum of our scientific knowledge, but only the most available form of it for practical guidance; where we know, not only that most A have the attribute B, but also the causes of B, or some properties by which the portion of A which has that attribute is distinguished from the portion which has it not; we are rather more favourably situated than in the preceding case. For we have now a double mode of ascertaining whether it be true that most A are B; the direct mode, as before, and an indirect one, that of examining whether the proposition admits of being. deduced from the known cause, or from any known criterion, of B. Let the question, for example, be whether most Scotchmen can read? We may not have observed, or received the testimony of others respecting, a sufficient number and variety of Scotchmen to ascertain this fact; but when we consider that the cause of being able to read is the having been taught it, another mode of determining the question presents itself, namely, by inquiring whether most Scotchmen have been sent to schools where reading is effectually taught. Of these two modes, sometimes one and sometimes the other is the more available. In some cases, the frequency of the effect is the more accessible to that extensive and varied observation which is indispensable to the establishment of an empirical law; at other times, the frequency of the causes, or of some collateral indications. It commonly happens that neither is susceptible of so satisfactory an induction as could be desired, and that the grounds on which the conclusion is received are compounded of both. Thus a person may believe that most Scotchmen can read, because, so far as his information extends, most Scotchmen have been sent to school, and most Scotch schools teach reading effectually; and also because most of the Scotchmen whom he has known or heard of, could read; though neither of these two sets of observations may by itself fulfil the necessary conditions of extent and variety.

Although the approximate generalization may in most cases be indispensable for our guidance, even when we know the cause, or some certain mark, of the attribute predicated; it needs hardly be observed that we may always replace the uncertain indication by a certain one, in any case in which we can actually recognise the existence of the cause or mark. For example, an assertion is made by a witness, and the question is, whether to believe it. If we do not look to any of the individual circumstances of the case, we have nothing to direct us but the approximate generalization, that truth is more common than falsehood, or, in other words, that most persons, on most occasions, speak truth. But if we consider in what circumstances the cases where truth is spoken differ from those in which it is not, we find, for instance, the following: the witness's being an honest person or not; his being an accurate observer or not; his having an interest to serve in the matter or not. Now, not only may we be able to obtain other approximate generalizations respecting the degree of frequency of these various possibilities, but we may know which of them is positively realized in the individual case. That the witness has or has not an interest to serve, we perhaps know directly; and the other two points indirectly, by means of marks; as, for example, from his conduct on some former occasion; or from his reputation, which, though a very uncertain mark, affords an approximate generalization (as, for instance, Most persons who are believed to be honest by those with whom they have had frequent dealings, are really so) which approaches nearer to an universal truth than the approximate general proposition with which we set out, viz. Most persons on most occasions speak truth.

As it seems unnecessary to dwell further on the question of the evidence of approximate generalizations, we shall proceed to a not less important topic, that of the cautions to be observed in arguing from these incompletely universal propositions to particular cases.

§ 5. So far as regards the direct application of an approximate generalization to an individual instance, this question presents no difficulty. If the proposition, Most A are B, has been established, by a sufficient induction, as an empirical law, we may conclude that any particular A is B with a probability proportioned to the preponderance of the number of affirmative instances over the number of exceptions. If it has been found practicable to attain numerical precision in the data, a corresponding degree of precision may be given to the evaluation of the chances of error in the conclusion. If it can be established as an empirical law that nine out of every ten A are B, there will be one chance in ten of error in assuming that any A, not individually known to us, is a B: but this of course holds only within the limits of time, place, and circumstance, embraced in the observations, and therefore cannot be counted on for any sub-class or variety of A (or for A in any set of external circumstances) which were not included in the average. It must be added, that we can guide ourselves by the proposition, Nine out of every ten A are B, only in cases of which we know nothing except that they fall within the class A. For if we know, of any particular instance i, not only that it falls under A, but to what species or variety of A it belongs, we shall generally err in applying to i the average struck for the whole genus, from which the average corresponding to that species alone would, in all probability, materially differ. And so if i, instead of being a particular sort of instance, is an instance known to be under the influence of a particular set of circumstances. The presumption drawn from the numerical proportions in the whole genus would probably, in such a case, only mislead. A general average should only be applied to cases which are neither known, nor can be presumed, to be other than average cases. Such averages, therefore, are commonly of little use for the practical guidance of any affairs but those which concern large numbers. Tables of the chances of life are useful to insurance offices, but they go a very little way towards informing any one of the chances of his own life, or any other life in which he is interested, since almost every life is either better or worse than the average. Such averages can only be considered as supplying the first term in a series of approximations; the subsequent terms proceeding on an appreciation of the circumstances belonging to the particular case.

§ 6. From the application of a single approximate generalization to individual cases, we proceed to the application of two or more of them together to the same case.

When a judgment applied to an individual instance is grounded on two approximate generalizations taken in conjunction, the propositions may co-operate towards the result in two different ways. In the one, each proposition is separately applicable to the case in hand, and our object in combining them is to give to the conclusion in that particular case the double probability arising from the two propositions separately. This may be called joining two probabilities by way of Addition; and the result is a probability greater than either. The other mode is, when only one of the propositions is directly applicable to the case, the second being only applicable to it by virtue of the application of the first. This is joining two probabilities by way of Ratiocination or Deduction; the result of which is a less probability than either. The type of the first argument is, Most A are B; most C are B; this thing is both an A and a C; therefore it is probably a B. The type of the second is, Most A are B; most C are A; this is a C; therefore it is probably an A, therefore it is probably a B. The first is exemplified when we prove a fact by the testimony of two unconnected witnesses; the second, when we adduce only the testimony of one witness that he has heard the thing asserted by another. Or again, in the first mode it may be argued that the accused committed the crime, because he concealed himself, and because his clothes were stained with blood; in the second, that he committed it because he washed or destroyed his clothes, which is supposed to render it probable that they were stained with blood. Instead of only two links, as in these instances, we may suppose chains of any length. A chain of the former kind was termed by Bentham* a self-corroborative chain of evidence; the second, a self-infirmative chain.

When approximate generalizations are joined by way of addition, we may deduce from the theory of probabilities laid down in a former chapter, in what manner each of them adds to the probability of a conclusion which has the warrant of them all.

In the early editions of this treatise, the joint probability arising from the sum of two independent probabilities was estimated in the following manner. If, on an average, two of

---

* *Rationale of Judicial Evidence*, vol. iii. p. 224.

every three As are Bs, and three of every four Cs are Bs, the probability that something which is both an A and a C is a B, will be more than two in three, or than three in four. Of every twelve things which are As, all except four are Bs by the supposition; and if the whole twelve, and consequently those four, have the characters of C likewise, three of these will be Bs on that ground. Therefore, out of twelve which are both As and Cs, eleven are Bs. To state the argument in another way; a thing which is both an A and a C, but which is not a B, is found in only one of three sections of the class A, and in only one of four sections of the class C; but this fourth of C being spread over the whole of A indiscriminately, only one-third part of it (or one-twelfth of the whole number) belongs to the third section of A; therefore a thing which is not a B occurs only once, among twelve things which are both As and Cs. The argument would in the language of the doctrine of chances, be thus expressed: the chance that an A is not a B is $\frac{1}{3}$, the chance that a C is not a B is $\frac{1}{4}$; hence if the thing be both an A and a C, the chance is $\frac{1}{3}$ of $\frac{1}{4} = \frac{1}{12}$.

It has, however, been pointed out to me by a mathematical friend, that in this statement the evaluation of the chances is erroneous. The correct mode of setting out the possibilities is as follows. If the thing (let us call it T) which is both an A and a C, is a B, something is true which is only true twice in every thrice, and something else which is only true thrice in every four times. The first fact being true eight times in twelve, and the second being true six times in every eight, and consequently six times in those eight; both facts will be true only six times in twelve. On the other hand if T, although it is both an A and a C, is not a B, something is true which is only true once in every thrice, and something else which is only true once in every four times. The former being true four times out of twelve, and the latter once in every four, and therefore once in those four; both are only true in one case out of twelve. So that T is a B six times in twelve, and T is not a B, only once: making the comparative probabilities, not eleven to one, as I had previously made them, but six to one.

It may be asked, what happens in the remaining cases?

since in this calculation seven out of twelve cases seem to have exhausted the possibilities. If T is a B in only six cases of every twelve, and a not-B in only one, what is it in the other five? The only supposition remaining for those cases is that it is neither a B nor not a B, which is impossible. But this impossibility merely proves that the state of things supposed in the hypothesis does not exist in those cases. They are cases that do not furnish anything which is both an A and a C.

To make this intelligible, we will substitute for our symbols a concrete case. Let there be two witnesses, M and N, whose probabilities of veracity correspond with the ratios of the preceding example: M speaks truth twice in every thrice, N thrice in every four times. The question is, what is the probability that a statement, in which they both concur, will be true. The cases may be classed as follows. Both the witnesses will speak truly six in every twelve times; both falsely once in twelve times. Therefore, if they both agree in an assertion, it will be true six times, for once that it will be false. What happens in the remaining cases is here evident; there will be five cases in every twelve in which the witnesses will not agree. M will speak truth and N falsehood in two cases of every twelve; N will speak truth and M falsehood in three cases, making in all five. In these cases, however, the witnesses will not agree in their testimony. But disagreement between them is excluded by the supposition. There are, therefore, only seven cases which are within the conditions of the hypothesis; of which seven, veracity exists in six, and falsehood in one. Resuming our former symbols, in five cases out of twelve T is not both an A and a C, but an A only, or a C only. The cases in which it is both are only seven, in six of which it is a B, in one not a B, making the chance six to one, or $\frac{6}{7}$ and $\frac{1}{7}$ respectively.

In this correct, as in the former incorrect computation, it is of course presupposed that the probabilities arising from A and C are independent of each other. There must not be any such connexion between A and C, that when a thing belongs to the one class it will therefore belong to the other, or even have a greater chance of doing so.   Otherwise the not-Bs which are Cs may be, most or even all of them, identical with the not-Bs which are As; in which last case the probability arising from A and C together will be no greater than that arising from A alone.

When approximate generalizations are joined together in the other mode, that of deduction, the degree of probability of the inference, instead of increasing, diminishes at each step. From two such premises as Most A are B, Most B are C, we cannot with certainty conclude that even a single A is C; for the whole of the portion of A which in any way falls under B, may perhaps be comprised in the exceptional part of it. Still, the two propositions in question afford an appreciable probability that any given A is C, provided the average on which the second proposition is grounded, was taken fairly with reference to the first; provided the proposition, Most B are C, was arrived at in a manner leaving no suspicion that the probability arising from it is otherwise than fairly distributed over the section of B which belongs to A.   For though the instances which are A *may* be all in the minority, they may, also, be all in the majority; and the one possibility is to be set against the other.   On the whole, the probability arising from the two propositions taken together, will be correctly measured by the probability arising from the one, abated in the ratio of that arising from the other.   If nine out of ten Swedes have light hair, and eight out of nine inhabitants of Stockholm are Swedes, the probability arising from these two propositions, that any given inhabitant of Stockholm is light-haired, will amount to eight in ten; though it is rigorously possible that the whole Swedish population of Stockholm might belong to that tenth section of the people of Sweden who are an exception to the rest.

If the premises are known to be true not of a bare majority, but of nearly the whole, of their respective subjects, we may go on joining one such proposition to another for several steps, before we reach a conclusion not presumably true even of a majority.   The error of the conclusion will amount to the aggregate of the errors of all the premises.

Let the proposition, Most A are B, be true of nine in ten; Most B are C, of eight in nine: then not only will one A in ten not be C, because not B, but even of the nine-tenths which are B, only eight-ninths will be C: that is, the cases of A which are C will be only $\frac{8}{9}$ of $\frac{9}{10}$, or four-fifths. Let us now add Most C are D, and suppose this to be true of seven cases out of eight; the proportion of A which is D will be only $\frac{7}{8}$ of $\frac{8}{9}$ of $\frac{9}{10}$, or $\frac{7}{10}$. Thus the probability progressively dwindles. The experience, however, on which our approximate generalizations are grounded, has so rarely been subjected to, or admits of, accurate numerical estimation, that we cannot in general apply any measurement to the diminution of probability which takes place at each illation; but must be content with remembering that it does diminish at every step, and that unless the premises approach very nearly indeed to being universally true, the conclusion after a very few steps is worth nothing. A hearsay of a hearsay, or an argument from presumptive evidence depending not on immediate marks but on marks of marks, is worthless at a very few removes from the first stage.

§ 7. There are, however, two cases in which reasonings depending on approximate generalizations may be carried to any length we please with as much assurance, and are as strictly scientific, as if they were composed of universal laws of nature. But these cases are exceptions of the sort which are currently said to prove the rule. The approximate generalizations are as suitable, in the cases in question, for purposes of ratiocination, as if they were complete generalizations, because they are capable of being transformed into complete generalizations exactly equivalent.

First: If the approximate generalization is of the class in which our reason for stopping at the approximation is not the impossibility, but only the inconvenience, of going further; if we are cognizant of the character which distinguishes the cases that accord with the generalization from those which are exceptions to it; we may then substitute for the approximate proposition, an universal proposition with a proviso. The proposition, Most persons who have uncontrolled power employ it ill, is a generalization of this class, and may be transformed into the following:—All persons who have uncontrolled power employ it ill, provided they are not persons of unusual strength of judgment and rectitude of purpose. The proposition, carrying the hypothesis or proviso with it, may then be dealt with no longer as an approximate, but as an universal proposition; and to whatever number of steps the reasoning may reach, the hypothesis, being carried forward to the conclusion, will exactly indicate how far that conclusion is from being applicable universally. If in the course of the argument other approximate generalizations are introduced, each of them being in like manner expressed as an universal proposition with a condition annexed, the sum of all the conditions will appear at the end as the sum of all the errors which affect the conclusion. Thus, to the proposition last cited, let us add the following:—All absolute monarchs have uncontrolled power, unless their position is such that they need the active support of their subjects (as was the case with Queen Elizabeth, Frederick of Prussia, and others). Combining these two propositions, we can deduce from them an universal conclusion, which will be subject to both the hypotheses in the premises; All absolute monarchs employ their power ill, unless their position makes them need the active support of their subjects, or unless they are persons of unusual strength of judgment and rectitude of purpose. It is of no consequence how rapidly the errors in our premises accumulate, if we are able, in this manner to record each error, and keep an account of the aggregate as it swells up.

Secondly: there is a case in which approximate propositions, even without our taking note of the conditions under which they are not true of individual cases, are yet, for the purposes of science, universal ones; namely, in the inquiries which relate to the properties not of individuals, but of multitudes. The principal of these is the science of politics, or of human society. This science is principally concerned with the actions not of solitary individuals, but of masses; with the fortunes not of single persons, but of communities.

For the statesman, therefore, it is generally enough to know that *most* persons act or are acted upon in a particular way; since his speculations and his practical arrangements refer almost exclusively to cases in which the whole community, or some large portion of it, is acted upon at once, and in which, therefore, what is done or felt by *most* persons determines the result produced by or upon the body at large. He can get on well enough with approximate generalizations on human nature, since what is true approximately of all individuals is true absolutely of all masses. And even when the operations of individual men have a part to play in his deductions, as when he is reasoning of kings, or other single rulers, still, as he is providing for indefinite duration, involving an indefinite succession of such individuals, he must in general both reason and act as if what is true of most persons were true of all.

The two kinds of considerations above adduced are a sufficient refutation of the popular error, that speculations on society and government, as resting on merely probable evidence, must be inferior in certainty and scientific accuracy to the conclusions of what are called the exact sciences, and less to be relied on in practice. There are reasons enough why the moral sciences must remain inferior to at least the more perfect of the physical: why the laws of their more complicated phenomena cannot be so completely deciphered, nor the phenomena predicted with the same degree of assurance. But though we cannot attain to so many truths, there is no reason that those we can attain should deserve less reliance, or have less of a scientific character. Of this topic, however, I shall treat more systematically in the concluding Book, to which place any further consideration of it must be deferred.

CHAPTER XXIV.

OF THE REMAINING LAWS OF NATURE.

§ 1. In the First Book, we found that all the assertions which can be conveyed by language, express some one or more of five different things: Existence; Order in Place; Order in Time; Causation; and Resemblance.* Of these, Causation, in our view of the subject, not being fundamentally different from Order in Time, the five species of possible assertions are reduced to four. The propositions which affirm Order in Time, in either of its two modes, Coexistence and Succession, have formed, thus far, the subject of the present Book. And we have now concluded the exposition, so far as it falls within the limits assigned to this work, of the nature of the evidence on which these propositions rest, and the processes of investigation by which they are ascertained and proved. There remain three classes of facts: Existence, Order in Place, and Resemblance; in regard to which the same questions are now to be resolved.

Regarding the first of these, very little needs be said. Existence in general, is a subject not for our science, but for metaphysics. To determine what things can be recognised as really existing, independently of our own sensible or other impressions, and in what meaning the term is, in that case, predicated of them, belongs to the consideration of "Things in themselves," from which, throughout this work, we have as much as possible kept aloof. Existence, so far as Logic is concerned about it, has reference only to phenomena; to actual, or possible, states of external or internal consciousness, in ourselves or others. Feelings of sensitive beings, or possibilities of having such feelings, are the only things the exist-

---

* Supra, vol. i. p. 115.

ence of which can be a subject of logical induction, because the only things of which the existence in individual cases can be a subject of experience.

It is true that a thing is said by us to exist, even when it is absent, and therefore is not and cannot be perceived. But even then, its existence is to us only another word for our conviction that we should perceive it on a certain supposition; namely, if we were in the needful circumstances of time and place, and endowed with the needful perfection of organs. My belief that the Emperor of China exists, is simply my belief that if I were transported to the imperial palace or some other locality in Pekin, I should see him. My belief that Julius Cæsar existed, is my belief that I should have seen him if I had been present in the field of Pharsalia, or in the senate-house at Rome. When I believe that stars exist beyond the utmost range of my vision, though assisted by the most powerful telescopes yet invented, my belief, philosophically expressed, is, that with still better telescopes, if such existed, I could see them, or that they may be perceived by beings less remote from them in space, or whose capacities of perception are superior to mine.

The existence, therefore, of a phenomenon, is but another word for its being perceived, or for the inferred possibility of perceiving it. When the phenomenon is within the range of present observation, by present observation we assure ourselves of its existence; when it is beyond that range, and is therefore said to be absent, we infer its existence from marks or evidences. But what can these evidences be? Other phenomena; ascertained by induction to be connected with the given phenomenon, either in the way of succession or of coexistence. The simple existence, therefore, of an individual phenomenon, when not directly perceived, is inferred from some inductive law of succession or coexistence: and is consequently not amenable to any peculiar inductive principles. We prove the existence of a thing, by proving that it is connected by succession or coexistence with some known thing.

With respect to *general* propositions of this class, that is, which affirm the bare fact of existence, they have a peculiarity which renders the logical treatment of them a very easy matter; they are generalizations which are sufficiently proved by a single instance. That ghosts, or unicorns, or sea-serpents exist, would be fully established if it could be ascertained positively that such things had been even once seen. Whatever has once happened, is capable of happening again; the only question relates to the conditions under which it happens.

So far, therefore, as relates to simple existence, the Inductive Logic has no knots to untie. And we may proceed to the remaining two of the great classes into which facts have been divided; Resemblance, and Order in Space.

§ 2. Resemblance and its opposite, except in the case in which they assume the names of Equality and Inequality, are seldom regarded as subjects of science; they are supposed to be perceived by simple apprehension; by merely applying our senses or directing our attention to the two objects at once, or in immediate succession. And this simultaneous, or virtually simultaneous, application of our faculties to the two things which are to be compared, does necessarily constitute the ultimate appeal, wherever such application is practicable. But, in most cases, it is not practicable: the objects cannot be brought so close together that the feeling of their resemblance (at least a complete feeling of it) directly arises in the mind. We can only compare each of them with some third object, capable of being transported from one to the other. And besides, even when the objects can be brought into immediate juxtaposition, their resemblance or difference is but imperfectly known to us, unless we have compared them minutely, part by part. Until this has been done, things in reality very dissimilar often appear undistinguishably alike. Two lines of very unequal length will appear about equal when lying in different directions; but place them parallel, with their farther extremities even, and if we look at the nearer extremities, their inequality becomes a matter of direct perception.

To ascertain whether, and in what, two phenomena resemble or differ, is not always, therefore, so easy a thing as it might at first appear. When the two cannot be brought into juxtaposition, or not so that the observer is able to compare their several parts in detail, he must employ the indirect means of reasoning and general propositions. When we cannot bring two straight lines together, to determine whether they are equal, we do it by the physical aid of a foot rule applied first to one and then to the other, and the logical aid of the general proposition or formula, "Things which are equal to the same thing are equal to one another." The comparison of two things through the intervention of a third thing, when their direct comparison is impossible, is the appropriate scientific process for ascertaining resemblances and dissimilarities, and is the sum total of what Logic has to teach on the subject.

An undue extension of this remark induced Locke to consider reasoning itself as nothing but the comparison of two ideas through the medium of a third, and knowledge as the perception of the agreement or disagreement of two ideas: doctrines which the Condillac school blindly adopted, without the qualifications and distinctions with which they were studiously guarded by their illustrious author. Where, indeed, the agreement or disagreement (otherwise called resemblance or dissimilarity) of any two things is the very matter to be determined, as is the case particularly in the sciences of quantity and extension; there, the process by which a solution, if not attainable by direct perception, must be indirectly sought, consists in comparing these two things through the medium of a third. But this is far from being true of all inquiries. The knowledge that bodies fall to the ground is not a perception of agreement or disagreement, but of a series of physical occurrences, a succession of sensations. Locke's definitions of knowledge and of reasoning required to be limited to our knowledge of, and reasoning about, resemblances. Nor, even when thus restricted, are the propositions strictly correct; since the comparison is not made, as he represents, between the ideas of the two phenomena, but between the phenomena themselves. This mistake has been pointed out in an earlier part of our inquiry,* and we traced it to an imperfect conception of what takes place in mathematics, where very often the comparison is really made between the ideas, without any appeal to the outward senses; only, however, because in mathematics a comparison of the ideas is strictly equivalent to a comparison of the phenomena themselves. Where, as in the case of numbers, lines, and figures, our idea of an object is a complete picture of the object, so far as respects the matter in hand; we can, of course, learn from the picture, whatever could be learnt from the object itself by mere contemplation of it as it exists at the particular instant when the picture is taken. No mere contemplation of gunpowder would ever teach us that a spark would make it explode, nor, consequently, would the contemplation of the idea of gun-powder do so: but the mere contemplation of a straight line shows that it cannot inclose a space: accordingly the contemplation of the idea of it will show the same. What takes place in mathematics is thus no argument that the comparison is between the ideas only. It is always, either indirectly or directly, a comparison of the phenomena.

In cases in which we cannot bring the phenomena to the test of direct inspection at all, or not in a manner sufficiently precise, but must judge of their resemblance by inference from other resemblances or dissimilarities more accessible to observation, we of course require, as in all cases of ratiocination, generalizations or formulæ applicable to the subject. We must reason from laws of nature; from the uniformities which are observable in the fact of likeness or unlikeness.

§ 3. Of these laws or uniformities, the most comprehensive are those supplied by mathematics; the axioms relating to equality, inequality, and proportionality, and the various theorems thereon founded. And these are the only Laws of Resemblance which require to be, or which can be, treated apart. It is true there are innumerable other

---

* Supra, book i. ch. v. § 1, and book ii. ch. v. § 5.

theorems which affirm resemblances among phenomena; as that the angle of the reflection of light is *equal* to its angle of incidence (equality being merely exact resemblance in magnitude). Again, that the heavenly bodies describe *equal* areas in equal times; and that their periods of revolution are *proportional* (another species of resemblance) to the sesquiplicate powers of their distances from the centre of force. These and similar propositions affirm resemblances, of the same nature with those asserted in the theorems of mathematics; but the distinction is, that the propositions of mathematics are true of all phenomena whatever, or at least without distinction of origin; while the truths in question are affirmed only of special phenomena, which originate in a certain way; and the equalities, proportionalities, or other resemblances, which exist between such phenomena, must necessarily be either derived from, or identical with, the law of their origin—the law of causation on which they depend. The equality of the areas described in equal times by the planets, is *derived* from the laws of the causes; and, until its derivation was shown, it was an empirical law. The equality of the angles of reflexion and incidence is *identical* with the law of the cause; for the cause is the incidence of a ray of light upon a reflecting surface, and the equality in question is the very law according to which that cause produces its effects. This class, therefore, of the uniformities of resemblance between phenomena, are inseparable, in fact and in thought, from the laws of the production of those phenomena: and the principles of induction applicable to them are no other than those of which we have treated in the preceding chapters of this Book.

It is otherwise with the truths of mathematics. The laws of equality and inequality between spaces, or between numbers, have no connexion with laws of causation. That the angle of reflexion is equal to the angle of incidence, is a statement of the mode of action of a particular cause; but that when two straight lines intersect each other the opposite angles are equal, is true of all such lines and angles, by whatever cause produced. That the squares of the periodic times of the planets are proportional to the cubes of their distances from

the sun, is an uniformity derived from the laws of the causes
(or forces) which produce the planetary motions; but that the
square of any number is four times the square of half the
number, is true independently of any cause. The only laws
of resemblance, therefore, which we are called upon to consider
independently of causation, belong to the province of mathematics.

§ 4. The same thing is evident with respect to the only
one remaining of our five categories, Order in Place. The
order in place, of the effects of a cause, is (like everything else
belonging to the effects) a consequence of the laws of that
cause. The order in place, or, as we have termed it, the collocation, of the primeval causes, is (as well as their resemblance) in each instance an ultimate fact, in which no laws or
uniformities are traceable. The only remaining general propositions respecting order in place, and the only ones which
have nothing to do with causation, are some of the truths of
geometry; laws through which we are able, from the order in
place of certain points, lines, or spaces, to infer the order in
place of others which are connected with the former in some
known mode; quite independently of the particular nature of
those points, lines, or spaces, in any other respect than position or magnitude, as well as independently of the physical
cause from which in any particular case they happen to derive
their origin.

It thus appears that mathematics is the only department
of science into the methods of which it still remains to inquire.
And there is the less necessity that this inquiry should occupy
us long, as we have already, in the Second Book, made considerable progress in it. We there remarked, that the directly
inductive truths of mathematics are few in number; consisting
of the axioms, together with certain propositions concerning
existence, tacitly involved in most of the so-called definitions.
And we gave what appeared conclusive reasons for affirming
that these original premises, from which the remaining truths
of the science are deduced, are, notwithstanding all appearances
to the contrary, results of observation and experience; founded,

in short, on the evidence of the senses. That things equal to
the same thing are equal to one another, and that two straight
lines which have once intersected one another continue to
diverge, are inductive truths; resting, indeed, like the law of
universal causation, only on induction *per enumerationem
simplicem;* on the fact that they have been perpetually perceived to be true, and never once found to be false. But, as
we have seen in a recent chapter that this evidence, in the
case of a law so completely universal as the law of causation,
amounts to the fullest proof, so is this even more evidently
true of the general propositions to which we are now adverting; because, as a perception of their truth in any individual
case whatever, requires only the simple act of looking at the
objects in a proper position, there never could have been in
their case (what, for a long period, there were in the case of
the law of causation) instances which were apparently, though
not really, exceptions to them. Their infallible truth was
recognised from the very dawn of speculation; and as their
extreme familiarity made it impossible for the mind to conceive
the objects under any other law, they were, and still are,
generally considered as truths recognised by their own evidence, or by instinct.

§ 5. There is something which seems to require explanation, in the fact that the immense multitude of truths (a multitude still as far from being exhausted as ever) comprised in
the mathematical sciences, can be elicited from so small a
number of elementary laws. One sees not, at first, how it is
that there can be room for such an infinite variety of true
propositions, on subjects apparently so limited.

To begin with the science of number. The elementary or
ultimate truths of this science are the common axioms concerning equality, namely, "Things which are equal to the
same thing are equal to one another," and "Equals added to
equals make equal sums," (no other axioms are required,*)

---

* The axiom, "Equals subtracted from equals leave equal differences,"
may be demonstrated from the two axioms in the text. If A = a and B = b,

together with the definitions of the various numbers. Like other so-called definitions, these are composed of two things, the explanation of a name, and the assertion of a fact: of which the latter alone can form a first principle or premise of a science. The fact asserted in the definition of a number is a physical fact. Each of the numbers two, three, four, &c., denotes physical phenomena, and connotes a physical property of those phenomena. Two, for instance, denotes all pairs of things, and twelve all dozens of things, connoting what makes them pairs, or dozens; and that which makes them so is something physical; since it cannot be denied that two apples are physically distinguishable from three apples, two horses from one horse, and so forth: that they are a different visible and tangible phenomenon. I am not undertaking to say what the difference is; it is enough that there is a difference of which the senses can take cognizance. And although a hundred and two horses are not so easily distinguished from a hundred and three, as two horses are from three—though in most positions the senses do not perceive any difference—yet they may be so placed that a difference will be perceptible, or else we should never have distinguished them, and given them different names. Weight is confessedly a physical property of things; yet small differences between great weights are as imperceptible to the senses in most situations, as small differences between great numbers; and are only put in evidence by placing the two objects in a peculiar position—namely, in the opposite scales of a delicate balance.

$A - B = a - b$. For if not, let $A - B = a - b + c$. Then since $B = b$, adding equals to equals, $A = a + c$. But $A = a$. Therefore $a = a + c$, which is impossible.

This proposition having been demonstrated, we may, by means of it, demonstrate the following: "If equals be added to unequals, the sums are unequal." If $A = a$ and B not $= b$, $A + B$ is not $= a + b$. For suppose it be so. Then, since $A = a$ and $A + B = a + b$, subtracting equals from equals, $B = b$; which is contrary to the hypothesis.

So again, it may be proved that two things, one of which is equal and the other unequal to a third thing, are unequal to one another. If $A = a$ and A not $= B$, neither is $a = B$. For suppose it to be equal. Then since $A = a$ and $a = B$, and since things equal to the same thing are equal to one another, $A = B$: which is contrary to the hypothesis.

What, then, is that which is connoted by a name of number? Of course some property belonging to the agglomeration of things which we call by the name; and that property is, the characteristic manner in which the agglomeration is made up of, and may be separated into, parts. I will endeavour to make this more intelligible by a few explanations.

When we call a collection of objects *two*, *three*, or *four*, they are not two, three, or four in the abstract; they are two, three, or four things of some particular kind; pebbles, horses, inches, pounds weight. What the name of number connotes is, the manner in which single objects of the given kind must be put together, in order to produce that particular aggregate. If the aggregate be of pebbles, and we call it *two*, the name implies that, to compose the aggregate, one pebble must be joined to one pebble. If we call it *three*, one and one and one pebble must be brought together to produce it, or else one pebble must be joined to an aggregate of the kind called *two*, already existing. The aggregate which we call *four*, has a still greater number of characteristic modes of formation. One and one and one and one pebble may be brought together; or two aggregates of the kind called *two* may be united; or one pebble may be added to an aggregate of the kind called *three*. Every succeeding number in the ascending series, may be formed by the junction of smaller numbers in a progressively greater variety of ways. Even limiting the parts to two, the number may be formed, and consequently may be divided, in as many different ways as there are numbers smaller than itself; and, if we admit of threes, fours, &c., in a still greater variety. Other modes of arriving at the same aggregate present themselves, not by the union of smaller, but by the dismemberment of larger aggregates. Thus, *three pebbles* may be formed by taking away one pebble from an aggregate of four; *two pebbles*, by an equal division of a similar aggregate; and so on.

Every arithmetical proposition; every statement of the result of an arithmetical operation; is a statement of one of the modes of formation of a given number. It affirms that a certain aggregate might have been formed by putting together certain other aggregates, or by withdrawing certain portions of some aggregate; and that, by consequence, we might reproduce those aggregates from it, by reversing the process.

Thus, when we say that the cube of 12 is 1728, what we affirm is this: that if, having a sufficient number of pebbles or of any other objects, we put them together into the particular sort of parcels or aggregates called twelves; and put together these twelves again into similar collections; and, finally, make up twelve of these largest parcels; the aggregate thus formed will be such a one as we call 1728; namely, that which (to take the most familiar of its modes of formation) may be made by joining the parcel called a thousand pebbles, the parcel called seven hundred pebbles, the parcel called twenty pebbles, and the parcel called eight pebbles.

The converse proposition, that the cube root of 1728 is 12, asserts that this large aggregate may again be decomposed into the twelve twelves of twelves of pebbles which it consists of.

The modes of formation of any number are innumerable; but when we know one mode of formation of each, all the rest may be determined deductively. If we know that a is formed from b and c, b from a and e, c from d and f, and so forth, until we have included all the numbers of any scale we choose to select, (taking care that for each number the mode of formation be really a distinct one, not bringing us round again to the former numbers, but introducing a new number,) we have a set of propositions from which we may reason to all the other modes of formation of those numbers from one another. Having established a chain of inductive truths connecting together all the numbers of the scale, we can ascertain the formation of any one of those numbers from any other by merely travelling from one to the other along the chain. Suppose that we know only the following modes of formation: $6 = 4 + 2$, $4 = 7 - 3$, $7 = 5 + 2$, $5 = 9 - 4$. We could determine how 6 may be formed from 9. For $6 = 4 + 2 = 7 - 3 + 2 = 5 + 2 - 3 + 2 = 9 - 4 + 2 - 3 + 2$. It may therefore be formed by taking away 4 and 3, and adding 2 and 2. If we know besides that $2 + 2 = 4$, we obtain 6 from 9 in a simpler mode, by merely taking away 3.

It is sufficient, therefore, to select one of the various modes of formation of each number, as a means of ascertaining all the rest. And since things which are uniform, and therefore simple, are most easily received and retained by the understanding, there is an obvious advantage in selecting a mode of formation which shall be alike for all ; in fixing the connotation of names of number on one uniform principle. The mode in which our existing numerical nomenclature is contrived possesses this advantage, with the additional one, that it happily conveys to the mind two of the modes of formation of every number. Each number is considered as formed by the addition of an unit to the number next below it in magnitude, and this mode of formation is conveyed by the place which it occupies in the series. And each is also considered as formed by the addition of a number of units less than ten, and a number of aggregates each equal to one of the successive powers of ten ; and this mode of its formation is expressed by its spoken name, and by its numerical character.

What renders arithmetic the type of a deductive science, is the fortunate applicability to it of a law so comprehensive as " The sums of equals are equals:" or (to express the same principle in less familiar but more characteristic language), Whatever is made up of parts, is made up of the parts of those parts. This truth, obvious to the senses in all cases which can be fairly referred to their decision, and so general as to be coextensive with nature itself, being true of all sorts of phenomena, (for all admit of being numbered,) must be considered an inductive truth, or law of nature, of the highest order. And every arithmetical operation is an application of this law, or of other laws capable of being deduced from it. This is our warrant for all calculations. We believe that five and two are equal to seven, on the evidence of this inductive law, combined with the definitions of those numbers. We arrive at that conclusion (as all know who remember how they first learned it) by adding a single unit at a time: $5 + 1 = 6$, therefore $5 + 1 + 1 = 6 + 1 = 7$: and again $2 = 1 + 1$, therefore $5 + 2 = 5 + 1 + 1 = 7$.

§ 6. Innumerable as are the true propositions which can be formed concerning particular numbers, no adequate conception could be gained, from these alone, of the extent of the truths composing the science of number. Such propositions as we have spoken of are the least general of all numerical truths. It is true that even these are coextensive with all nature: the properties of the number four are true of all objects that are divisible into four equal parts, and all objects are either actually or ideally so divisible. But the propositions which compose the science of algebra are true, not of a particular number, but of all numbers; not of all things under the condition of being divided in a particular way, but of all things under the condition of being divided in any way—of being designated by a number at all.

Since it is impossible for different numbers to have any of their modes of formation completely in common, it is a kind of paradox to say, that all propositions which can be made concerning numbers relate to their modes of formation from other numbers, and yet that there are propositions which are true of all numbers. But this very paradox leads to the real principle of generalization concerning the properties of numbers. Two different numbers cannot be formed in the same manner from the same numbers; but they may be formed in the same manner from different numbers; as nine is formed from three by multiplying it into itself, and sixteen is formed from four by the same process. Thus there arises a classification of modes of formation, or in the language commonly used by mathematicians, a classification of Functions. Any number, considered as formed from any other number, is called a function of it; and there are as many kinds of functions as there are modes of formation. The simple functions are by no means numerous, most functions being formed by the combination of several of the operations which form simple functions, or by successive repetitions of some one of those operations. The simple functions of any number x are all reducible to the following forms: $x + a$, $x - a$, $a\,x$, $\dfrac{x}{a}$, x^a, $\sqrt[a]{x}$, log. x (to the base a), and the same expressions varied by putting x for a and a for x, wherever that substitution would alter the value: to which perhaps ought to be added sin x, and arc (sin$=x$). All other functions of x are formed by putting some one or more of the simple functions in the place of x or a, and subjecting them to the same elementary operations.

In order to carry on general reasonings on the subject of Functions, we require a nomenclature enabling us to express any two numbers by names which, without specifying what particular numbers they are, shall show what function each is of the other; or, in other words, shall put in evidence their mode of formation from one another. The system of general language called algebraical notation does this. The expressions a and $a^2 + 3a$ denote, the one any number, the other the number formed from it in a particular manner. The expressions a, b, n, and $(a + b)^n$, denote any three numbers, and a fourth which is formed from them in a certain mode.

The following may be stated as the general problem of the algebraical calculus : F being a certain function of a given number, to find what function F will be of any function of that number. For example, a binomial $a + b$ is a function of its two parts a and b, and the parts are, in their turn, functions of $a + b$: now $(a + b)^n$ is a certain function of the binomial; what function will this be of a and b, the two parts ? The answer to this question is the binomial theorem.

The formula $(a + b)^n = a^n + \dfrac{n}{1}\,a^{n-1}b + \dfrac{n.\overline{n-1}}{1.2}a^{n-2}b^2 + \&c.$, shows in what manner the number which is formed by multiplying $a + b$ into itself n times, might be formed without that process, directly from a, b, and n. And of this nature are all the theorems of the science of number. They assert the identity of the result of different modes of formation. They affirm that some mode of formation from x, and some mode of formation from a certain function of x, produce the same number.

Besides these general theorems of formulæ, what remains in the algebraical calculus is the resolution of equations. But the resolution of an equation is also a theorem. If the equation be $x^2 + ax = b$, the resolution of this equation, viz. $x = -\frac{1}{2} a + \sqrt{\frac{1}{4} a^2 + b}$, is a general proposition, which may be regarded as an answer to the question, If b is a certain function of x and a (namely $x^2 + ax$), what function is x of b and a? The resolution of equations is, therefore, a mere variety of the general problem as above stated. The problem is—Given a function, what function is it of some other function? And in the resolution of an equation, the question is, to find what function of one of its own functions the number itself is.

Such as above described, is the aim and end of the calculus. As for its processes, every one knows that they are simply deductive. In demonstrating an algebraical theorem, or in resolving an equation, we travel from the *datum* to the *quæsitum* by pure ratiocination; in which the only premises introduced, besides the original hypotheses, are the fundamental axioms already mentioned—that things equal to the same thing are equal to one another, and that the sums of equal things are equal. At each step in the demonstration or in the calculation, we apply one or other of these truths, or truths deducible from them, as, that the differences, products, &c., of equal numbers are equal.

It would be inconsistent with the scale of this work, and not necessary to its design, to carry the analysis of the truths and processes of algebra any farther; which is also the less needful, as the task has been, to a very great extent, performed by other writers. Peacock's Algebra, and Dr. Whewell's *Doctrine of Limits*, are full of instruction on the subject. The profound treatises of a truly philosophical mathematician, Professor De Morgan, should be studied by every one who desires to comprehend the evidence of mathematical truths, and the meaning of the obscurer processes of the calculus; and the speculations of M. Comte, in his *Cours de Philosophie*

Positive, on the philosophy of the higher branches of mathematics, are among the many valuable gifts for which philosophy is indebted to that eminent thinker.

§ 7. If the extreme generality, and remoteness not so much from sense as from the visual and tactual imagination, of the laws of number, renders it a somewhat difficult effort of abstraction to conceive those laws as being in reality physical truths obtained by observation; the same difficulty does not exist with regard to the laws of extension. The facts of which those laws are expressions, are of a kind peculiarly accessible to the senses, and suggesting eminently distinct images to the fancy. That geometry is a strictly physical science would doubtless have been recognised in all ages, had it not been for the illusions produced by two circumstances. One of these is the characteristic property, already noticed, of the facts of geometry, that they may be collected from our ideas or mental pictures of objects as effectually as from the objects themselves. The other is, the demonstrative character of geometrical truths; which was at one time supposed to constitute a radical distinction between them and physical truths, the latter, as resting on merely probable evidence, being deemed essentially uncertain and unprecise. The advance of knowledge has, however, made it manifest that physical science, in its better understood branches, is quite as demonstrative as geometry. The task of deducing its details from a few comparatively simple principles is found to be anything but the impossibility it was once supposed to be; and the notion of the superior certainty of geometry is an illusion, arising from the ancient prejudice which, in that science, mistakes the ideal data from which we reason, for a peculiar class of realities, while the corresponding ideal data of any deductive physical science are recognised as what they really are, mere hypotheses.

Every theorem in geometry is a law of external nature, and might have been ascertained by generalizing from observation and experiment, which in this case resolve themselves into comparison and measurement. But it was found prac-

ticable, and being practicable, was desirable, to deduce these
truths by ratiocination from a small number of general laws
of nature, the certainty and universality of which are obvious
to the most careless observer, and which compose the first
principles and ultimate premises of the science. Among
these general laws must be included the same two which we
have noticed as ultimate principles of the Science of Number
also, and which are applicable to every description of quantity;
viz. The sums of equals are equal, and Things which are equal
to the same thing are equal to one another; the latter of
which may be expressed in a manner more suggestive of the
inexhaustible multitude of its consequences, by the following
terms: Whatever is equal to any one of a number of equal
magnitudes, is equal to any other of them. To these two must
be added, in geometry, a third law of equality, namely, that
lines, surfaces, or solid spaces, which can be so applied to one
another as to coincide, are equal. Some writers have asserted
that this law of nature is a mere verbal definition; that the
expression "equal magnitudes" *means* nothing but magnitudes which can be so applied to one another as to coincide.
But in this opinion I cannot agree. The equality of two geometrical magnitudes cannot differ fundamentally in its nature
from the equality of two weights, two degrees of heat, or
two portions of duration, to none of which would this pretended definition of equality be suitable. None of these things
can be so applied to one another as to coincide, yet we perfectly understand what we mean when we call them equal.
Things are equal in magnitude, as things are equal in weight,
when they are felt to be exactly similar in respect of the attribute in which we compare them: and the application of the
objects to each other in the one case, like the balancing them
with a pair of scales in the other, is but a mode of bringing
them into a position in which our senses can recognise deficiencies of exact resemblance that would otherwise escape our
notice.

Along with these three general principles or axioms, the
remainder of the premises of geometry consists of the so-called
definitions, that is to say, propositions asserting the real

existence of the various objects therein designated, together
with some one property of each. In some cases more than
one property is commonly assumed, but in no case is more
than one necessary. It is assumed that there are such things
in nature as straight lines, and that any two of them setting
out from the same point, diverge more and more without
limit. This assumption, (which includes and goes beyond
Euclid's axiom that two straight lines cannot inclose a space,)
is as indispensable in geometry, and as evident, resting on as
simple, familiar, and universal observation, as any of the other
axioms. It is also assumed that straight lines diverge from
one another in different degrees; in other words, that there
are such things as angles, and that they are capable of being
equal or unequal. It is assumed that there is such a thing as
a circle, and that all its radii are equal; such things as ellipses,
and that the sums of the focal distances are equal for every
point in an ellipse; such things as parallel lines, and that
those lines are everywhere equally distant.*

§ 8. It is a matter of more than curiosity to consider, to
what peculiarity of the physical truths which are the subject
of geometry, it is owing that they can all be deduced from so

---

* Geometers have usually preferred to define parallel lines by the property
of being in the same plane and never meeting. This, however, has rendered it
necessary for them to assume, as an additional axiom, some other property of
parallel lines; and the unsatisfactory manner in which properties for that
purpose have been selected by Euclid and others has always been deemed the
opprobrium of elementary geometry. Even as a verbal definition, equidistance
is a fitter property to characterize parallels by, since it is the attribute really
involved in the signification of the name. If to be in the same plane and never
to meet were all that is meant by being parallel, we should feel no incongruity
in speaking of a curve as parallel to its asymptote. The meaning of parallel
lines is, lines which pursue exactly the same direction, and which, therefore,
neither draw nearer nor go farther from one another; a conception suggested
at once by the contemplation of nature. That the lines will never meet is of
course included in the more comprehensive proposition that they are everywhere equally distant. And that any straight lines which are in the same
plane and not equidistant will certainly meet, may be demonstrated in the most
rigorous manner from the fundamental property of straight lines assumed in
the text, viz. that if they set out from the same point, they diverge more and
more without limit.

small a number of original premises : why it is that we can set
out from only one characteristic property of each kind of
phenomenon, and with that and two or three general truths
relating to equality, can travel from mark to mark until we
obtain a vast body of derivative truths, to all appearance
extremely unlike those elementary ones.

The explanation of this remarkable fact seems to lie in
the following circumstances. In the first place, all questions
of position and figure may be resolved into questions of
magnitude. The position and figure of any object are determined, by determining the position of a sufficient number of
points in it; and the position of any point may be determined by the magnitude of three rectangular co-ordinates,
that is, of the perpendiculars drawn from the point to three
planes at right angles to one another, arbitrarily selected. By
this transformation of all questions of quality into questions
only of quantity, geometry is reduced to the single problem
of the measurement of magnitudes, that is, the ascertainment
of the equalities which exist between them. Now when we
consider that by one of the general axioms, any equality,
when ascertained, is proof of as many other equalities as
there are other things equal to either of the two equals; and
that by another of those axioms, any ascertained equality is
proof of the equality of as many pairs of magnitudes as
can be formed by the numerous operations which resolve
themselves into the addition of the equals to themselves or
to other equals; we cease to wonder that in proportion as
a science is conversant about equality, it should afford a
more copious supply of marks of marks; and that the sciences
of number and extension, which are conversant with little
else than equality, should be the most deductive of all the
sciences.

There are also two or three of the principal laws of space
or extension which are unusually fitted for rendering one
position or magnitude a mark of another, and thereby con-
tributing to render the science largely deductive. First; the
magnitudes of inclosed spaces, whether superficial or solid,
are completely determined by the magnitudes of the lines

and angles which bound them. Secondly, the length of any line, whether straight or curve, is measured (certain other things being given) by the angle which it subtends, and *vice versâ*. Lastly, the angle which any two straight lines make with each other at an inaccessible point, is measured by the angles they severally make with any third line we choose to select. By means of these general laws, the measurement of all lines, angles, and spaces whatsoever might be accomplished by measuring a single straight line and a sufficient number of angles ; which is the plan actually pursued in the trigonometrical survey of a country; and fortunate it is that this is practicable, the exact measurement of long straight lines being always difficult, and often impossible, but that of angles very easy. Three such generalizations as the foregoing afford such facilities for the indirect measurement of magnitudes, (by supplying us with known lines or angles which are marks of the magnitude of unknown ones, and thereby of the spaces which they inclose,) that it is easily intelligible how from a few data we can go on to ascertain the magnitude of an indefinite multitude of lines, angles, and spaces, which we could not easily, or could not at all, measure by any more direct process.

§ 9. Such are the few remarks which it seemed necessary to make in this place, respecting the laws of nature which are the peculiar subject of the sciences of number and extension. The immense part which those laws take in giving a deductive character to the other departments of physical science, is well known ; and is not surprising, when we consider that all causes operate according to mathematical laws. The effect is always dependent on, or is a function of, the quantity of the agent ; and generally of its position also. We cannot, therefore, reason respecting causation, without introducing considerations of quantity and extension at every step ; and if the nature of the phenomena admits of our obtaining numerical data of sufficient accuracy, the laws of quantity become the grand instrument for calculating forward to an effect, or backward to a cause. That in all other sciences, as well as in geometry, questions of quality are scarcely ever independent of questions of quantity, may be seen from the most familiar phenomena. Even when several colours are mixed on a painter's palette, the comparative quantity of each entirely determines the colour of the mixture.

With this mere suggestion of the general causes which render mathematical principles and processes so predominant in those deductive sciences which afford precise numerical data, I must, on the present occasion, content myself: referring the reader who desires a more thorough acquaintance with the subject, to the first two volumes of M. Comte's systematic work.

In the same work, and more particularly in the third volume, are also fully discussed the limits of the applicability of mathematical principles to the improvement of other sciences. Such principles are manifestly inapplicable, where the causes on which any class of phenomena depend are so imperfectly accessible to our observation, that we cannot ascertain, by a proper induction, their numerical laws; or where the causes are so numerous, and intermixed in so complex a manner with one another, that even supposing their laws known, the computation of the aggregate effect transcends the powers of the calculus as it is, or is likely to be; or lastly, where the causes themselves are in a state of perpetual fluctuation; as in physiology, and still more, if possible, in the social science. The mathematical solutions of physical questions become progressively more difficult and imperfect, in proportion as the questions divest themselves of their abstract and hypothetical character, and approach nearer to the degree of complication actually existing in nature; insomuch that beyond the limits of astronomical phenomena, and of those most nearly analogous to them, mathematical accuracy is generally obtained "at the expense of the reality of the inquiry:" while even in astronomical questions, "notwithstanding the admirable simplicity of their mathematical elements, our feeble intelligence becomes incapable of following out effectually the logical combinations of the laws on which the phenomena are dependent, as soon as we attempt to take into simultaneous

consideration more than two or three essential influences."[*]
Of this, the problem of the Three Bodies has already been
cited, more than once, as a remarkable instance; the complete
solution of so comparatively simple a question having vainly
tried the skill of the most profound mathematicians. We
may conceive, then, how chimerical would be the hope that
mathematical principles could be advantageously applied to
phenomena dependent on the mutual action of the innumerable minute particles of bodies, as those of chemistry, and
still more, of physiology; and for similar reasons those
principles remain inapplicable to the still more complex inquiries, the subjects of which are phenomena of society and
government.

The value of mathematical instruction as a preparation for
those more difficult investigations, consists in the applicability
not of its doctrines, but of its method. Mathematics will ever
remain the most perfect type of the Deductive Method in
general; and the applications of mathematics to the deductive
branches of physics, furnish the only school in which philosophers can effectually learn the most difficult and important
portion of their art, the employment of the laws of simpler
phenomena for explaining and predicting those of the more
complex. These grounds are quite sufficient for deeming
mathematical training an indispensable basis of real scientific
education, and regarding (according to the *dictum* which an
old but unauthentic tradition ascribes to Plato) one who is
ἀγεωμέτρητος, as wanting in one of the most essential qualifications for the successful cultivation of the higher branches of
philosophy.

---

* *Philosophie Positive*, iii. 414–416.

CHAPTER XXV.

OF THE GROUNDS OF DISBELIEF.

§ 1. THE method of arriving at general truths, or general propositions fit to be believed, and the nature of the evidence on which they are grounded, have been discussed, as far as space and the writer's faculties permitted, in the twenty-four preceding chapters. But the result of the examination of evidence is not always belief, nor even suspension of judgment; it is sometimes disbelief. The philosophy, therefore, of induction and experimental inquiry is incomplete, unless the grounds not only of belief, but of disbelief, are treated of; and to this topic we shall devote one, and the final, chapter.

By disbelief is not here to be understood the mere absence of belief. The ground for abstaining from belief is simply the absence or insufficiency of proof; and in considering what is sufficient evidence to support any given conclusion, we have already, by implication, considered what evidence is not sufficient for the same purpose. By disbelief is here meant, not the state of mind in which we form no opinion concerning a subject, but that in which we are fully persuaded that some opinion is not true; insomuch that if evidence, even of great apparent strength, (whether grounded on the testimony of others or on our own supposed perceptions,) were produced in favour of the opinion, we should believe that the witnesses spoke falsely, or that they, or we ourselves if we were the direct percipients, were mistaken.

That there are such cases, no one is likely to dispute. Assertions for which there is abundant positive evidence are often disbelieved, on account of what is called their improbability, or impossibility. And the question for consideration is what, in the present case, these words mean, and how far and in what circumstances the properties which they express are sufficient grounds for disbelief.

§ 2. It is to be remarked in the first place, that the positive evidence produced in support of an assertion which is nevertheless rejected on the score of impossibility or improbability, is never such as amounts to full proof. It is always grounded on some approximate generalization. The fact may have been asserted by a hundred witnesses; but there are many exceptions to the universality of the generalization that what a hundred witnesses affirm is true. We may seem to ourselves to have actually seen the fact: but, that we really see what we think we see, is by no means an universal truth; our organs may have been in a morbid state; or we may have inferred something, and imagined that we perceived it. The evidence, then, in the affirmative being never more than an approximate generalization, all will depend on what the evidence in the negative is. If that also rests on an approximate generalization, it is a case for comparison of probabilities. If the approximate generalizations leading to the affirmative are, when added together, less strong, or in other words, farther from being universal, than the approximate generalizations which support the negative side of the question, the proposition is said to be improbable, and is to be disbelieved provisionally. If however an alleged fact be in contradiction, not to any number of approximate generalizations, but to a completed generalization grounded on a rigorous induction, it is said to be impossible, and is to be disbelieved totally.

This last principle, simple and evident as it appears, is the doctrine which, on the occasion of an attempt to apply it to the question of the credibility of miracles, excited so violent a controversy. Hume's celebrated doctrine, that nothing is credible which is contradictory to experience, or at variance with laws of nature, is merely this very plain and harmless proposition, that whatever is contradictory to a complete induction is incredible. That such a maxim as this should either be accounted a dangerous heresy, or mistaken for a great and

recondite truth, speaks ill for the state of philosophical speculation on such subjects.

But does not (it may be asked) the very statement of the
proposition imply a contradiction ? An alleged fact, according
to this theory, is not to be believed if it contradict a complete
induction. But it is essential to the completeness of an induction that it shall not contradict any known fact. Is it not
then a *petitio principii* to say, that the fact ought to be disbelieved because the induction opposed to it is complete ? How
can we have a right to declare the induction complete, while
facts, supported by credible evidence, present themselves in
opposition to it ?

I answer, we have that right whenever the scientific canons
of induction give it to us ; that is, whenever the induction *can*
be complete. We have it, for example, in a case of causation
in which there has been an *experimentum crucis*. If an antecedent A, superadded to a set of antecedents in all other
respects unaltered, is followed by an effect B which did not
exist before, A is, in that instance at least, the cause of B, or
an indispensable part of its cause ; and if A be tried again
with many totally different sets of antecedents and B still
follows, then it is the whole cause. If these observations or
experiments have been repeated so often, and by so many persons,
as to exclude all supposition of error in the observer, a law of
nature is established ; and so long as this law is received as
such, the assertion that on any particular occasion A took place,
and yet B did not follow, *without any counteracting cause*, must
be disbelieved. Such an assertion is not to be credited on any
less evidence than what would suffice to overturn the law.
The general truths, that whatever has a beginning has a cause,
and that when none but the same causes exist, the same effects
follow, rest on the strongest inductive evidence possible ; the
proposition that things affirmed by even a crowd of respectable
witnesses are true, is but an approximate generalization ; and
—even if we fancy we actually saw or felt the fact which is in
contradiction to the law—what a human being can see is no
more than a set of appearances ; from which the real nature
of the phenomenon is merely an inference, and in this infe-

rence approximate generalizations usually have a large share.
If, therefore, we make our election to hold by the law, no
quantity of evidence whatever ought to persuade us that there
has occurred anything in contradiction to it. If, indeed, the
evidence produced is such that it is more likely that the set of
observations and experiments on which the law rests should
have been inaccurately performed or incorrectly interpreted,
than that the evidence in question should be false, we may
believe the evidence; but then we must abandon the law.
And since the law was received on what seemed a complete
induction, it can only be rejected on evidence equivalent;
namely, as being inconsistent not with any number of approximate generalizations, but with some other and better established law of nature. This extreme case, of a conflict between
two supposed laws of nature, has probably never actually
occurred where, in the process of investigating both the laws,
the true canons of scientific induction had been kept in view;
but if it did occur, it must terminate in the total rejection of
one of the supposed laws. It would prove that there must be
a flaw in the logical process by which either one or the other
was established: and if there be so, that supposed general truth
is no truth at all. We cannot admit a proposition as a law of
nature, and yet believe a fact in real contradiction to it. We
must disbelieve the alleged fact, or believe that we were mistaken in admitting the supposed law.

But in order that any alleged fact should be contradictory
to a law of causation, the allegation must be, not simply that
the cause existed without being followed by the effect, for that
would be no uncommon occurrence; but that this happened in
the absence of any adequate counteracting cause. Now in the
case of an alleged miracle, the assertion is the exact opposite of
this. It is, that the effect was defeated, not in the absence,
but in consequence of a counteracting cause, namely, a direct
interposition of an act of the will of some being who has
power over nature; and in particular of a Being, whose will
being assumed to have endowed all the causes with the
powers by which they produce their effects, may well be
supposed able to counteract them. A miracle (as was justly

remarked by Brown*) is no contradiction to the law of cause and effect; it is a new effect, supposed to be produced by the introduction of a new cause. Of the adequacy of that cause, if present, there can be no doubt; and the only antecedent improbability which can be ascribed to the miracle, is the improbability that any such cause existed.

All, therefore, which Hume has made out, and this he must be considered to have made out, is, that (at least in the imperfect state of our knowledge of natural agencies, which leaves it always possible that some of the physical antecedents may have been hidden from us,) no evidence can prove a miracle to any one who did not previously believe the existence of a being or beings with supernatural power; or who believes himself to have full proof that the character of the Being whom he recognises, is inconsistent with his having seen fit to interfere on the occasion in question.

If we do not already believe in supernatural agencies, no miracle can prove to us their existence. The miracle itself, considered merely as an extraordinary fact, may be satisfactorily certified by our senses or by testimony; but nothing can ever prove that it is a miracle: there is still another possible hypothesis, that of its being the result of some unknown natural cause: and this possibility cannot be so completely shut out, as to leave no alternative but that of admitting the existence and intervention of a being superior to nature. Those, however, who already believe in such a being, have two hypotheses to choose from, a supernatural and an unknown natural agency; and they have to judge which of the two is the most probable in the particular case. In forming this judgment, an important element of the question will be the conformity of the result to the laws of the supposed agent, that is, to the character of the Deity as they conceive it. But, with the knowledge which we now possess of the general uniformity of the course of nature, religion, following in the wake of science, has been compelled to acknowledge the government of the

---

* See the two remarkable notes (A) and (F), appended to his *Inquiry into the Relation of Cause and Effect.*

universe as being on the whole carried on by general laws, and
not by special interpositions. To whoever holds this belief,
there is a general presumption against any supposition of divine
agency not operating through general laws, or in other words,
there is an antecedent improbability in every miracle, which,
in order to outweigh it, requires an extraordinary strength of
antecedent probability derived from the special circumstances
of the case.

§ 3. It appears from what has been said, that the assertion that a cause has been defeated of an effect which is connected with it by a completely ascertained law of causation,
is to be disbelieved or not, according to the probability or
improbability that there existed in the particular instance an
adequate counteracting cause. To form an estimate of this,
is not more difficult than of other probabilities. With regard to all *known* causes capable of counteracting the given
causes, we have generally some previous knowledge of the
frequency or rarity of their occurrence, from which we may
draw an inference as to the antecedent improbability of their
having been present in any particular case. And neither in
respect to known or unknown causes are we required to pronounce on the probability of their existing in nature, but only
of their having existed at the time and place at which the
transaction is alleged to have happened. We are seldom,
therefore, without the means (when the circumstances of the
case are at all known to us) of judging how far it is likely
that such a cause should have existed at that time and place
without manifesting its presence by some other marks, and (in
the case of an unknown cause) without having hitherto manifested its existence in any other instance. According as this
circumstance, or the falsity of the testimony, appears more
improbable, that is, conflicts with an approximate generalization of a higher order, we believe the testimony, or disbelieve
it; with a stronger or a weaker degree of conviction, according to the preponderance: at least until we have sifted the
matter further.

So much, then, for the case in which the alleged fact conflicts, or appears to conflict, with a real law of causation. But a more common case, perhaps, is that of its conflicting with uniformities of mere coexistence, not proved to be dependent on causation : in other words, with the properties of Kinds. It is with these uniformities principally, that the marvellous stories related by travellers are apt to be at variance : as of men with tails, or with wings, and (until confirmed by experience) of flying fish; or of ice, in the celebrated anecdote of the Dutch travellers and the King of Siam. Facts of this description, facts previously unheard of but which could not from any known law of causation be pronounced impossible, are what Hume characterizes as not contrary to experience, but merely unconformable to it ; and Bentham, in his treatise on Evidence, denominates them facts disconformable *in specie*, as distinguished from such as are disconformable *in toto* or in *degree*.

In a case of this description, the fact asserted is the existence of a new Kind; which in itself is not in the slightest degree incredible, and only to be rejected if the improbability that any variety of object existing at the particular place and time should not have been discovered sooner, be greater than that of error or mendacity in the witnesses. Accordingly, such assertions, when made by credible persons, and of unexplored places, are not disbelieved, but at most regarded as requiring confirmation from subsequent observers ; unless the alleged properties of the supposed new Kind are at variance with known properties of some larger kind which includes it ; or in other words, unless, in the new Kind which is asserted to exist, some properties are said to have been found disjoined from others which have always been known to accompany them ; as in the case of Pliny's men, or any other kind of animal of a structure different from that which has always been found to coexist with animal life. On the mode of dealing with any such case, little needs be added to what has been said on the same topic in the twenty-second chapter.* When the uniformities of coexistence which the alleged fact

---

* Supra, pp. 119, 120.

would violate, are such as to raise a strong presumption of
their being the result of causation, the fact which conflicts
with them. is to be disbelieved; at least provisionally, and
subject to further investigation. When the presumption
amounts to a virtual certainty, as in the case of the general
structure of organized beings, the only question requiring consideration is whether, in phenomena so little understood, there
may not be liabilities to counteraction from causes hitherto
unknown; or whether the phenomena may not be capable of
originating in some other way, which would produce a different set of derivative uniformities. Where (as in the case
of the flying fish, or the ornithorhynchus) the generalization
to which the alleged fact would be an exception is very special
and of limited range, neither of the above suppositions can be
deemed very improbable; and it is generally, in the case of
such alleged anomalies, wise to suspend our judgment, pend-
ing the subsequent inquiries which will not fail to confirm the
assertion if it be true. But when the generalization is very
comprehensive, embracing a vast number and variety of observations, and covering a considerable province of the domain
of nature; then, for reasons which have been fully explained,
such an empirical law comes near to the certainty of an ascertained law of causation: and any alleged exception to it cannot
be admitted, unless on the evidence of some law of causation
proved by a still more complete induction.

Such uniformities in the course of nature as do not bear
marks of being the results of causation, are, as we have already
seen, admissible as universal truths with a degree of credence
proportioned to their generality. Those which are true of all
things whatever, or at least which are totally independent of
the varieties of Kinds, namely, the laws of number and extension, to which we may add the law of causation itself, are
probably the only ones, an exception to which is absolutely
and permanently incredible. Accordingly, it is to assertions
supposed to be contradictory to these laws, or to some others
coming near to them in generality, that the word impossibility (at least *total* impossibility) seems to be generally confined. Violations of other laws, of special laws of causation

for instance, are said, by persons studious of accuracy in expression, to be impossible *in the circumstances of the case;* or impossible unless some cause had existed which did not exist in the particular case.* Of no assertion, not in contradiction to some of these very general laws, will more than improbability be asserted by any cautious person; and improbability not of the highest degree, unless the time and place in which the fact is said to have occurred, render it almost certain that the anomaly, if real, could not have been overlooked by other observers. Suspension of judgment is in all other cases the resource of the judicious inquirer; provided the testimony in favour of the anomaly presents, when well sifted, no suspicious circumstances.

But the testimony is scarcely ever found to stand that test, in cases in which the anomaly is not real. In the instances on record in which a great number of witnesses, of good reputation and scientific acquirements, have testified to the truth of something which has turned out untrue, there have almost always been circumstances which, to a keen observer who had taken due pains to sift the matter, would have rendered the testimony untrustworthy. There have generally been means of accounting for the impression on the senses or minds of the alleged percipients, by fallacious appearances; or some epidemic delusion, propagated by the contagious influence of popular feeling, has been concerned in the case; or some strong

---

* A writer to whom I have several times referred, gives as the definition of an impossibility, that which there exists in the world no cause adequate to produce. This definition does not take in such impossibilities as these—that two and two should make five; that two straight lines should inclose a space; or that anything should begin to exist without a cause. I can think of no definition of impossibility comprehensive enough to include all its varieties, except the one which I have given: viz. An impossibility is that, the truth of which would conflict with a complete induction, that is, with the most conclusive evidence which we possess of universal truth.

As to the reputed impossibilities which rest on no other grounds than our ignorance of any cause capable of producing the supposed effects; very few of them are certainly impossible, or permanently incredible. The facts of travelling seventy miles an hour, painless surgical operations, and conversing by instantaneous signals between London and New York, held a high place, not many years ago, among such impossibilities.

interest has been implicated—religious zeal, party feeling, vanity, or at least the passion for the marvellous, in persons strongly susceptible of it. When none of these or similar circumstances exist to account for the apparent strength of the testimony; and where the assertion is not in contradiction either to those universal laws which know no counteraction or anomaly, or to the generalizations next in comprehensiveness to them, but would only amount, if admitted, to the existence of an unknown cause or an anomalous Kind, in circumstances not so thoroughly explored but that it is credible that things hitherto unknown may still come to light; a cautious person will neither admit nor reject the testimony, but will wait for confirmation at other times and from other unconnected sources. Such ought to have been the conduct of the King of Siam when the Dutch travellers affirmed to him the existence of ice. But an ignorant person is as obstinate in his contemptuous incredulity as he is unreasonably credulous. Anything unlike his own narrow experience he disbelieves, if it flatters no propensity; any nursery tale is swallowed implicitly by him if it does.

§ 4. I shall now advert to a very serious misapprehension of the principles of the subject, which has been committed by some of the writers against Hume's Essay on Miracles, and by Bishop Butler before them, in their anxiety to destroy what appeared to them a formidable weapon of assault against the Christian religion; and the effect of which is entirely to confound the doctrine of the Grounds of Disbelief. The mistake consists in overlooking the distinction between (what may be called) improbability before the fact, and improbability after it; or (since, as Mr. Venn remarks, the distinction of past and future is not the material circumstance) between the improbability of a mere guess being right, and the improbability of an alleged fact being true.

Many events are altogether improbable to us, before they have happened, or before we are informed of their happening, which are not in the least incredible when we are informed of them, because not contrary to any, even approximate, induction. In the cast of a perfectly fair die, the chances are five to one against throwing ace, that is, ace will be thrown on an average only once in six throws. But this is no reason against believing that ace was thrown on a given occasion, if any credible witness asserts it; since though ace is only thrown once in six times, *some* number which is only thrown once in six times must have been thrown if the die was thrown at all. The improbability, then, or in other words, the unusualness, of any fact, is no reason for disbelieving it, if the nature of the case renders it certain that either that or something equally improbable, that is, equally unusual, did happen. Nor is this all: for even if the other five sides of the die were all twos, or all threes, yet as ace would still on the average come up once in every six throws, its coming up in a given throw would be not in any way contradictory to experience. If we disbelieved all facts which had the chances against them beforehand, we should believe hardly anything. We are told that A. B. died yesterday: the moment before we were so told, the chances against his having died on that day may have been ten thousand to one; but since he was certain to die at some time or other, and when he died must necessarily die on some particular day, while the preponderance of chances is very great against every day in particular, experience affords no ground for discrediting any testimony which may be produced to the event's having taken place on a given day.

Yet it has been considered, by Dr. Campbell and others, as a complete answer to Hume's doctrine (that things are incredible which are *contrary* to the uniform course of experience), that we do not disbelieve, merely because the chances were against them, things in strict *conformity* to the uniform course of experience; that we do not disbelieve an alleged fact merely because the combination of causes on which it depends occurs only once in a certain number of times. It is evident that whatever is shown by observation, or can be proved from laws of nature, to occur in a certain proportion (however small) of the whole number of possible cases, is not contrary to experience; though we are right in disbelieving it, if some other supposition respecting the matter in question involves on the whole a less departure from the ordinary course of events. Yet, on such grounds as this have able writers been led to the extraordinary conclusion, that nothing supported by credible testimony ought ever to be disbelieved.

§ 5. We have considered two species of events, commonly said to be improbable; one kind which are in no way extraordinary, but which, having an immense preponderance of chances against them, are improbable until they are affirmed, but no longer; another kind which, being contrary to some recognised law of nature, are incredible on any amount of testimony except such as would be sufficient to shake our belief in the law itself. But between these two classes of events, there is an intermediate class, consisting of what are commonly termed Coincidences: in other words, those combinations of chances which present some peculiar and unexpected regularity, assimilating them, in so far, to the results of law. As if, for example, in a lottery of a thousand tickets, the numbers should be drawn in the exact order of what are called the natural numbers, 1, 2, 3, &c. We have still to consider the principles of evidence applicable to this case: whether there is any difference between coincidences and ordinary events, in the amount of testimony or other evidence necessary to render them credible.

It is certain, that on every rational principle of expectation, a combination of this peculiar sort may be expected quite as often as any other given series of a thousand numbers; that with perfectly fair dice, sixes will be thrown twice, thrice, or any number of times in succession, quite as often in a thousand or a million throws, as any other succession of numbers fixed upon beforehand; and that no judicious player would give greater odds against the one series than against the other. Notwithstanding this, there is a general disposition to regard the one as much more improbable than the other, and as requiring much stronger evidence to make it credible. Such is the force of this impression, that it has led some thinkers to the conclusion, that nature has greater difficulty in producing regular combinations than irregular ones; or in other words, that there is some general tendency of things, some law, which prevents regular combinations from occurring, or at least from occurring so often as others. Among these thinkers may be numbered D'Alembert; who, in an Essay on Probabilities to be found in the fifth volume of his *Mélanges*, contends that regular combinations, though equally probable according to the mathematical theory with any others, are physically less probable. He appeals to common sense, or in other words, to common impressions; saying, if dice thrown repeatedly in our presence gave sixes every time, should we not, before the number of throws had reached ten, (not to speak of thousands of millions,) be ready to affirm, with the most positive conviction, that the dice were false?

The common and natural impression is in favour of D'Alembert: the regular series would be thought much more unlikely than an irregular. But this common impression is, I apprehend, merely grounded on the fact, that scarcely anybody remembers to have ever seen one of these peculiar coincidences: the reason of which is simply that no one's experience extends to anything like the number of trials, within which that or any other given combination of events can be expected to happen. The chance of sixes on a single throw of two dice being $\frac{1}{36}$, the chance of sixes ten times in succession is 1 divided by the tenth power of 36; in other words, such a concurrence is only likely to happen once in 3,656,158,440,062,976 trials, a number which no dice-player's experience comes up to a millionth part of. But if, instead of sixes ten times, any other given succession of ten throws had been fixed upon, it would have been exactly as unlikely that in any individual's experience that particular succession had ever occurred; although this does not *seem* equally improbable, because no one could possibly have remembered whether it had occurred or not, and because the comparison is tacitly made, not between sixes ten times and any one particular series of throws, but between all regular and all irregular successions taken together.

That (as D'Alembert says) if the succession of sixes was actually thrown before our eyes, we should ascribe it not to

chance, but to unfairness in the dice, is unquestionably true.
But this arises from a totally different principle. We should
then be considering, not the probability of the fact in itself,
but the comparative probability with which, when it is known
to have happened, it may be referred to one or to another
cause. The regular series is not at all less likely than the
irregular one to be brought about by chance, but it is much
more likely than the irregular one to be produced by design;
or by some general cause operating through the structure of
the dice. It is the nature of casual combinations to produce
a repetition of the same event, as often and no oftener than
any other series of events. But it is the nature of general
causes to reproduce, in the same circumstances, always the
same event. Common sense and science alike dictate that,
all other things being the same, we should rather attribute
the effect to a cause which if real would be very likely to
produce it, than to a cause which would be very unlikely to
produce it. According to Laplace's sixth theorem, which we
demonstrated in a former chapter, the difference of probability
arising from the superior *efficacy* of the constant cause, unfairness in the dice, would after a very few throws far outweigh
any antecedent probability which there could be against its
existence.

D'Alembert should have put the question in another
manner. He should have supposed that we had ourselves
previously tried the dice, and knew by ample experience that
they were fair. Another person then tries them in our
absence, and assures us that he threw sixes ten times in
succession. Is the assertion credible or not? Here the effect
to be accounted for is not the occurrence itself, but the fact of
the witness's asserting it. This may arise either from its
having really happened, or from some other cause. What we
have to estimate is the comparative probability of these two
suppositions.

If the witness affirmed that he had thrown any other
series of numbers, supposing him to be a person of veracity,
and tolerable accuracy, and to profess that he took particular
notice, we should believe him. But the ten sixes are exactly

as likely to have been really thrown as the other series. If, therefore, this assertion is less credible than the other, the reason must be, not that it is less likely than the other to be made truly, but that it is more likely than the other to be made falsely.

One reason obviously presents itself why what is called a coincidence, should be oftener asserted falsely than an ordinary combination. It excites wonder. It gratifies the love of the marvellous. The motives, therefore, to falsehood, one of the most frequent of which is the desire to astonish, operate more strongly in favour of this kind of assertion than of the other kind. Thus far there is evidently more reason for discrediting an alleged coincidence, than a statement in itself not more probable, but which if made would not be thought remarkable. There are cases, however, in which the presumption on this ground would be the other way. There are some witnesses who, the more extraordinary an occurrence might appear, would be the more anxious to verify it by the utmost carefulness of observation before they would venture to believe it, and still more before they would assert it to others.

§ 6. Independently, however, of any peculiar chances of mendacity arising from the nature of the assertion, Laplace contends, that merely on the general ground of the fallibility of testimony, a coincidence is not credible on the same amount of testimony on which we should be warranted in believing an ordinary combination of events. In order to do justice to his argument, it is necessary to illustrate it by the example chosen by himself.

If, says Laplace, there were one thousand tickets in a box, and one only has been drawn out, then if an eye-witness affirms that the number drawn was 79, this, though the chances were 999 in 1000 against it, is not on that account the less credible; its credibility is equal to the antecedent probability of the witness's veracity. But if there were in the box 999 black balls and only one white, and the witness affirms that the white ball was drawn, the case according to Laplace is very different : the credibility of his assertion is but a small fraction of what it was in the former case ; the reason of the difference being as follows.

The witnesses of whom we are speaking must, from the nature of the case, be of a kind whose credibility falls materially short of certainty : let us suppose, then, the credibility of the witness in the case in question to be $\frac{9}{10}$; that is, let us suppose that in every ten statements which the witness makes, nine on an average are correct, and one incorrect. Let us now suppose that there have taken place a sufficient number of drawings to exhaust all the possible combinations, the witness deposing in every one. In one case out of every ten in all these drawings he will actually have made a false announcement. But in the case of the thousand tickets these false announcements will have been distributed impartially over all the numbers, and of the 999 cases in which No. 79 was not drawn, there will have been only one case in which it was announced. On the contrary, in the case of the thousand balls, (the announcement being always either " black" or " white,") if white was not drawn, and there was a false announcement, that false announcement *must* have been white ; and since by the supposition there was a false announcement once in every ten times, white will have been announced falsely in one tenth part of all the cases in which it was not drawn, that is, in one tenth part of 999 cases out of every thousand. White, then, is drawn, on an average, exactly as often as No. 79, but it is announced, without having been really drawn, 999 times as often as No. 79 ; the announcement therefore requires a much greater amount of testimony to render it credible.*

---

* Not, however, as might at first sight appear, 999 times as much. A complete analysis of the cases shows that (always assuming the veracity of the witness to be $\frac{9}{10}$) in 10,000 drawings, the drawing of No. 79 will occur nine times, and be announced incorrectly once ; the credibility therefore of the announcement of No. 79 is $\frac{9}{10}$; while the drawing of a white ball will occur nine times, and be announced incorrectly 999 times. The credibility therefore of the announcement of white is $\frac{9}{1008}$, and the ratio of the two 1008 : 10 the one announcement being thus only about a hundred times more credible than the other, instead of 999 times.

To make this argument valid it must of course be supposed, that the announcements made by the witness are average specimens of his general veracity and accuracy ; or, at least, that they are neither more nor less so in the case of the black and white balls, than in the case of the thousand tickets. This assumption, however, is not warranted. A person is far less likely to mistake, who has only one form of error to guard against, than if he had 999 different errors to avoid. For instance, in the example chosen, a messenger who might make a mistake once in ten times in reporting the number drawn in a lottery, might not err once in a thousand times if sent simply to observe whether a ball was black or white. Laplace's argument therefore is faulty even as applied to his own case. Still less can that case be received as completely representing all cases of coincidence. Laplace has so contrived his example, that though black answers to 999 distinct possibilities, and white only to one, the witness has nevertheless no bias which can make him prefer black to white. The witness did not know that there were 999 black balls in the box and only one white ; or if he did, Laplace has taken care to make all the 999 cases so undistinguishably alike, that there is hardly a possibility of any cause of falsehood or error operating in favour of any of them, which would not operate in the same manner if there were only one. Alter this supposition, and the whole argument falls to the ground. Let the balls, for instance, be numbered, and let the white ball be No. 79. Considered in respect of their colour, there are but two things which the witness can be interested in asserting, or can have dreamt or hallucinated, or has to choose from if he answers at random, viz. black and white : but considered in respect of the numbers attached to them, there are a thousand : and if his interest or error happens to be connected with the numbers, though the only assertion he makes is about the colour; the case becomes precisely assimilated to that of the thousand tickets. Or instead of the balls suppose a lottery, with 1000 tickets and but one prize, and that I hold No. 79, and being interested only in that, ask the witness not what was the number drawn, but whether it was 79 or some other. There are now only

two cases, as in Laplace's example; yet he surely would
not say that if the witness answered 79, the assertion would be
in an enormous proportion less credible, than if he made the
same answer to the same question asked in the other way.
If, for instance, (to put a case supposed by Laplace himself,)
he has staked a large sum on one of the chances, and thinks
that by announcing its occurrence he shall increase his credit:
he is equally likely to have betted on any one of the 999
numbers which are attached to black balls, and so far as the
chances of mendacity from this cause are concerned, there
will be 999 times as many chances of his announcing black
falsely, as white.

Or suppose a regiment of 1000 men, 999 Englishmen and
one Frenchman, and that of these one man has been killed,
and it is not known which. I ask the question, and the
witness answers, the Frenchman. This was not only as improbable *à priori*, but is in itself as singular a circumstance, as
remarkable a coincidence, as the drawing of the white ball:
yet we should believe the statement as readily, as if the
answer had been John Thompson. Because though the 999
Englishmen were all alike in the point in which they differed
from the Frenchman, they were not, like the 999 black
balls, undistinguishable in every other respect; but being
all different, they admitted as many chances of interest or
error, as if each man had been of a different nation; and if
a lie was told or a mistake made, the misstatement was as
likely to fall on any Jones or Thompson of the set, as on the
Frenchman.

The example of a coincidence selected by D'Alembert, that
of sixes thrown on a pair of dice ten times in succession, belongs
to this sort of cases rather than to such as Laplace's. The
coincidence is here far more remarkable, because of far rarer
occurrence, than the drawing of the white ball. But though
the improbability of its really occurring is greater, the superior
probability of its being announced falsely cannot be established
with the same evidence. The announcement "black" represented 999 cases, but the witness may not have known this,
and if he did, the 999 cases are so exactly alike, that there is

really only one set of possible causes of mendacity corresponding to the whole. The announcement " sixes *not* drawn ten times," represents, and is known by the witness to represent, a great multitude of contingencies, every one of which being unlike every other, there may be a different and a fresh set of causes of mendacity corresponding to each.

It appears to me, therefore, that Laplace's doctrine is not strictly true of any coincidences, and is wholly inapplicable to most: and that to know whether a coincidence does or does not require more evidence to render it credible than an ordinary event, we must refer, in every instance, to first principles, and estimate afresh what is the probability that the given testimony would have been delivered in that instance, supposing the fact which it asserts not to be true.

With these remarks we close the discussion of the Grounds of Disbelief; and along with it, such exposition as space admits, and as the writer has it in his power to furnish, of the Logic of Induction.

BOOK IV.

—

OF OPERATIONS SUBSIDIARY TO INDUCTION.

"Clear and distinct ideas are terms which, though familiar and frequent in men's mouths, I have reason to think every one who uses does not perfectly understand. And possibly it is but here and there one who gives himself the trouble to consider them so far as to know what he himself or others precisely mean by them; I have, therefore, in most places, chose to put determinate or determined, instead of clear and distinct, as more likely to direct men's thoughts to my meaning in this matter."—LOCKE'S *Essay on the Human Understanding*; Epistle to the Reader.

"Il ne peut y avoir qu'une méthode parfaite, qui est la *méthode naturelle*; on nomme ainsi un arrangement dans lequel les êtres du même genre seraient plus voisins entre eux que ceux de tous les autres genres; les genres du même ordre, plus que ceux de tous les autres ordres; et ainsi de suite. Cette méthode est l'idéal auquel l'histoire naturelle doit tendre; car il est évident que si l'on y parvenait, l'on aurait l'expression exacte et complète de la nature entière."—CUVIER, *Règne Animal*, Introduction.

"Deux grandes notions philosophiques dominent la théorie fondamentale de la méthode naturelle proprement dite, savoir la formation des groupes naturels, et ensuite leur succession hiérarchique."—COMTE, *Cours de Philosophie Positive*, 42me leçon.

CHAPTER I.

OF OBSERVATION AND DESCRIPTION.

§ 1. THE inquiry which occupied us in the two preceding books, has conducted us to what appears a satisfactory solution of the principal problem of Logic, according to the conception I have formed of the science. We have found, that the mental process with which Logic is conversant, the operation of ascertaining truths by means of evidence, is always, even when appearances point to a different theory of it, a process of induction. And we have particularized the various modes of induction, and obtained a clear view of the principles to which it must conform, in order to lead to results which can be relied on.

The consideration of Induction, however, does not end with the direct rules for its performance. Something must be said of those other operations of the mind, which are either necessarily presupposed in all induction, or are instrumental to the more difficult and complicated inductive processes. The present book will be devoted to the consideration of these subsidiary operations: among which our attention must first be given to those, which are indispensable preliminaries to all induction whatsoever.

Induction being merely the extension to a class of cases, of something which has been observed to be true in certain individual instances of the class; the first place among the operations subsidiary to induction, is claimed by Observation. This is not, however, the place to lay down rules for making good observers; nor is it within the competence of Logic to do so, but of the art of intellectual Education. Our business with observation is only in its connexion with the appropriate problem of logic, the estimation of evidence. We have to consider, not how or what to observe, but under what conditions observation is to be relied on ; what is needful, in order that the fact, supposed to be observed, may safely be received as true.

§ 2.   The answer to this question is very simple, at least in its first aspect.   The sole condition is, that what is supposed to have been observed shall really have been observed ; that it be an observation, not an inference.   For in almost every act of our perceiving faculties, observation and inference are intimately blended.   What we are said to observe is usually a compound result, of which one-tenth may be observation, and the remaining nine-tenths inference.

I affirm, for example, that I hear a man's voice. This would pass, in common language, for a direct perception. All, however, which is really perception, is that I hear a sound. That the sound is a voice, and that voice the voice of a man, are not perceptions but inferences. I affirm, again, that I saw my brother at a certain hour this morning. If any proposition concerning a matter of fact would commonly be said to be known by the direct testimony of the senses, this surely would be so. The truth, however, is far otherwise. I only saw a certain coloured surface ; or rather I had the kind of visual sensations which are usually produced by a coloured surface ; and from these as marks, known to be such by previous experience, I concluded that I saw my brother. I might have had sensations precisely similar, when my brother was not there. I might have seen some other person so nearly resembling him in appearance, as, at the distance, and with the degree of attention which I bestowed, to be mistaken for him. I might have been asleep, and have dreamed that I saw him ; or in a state of nervous disorder, which brought his image before me in a waking hallucination. In all these modes, many have been led to believe that they saw persons well known to them, who were dead or far distant. If any of these suppositions had been true, the affirmation that I saw my brother would have been erroneous ; but whatever was matter of direct perception, namely the visual sensations, would have been real. The inference only would have been ill grounded; I should have ascribed those sensations to a wrong cause.

Innumerable instances might be given, and analysed in the same manner, of what are vulgarly called errors of sense. There are none of them properly errors of sense; they are erroneous inferences from sense. When I look at a candle through a multiplying glass, I see what seems a dozen candles instead of one: and if the real circumstances of the case were skilfully disguised, I might suppose that there were really that number; there would be what is called an optical deception. In the kaleidoscope there really is that deception: when I look through the instrument, instead of what is actually there, namely a casual arrangement of coloured fragments, the appearance presented is that of the same combination several times repeated in symmetrical arrangement round a point. The delusion is of course effected by giving me the same sensations which I should have had if such a symmetrical combination had really been presented to me. If I cross two of my fingers, and bring any small object, a marble for instance, into contact with both, at points not usually touched simultaneously by one object, I can hardly, if my eyes are shut, help believing that there are two marbles instead of one. But it is not my touch in this case, nor my sight in the other, which is deceived; the deception, whether durable or only momentary, is in my judgment. From my senses I have only the sensations, and those are genuine. Being accustomed to have those or similar sensations when, and only when, a certain arrangement of outward objects is present to my organs, I have the habit of instantly, when I experience the sensations, inferring the existence of that state of outward things. This habit has become so powerful, that the inference, performed with the speed and certainty of an instinct, is confounded with intuitive perceptions. When it is correct, I am unconscious that it ever needed proof; even when I know it to be incorrect, I cannot without considerable effort abstain from making it. In order to be aware that it is not made by instinct but by an acquired habit, I am obliged to reflect on the slow process through which I learnt to judge by the eye of many things

which I now appear to perceive directly by sight ; and on the
reverse operation performed by persons learning to draw, who
with difficulty and labour divest themselves of their acquired
perceptions, and learn afresh to see things as they appear to
the eye.

It would be easy to prolong these illustrations, were
there any need to expatiate on a topic so copiously exemplified
in various popular works. From the examples already given,
it is seen sufficiently, that the individual facts from which we
collect our inductive generalizations are scarcely ever obtained
by observation alone. Observation extends only to the sensations by which we recognise objects ; but the propositions
which we make use of, either in science or in common life,
relate mostly to the objects themselves. In every act of what
is called observation, there is at least one inference—from the
sensations to the presence of the object ; from the marks or
diagnostics, to the entire phenomenon. And hence, among
other consequences, follows the seeming paradox, that a
general proposition collected from particulars is often more
certainly true than any one of the particular propositions from
which, by an act of induction, it was inferred. For, each of
those particular (or rather singular) propositions involved an
inference, from the impression on the senses to the fact which
caused that impression : and this inference may have been
erroneous in any one of the instances, but cannot well have
been erroneous in all of them, provided their number was
sufficient to eliminate chance. The conclusion, therefore, that
is, the general proposition, may deserve more complete reliance
tnan it would be safe to repose in any one of the inductive
premises.

The logic of observation, then, consists solely in a correct
discrimination between that, in a result of observation, which
has really been perceived, and that which is an inference from
the perception. Whatever portion is inference, is amenable to
the rules of induction already treated of, and requires no further notice here : the question for us in this place is, when
all which is inference is taken away, what remains. There
remains, in the first place, the mind's own feelings or states of

consciousness, namely, its outward feelings or sensations, and its inward feelings—its thoughts, emotions, and volitions. Whether anything else remains, or all else is inference from this; whether the mind is capable of directly perceiving or apprehending anything except states of its own consciousness— is a problem of metaphysics not to be discussed in this place. But after excluding all questions on which metaphysicians differ, it remains true, that for most purposes the discrimination we are called upon practically to exercise is that between sensations or other feelings, of our own or of other people, and inferences drawn from them. And on the theory of Observation this is all which seems necessary to be said for the purposes of the present work.

§ 8. If, in the simplest observation, or in what passes for such, there is a large part which is not observation but something else; so in the simplest description of an observation, there is, and must always be, much more asserted than is contained in the perception itself. We cannot describe a fact, without implying more than the fact. The perception is only of one individual thing; but to describe it is to affirm a connexion between it and every other thing which is either denoted or connoted by any of the terms used. To begin with an example, than which none can be conceived more elementary: I have a sensation of sight, and I endeavour to describe it by saying that I see something white. In saying this, I do not solely affirm my sensation; I also class it. I assert a resemblance between the thing I see, and all things which I and others are accustomed to call white. I assert that it resembles them in the circumstance in which they all resemble one another, in that which is the ground of their being called by the name. This is not merely one way of describing an observation, but the only way. If I would either register my observation for my own future use, or make it known for the benefit of others, I must assert a resemblance between the fact which I have observed and something else. It is inherent in a description, to be the statement of a resemblance, or resemblances.

We thus see that it is impossible to express in words any result of observation, without performing an act possessing what Dr. Whewell considers to be characteristic of Induction. There is always something introduced which was not included in the observation itself; some conception common to the phenomenon with other phenomena to which it is compared. An observation cannot be spoken of in language at all without declaring more than that one observation; without assimilating it to other phenomena already observed and classified. But this identification of an object—this recognition of it as possessing certain known characteristics—has never been confounded with Induction. It is an operation which precedes all induction, and supplies it with its materials. It is a perception of resemblances, obtained by comparison.

These resemblances are not always apprehended directly, by merely comparing the object observed with some other present object, or with our recollection of an object which is absent. They are often ascertained through intermediate marks, that is, deductively. In describing some new kind of animal, suppose me to say that it measures ten feet in length, from the forehead to the extremity of the tail. I did not ascertain this by the unassisted eye. I had a two-foot rule which I applied to the object, and, as we commonly say, measured it; an operation which was not wholly manual, but partly also mathematical, involving the two propositions, Five times two is ten, and Things which are equal to the same thing are equal to one another. Hence, the fact that the animal is ten feet long is not an immediate perception, but a conclusion from reasoning; the minor premises alone being furnished by observation of the object. Nevertheless, this is called an observation or a description of the animal, not an induction respecting it.

To pass at once from a very simple to a very complex example: I affirm that the earth is globular. The assertion is not grounded on direct perception; for the figure of the earth cannot, by us, be directly perceived, though the assertion would not be true unless circumstances could be supposed under which its truth could be so perceived. That the form of the earth is globular is inferred from certain marks, as for instance from this, that its shadow thrown upon the moon is circular; or this, that on the sea, or any extensive plain, our horizon is always a circle; either of which marks is incompatible with any other than a globular form. I assert further, that the earth is that particular kind of globe which is termed an oblate spheroid; because it is found by measurement in the direction of the meridian, that the length on the surface of the earth which subtends a given angle at its centre, diminishes as we recede from the equator and approach the poles. But these propositions, that the earth is globular, and that it is an oblate spheroid, assert, each of them, an individual fact; in its own nature capable of being perceived by the senses when the requisite organs and the necessary position are supposed, and only not actually perceived because those organs and that position are wanting. This identification of the earth, first as a globe, and next as an oblate spheroid, which, if the fact could have been seen, would have been called a description of the figure of the earth, may without impropriety be so called when, instead of being seen, it is inferred. But we could not without impropriety call either of these assertions an induction from facts respecting the earth. They are not general propositions collected from particular facts, but particular facts deduced from general propositions. They are conclusions obtained deductively, from premises originating in induction: but of these premises some were not obtained by observation of the earth, nor had any peculiar reference to it.

If, then, the truth respecting the figure of the earth is not an induction, why should the truth respecting the figure of the earth's orbit be so? The two cases only differ in this, that the form of the orbit was not, like the form of the earth itself, deduced by ratiocination from facts which were marks of ellipticity, but was got at by boldly guessing that the path was an ellipse, and finding afterwards, on examination, that the observations were in harmony with the hypothesis. According to Dr. Whewell, however, this process of guessing and verifying our guesses is not only induction, but the whole of induction: no other exposition can be given of that logical operation. That he is wrong in the latter assertion, the whole of the preceding book has, I hope, sufficiently proved; and that the process by which the ellipticity of the planetary orbits was ascertained, is not induction at all, was attempted to be shown in the second chapter of the same book.* We are now, however, prepared to go more into the heart of the matter than at that earlier period of our inquiry, and to show, not merely what the operation in question is not, but what it is.

§ 4. We observed, in the second chapter, that the proposition "the earth moves in an ellipse," so far as it only serves for the colligation or connecting together of actual observations, (that is, as it only affirms that the observed positions of the earth may be correctly represented by as many points in the circumference of an imaginary ellipse,) is not an induction, but a description: it is an induction, only when it affirms that the intermediate positions, of which there has been no direct observation, would be found to correspond to the remaining points of the same elliptic circumference. Now, though this real induction is one thing, and the description another, we are in a very different condition for making the induction before we have obtained the description, and after it. For inasmuch as the description, like all other descriptions, contains the assertion of a resemblance between the phenomenon described and something else; in pointing out something which the series of observed places of a planet resembles, it points out something in which the several places themselves agree. If the series of places correspond to as many points of an ellipse, the places themselves agree in being situated in that ellipse. We have, therefore, by the same process which gave us the description, obtained the requisites for an induction by the Method of Agreement. The successive observed places of the earth being considered as effects, and its motion as the cause which produces them, we find that those effects, that is, those places, agree in the circumstance of

---

* Supra, book iii. ch. ii. § 3, 4, 5.

being in an ellipse. We conclude that the remaining effects, the places which have not been observed, agree in the same circumstance, and that the *law* of the motion of the earth is motion in an ellipse.

The Colligation of Facts, therefore, by means of hypotheses, or, as Dr. Whewell prefers to say, by means of Conceptions, instead of being, as he supposes, Induction itself, takes its proper place among operations subsidiary to Induction. All Induction supposes that we have previously compared the requisite number of individual instances, and ascertained in what circumstances they agree. The Colligation of Facts is no other than this preliminary operation. When Kepler, after vainly endeavouring to connect the observed places of a planet by various hypotheses of circular motion, at last tried the hypothesis of an ellipse and found it answer to the phenomena; what he really attempted, first unsuccessfully and at last successfully, was to discover the circumstance in which all the observed positions of the planet agreed. And when he in like manner connected another set of observed facts, the periodic times of the different planets, by the proposition that the squares of the times are proportional to the cubes of the distances, what he did was simply to ascertain the property in which the periodic times of all the different planets agreed.

Since, therefore, all that is true and to the purpose in Dr. Whewell's doctrine of Conceptions might be fully expressed by the more familiar term Hypothesis; and since his Colligation of Facts by means of appropriate Conceptions, is but the ordinary process of finding by a comparison of phenomena, in what consists their agreement or resemblance; I would willingly have confined myself to those better understood expressions, and persevered to the end in the same abstinence which I have hitherto observed from ideological discussions; considering the mechanism of our thoughts to be a topic distinct from and irrelevant to the principles and rules by which the trustworthiness of the results of thinking is to be estimated. Since, however, a work of such high pretensions, and, it must also be said, of so much real merit, has rested the whole theory of Induction upon such ideological considerations, it seems necessary for others who follow, to claim for themselves and their doctrines whatever position may properly belong to them on the same metaphysical ground.　And this is the object of the succeeding chapter.

CHAPTER II.

§ 1. THE metaphysical inquiry into the nature and composition of what have been called Abstract Ideas, or in other words, of the notions which answer in the mind to classes and to general names, belongs not to Logíc, but to a different science, and our purpose does not require that we should enter upon it here. We are only concerned with the universally acknowledged fact, that such notions or conceptions do exist. The mind can conceive a multitude of individual things as one assemblage or class; and general names do really suggest to us certain ideas or mental representations, otherwise we could not use the names with consciousness of a meaning. Whether the idea called up by a general name is composed of the various circumstances in which all the individuals denoted by the name agree, and of no others, (which is the doctrine of Locke, Brown, and the Conceptualists;) or whether it be the idea of some one of those individuals, clothed in its individualizing peculiarities, but with the accompanying knowledge that those peculiarities are not properties of the class, (which is the doctrine of Berkeley, Mr. Bailey,* and the modern Nominalists;) or whether (as held by Mr. James

---

* Mr. Bailey has given by far the best statement of this theory. "The general name," he says, "raises up the image sometimes of one individual of the class formerly seen, sometimes of another, not unfrequently of many individuals in succession; and it sometimes suggests an image made up of elements from several different objects, by a latent process of which I am not conscious." (Letters on the Philosophy of the Human Mind, 1st series, letter 22.) But Mr. Bailey must allow that we carry on inductions and ratiocinations respecting the class, by means of this idea or conception of some one individual in it. This is all I require. The name of a class calls up some idea, through which we can, to all intents and purposes, think of the class as such, and not solely of an individual member of it.

Mill) the idea of the class is that of a miscellaneous assemblage of individuals belonging to the class; or whether, finally, (what appears to be the truest opinion,) it be any one or any other of all these, according to the accidental circumstances of the case; certain it is, that *some* idea or mental conception is suggested by a general name, whenever we either hear it or employ it with consciousness of a meaning. And this, which we may call if we please a general idea, *represents* in our minds the whole class of things to which the name is applied. Whenever we think or reason concerning the class, we do so by means of this idea. And the voluntary power which the mind has, of attending to one part of what is present to it at any moment, and neglecting another part, enables us to keep our reasonings and conclusions respecting the class unaffected by anything in the idea or mental image which is not really, or at least which we do not really believe to be, common to the whole class.*

There are, then, such things as general conceptions, or conceptions by means of which we can think generally: and when we form a set of phenomena into a class, that is, when we compare them with one another to ascertain in what they agree, some general conception is implied in this mental operation. And inasmuch as such a comparison is a necessary preliminary to Induction, it is most true that Induction could not go on without general conceptions.

§ 2. But it does not therefore follow that these general conceptions must have existed in the mind previously to the comparison. It is not a law of our intellect, that in comparing things with each other and taking note of their agreement we merely recognise as realized in the outward world something that we already had in our minds. The conception originally found its way to us as the *result* of such a comparison. It was obtained (in metaphysical phrase) by *abstraction* from individual things. These things may be things which we perceived or thought of on former occasions,

* I have entered rather fully into this question in chap. xvii. of *An Examination of Sir William Hamilton's Philosophy*, headed "The Doctrine of Concepts or General Notions," which contains my last views on the subject.

but they may also be the things which we are perceiving or
thinking of on the very occasion. When Kepler compared
the observed places of the planet Mars, and found that they
agreed in being points of an elliptic circumference, he
applied a general conception which was already in his mind,
having been derived from his former experience. But this
is by no means universally the case. When we compare
several objects and find them to agree in being white, or
when we compare the various species of ruminating animals
and find them to agree in being cloven-footed, we have just as
much a general conception in our minds as Kepler had in
his: we have the conception of "a white thing," or the conception of "a cloven-footed animal." But no one supposes that
we necessarily bring these conceptions with us, and *superinduce*
them (to adopt Dr. Whewell's expression) upon the facts:
because in these simple cases everybody sees that the very act
of comparison which ends in our connecting the facts by
means of the conception, may be the source from which we
derive the conception itself. If we had never seen any white
object or had never seen any cloven-footed animal before, we
should at the same time and by the same mental act acquire
the idea, and employ it for the colligation of the observed
phenomena. Kepler, on the contrary, really had to bring the
idea with him, and superinduce it upon the facts; he could
not evolve it out of them: if he had not already had the idea,
he would not have been able to acquire it by a comparison of
the planet's positions. But this inability was a mere accident:
the idea of an ellipse could have been acquired from the paths
of the planets as effectually as from anything else, if the paths
had not happened to be invisible. If the planet had left a
visible track, and we had been so placed that we could see it at
the proper angle, we might have abstracted our original idea
of an ellipse from the planetary orbit. Indeed, every conception which can be made the instrument for connecting a set of
facts, might have been originally evolved from those very facts.
The conception is a conception *of* something; and that which
it is a conception of, is really *in* the facts, and might, under
some supposable circumstances, or by some supposable exten-

sion of the faculties which we actually possess, have been
detected in them.   And not only is this always in itself
possible, but it actually happens, in almost all cases in which
the obtaining of the right conception is a matter of any considerable difficulty.   For if there be no new conception
required; if one of those already familiar to mankind will
serve the purpose, the accident of being the first to whom
the right one occurs, may happen to almost anybody; at
least in the case of a set of phenomena which the whole
scientific world are engaged in attempting to connect.   The
honour, in Kepler's case, was that of the accurate, patient,
and toilsome calculations by which he compared the results
that followed from his different guesses, with the observations
of Tycho Brahe; but the merit was very small of guessing
an ellipse; the only wonder is that men had not guessed
it before, nor could they have failed to do so if there had not
existed an obstinate *à priori* prejudice that the heavenly
bodies must move, if not in a circle, in some combination of
circles.

The really difficult cases are those in which the conception
destined to create light and order out of darkness and confusion, has to be sought for among the very phenomena which
it afterwards serves to arrange.   Why, according to Dr.
Whewell himself, did the ancients fail in discovering the laws
of mechanics, that is, of equilibrium and of the communication of motion?   Because they had not, or at least had not
clearly, the ideas or conceptions of pressure and resistance,
momentum, and uniform and accelerating force.   And whence
could they have obtained these ideas, except from the very
facts of equilibrium and motion?   The tardy development of
several of the physical sciences, for example of optics, electricity, magnetism, and the higher generalizations of chemistry,
he ascribes to the fact that mankind had not yet possessed
themselves of the Idea of Polarity, that is, the idea of opposite properties in opposite directions.   But what was there to
suggest such an idea, until, by a separate examination of
several of these different branches of knowledge, it was shown
that the facts of each of them did present, in some instances

at least, the curious phenomenon of opposite properties in opposite directions? The thing was superficially manifest only in two cases, those of the magnet, and of electrified bodies; and there the conception was encumbered with the circumstance of material poles, or fixed points in the body itself, in which points this opposition of properties seemed to be inherent. The first comparison and abstraction had led only to this conception of poles; and if anything corresponding to that conception had existed in the phenomena of chemistry or optics, the difficulty now justly considered so great, would have been extremely small. The obscurity rose from the fact, that the polarities in chemistry and optics were distinct species, though of the same genus, with the polarities in electricity and magnetism: and that in order to assimilate the phenomena to one another, it was necessary to compare a polarity without poles, such for instance as is exemplified in the polarization of light, and the polarity with (apparent) poles, which we see in the magnet; and to recognise that these polarities, while different in many other respects, agree in the one character which is expressed by the phrase, opposite properties in opposite directions. From the result of such a comparison it was that the minds of scientific men formed this new general conception: between which, and the first confused feeling of an analogy between some of the phenomena of light and those of electricity and magnetism, there is a long interval, filled up by the labours and more or less sagacious suggestions of many superior minds.

The conceptions, then, which we employ for the colligation and methodization of facts, do not develop themselves from within, but are impressed upon the mind from without; they are never obtained otherwise than by way of comparison and abstraction, and, in the most important and the most numerous cases, are evolved by abstraction from the very phenomena which it is their office to colligate. I am far, however, from wishing to imply that it is not often a very difficult thing to perform this process of abstraction well, or that the success of an inductive operation does not, in many cases, principally depend on the skill with which we perform it. Bacon was quite justified in designating as one of the principal obstacles to good induction, general conceptions wrongly formed, "notiones temerè à rebus abstractæ:" to which Dr. Whewell adds, that not only does bad abstraction make bad induction, but that in order to perform induction well, we must have abstracted well; our general conceptions must be "clear" and "appropriate" to the matter in hand.

§ 3. In attempting to show what the difficulty in this matter really is, and how it is surmounted, I must beg the reader, once for all, to bear this in mind; that although in discussing the opinions of a different school of philosophy, I am willing to adopt their language, and to speak, therefore, of connecting facts through the instrumentality of a conception, this technical phraseology means neither more nor less than what is commonly called comparing the facts with one another and determining in what they agree. Nor has the technical expression even the advantage of being metaphysically correct. The facts are not *connected*, except in a merely metaphorical acceptation of the term. The *ideas* of the facts may become connected, that is, we may be led to think of them together; but this consequence is no more than what may be produced by any casual association. What really takes place, is, I conceive, more philosophically expressed by the common word Comparison, than by the phrases "to connect" or "to superinduce." For, as the general conception is itself obtained by a comparison of particular phenomena, so, when obtained, the mode in which we apply it to other phenomena is again by comparison. We compare phenomena with each other to get the conception, and we then compare those and other phenomena *with* the conception. We get the conception of an animal (for instance) by comparing different animals, and when we afterwards see a creature resembling an animal, we compare it with our general conception of an animal; and if it agrees with that general conception, we include it in the class. The conception becomes the type of comparison.

And we need only consider what comparison is, to see that where the objects are more than two, and still more when they are an indefinite number, a type of some sort is an indispensable condition of the comparison. When we have to arrange and classify a great number of objects according to their agreements and differences, we do not make a confused attempt to compare all with all. We know that two things are as much as the mind can easily attend to at a time, and we therefore fix upon one of the objects, either at hazard or because it offers in a peculiarly striking manner some important character, and, taking this as our standard, compare it with one object after another. If we find a second object which presents a remarkable agreement with the first, inducing us to class them together, the question instantly arises, in what particular circumstances do they agree? and to take notice of these circumstances is already a first stage of abstraction, giving rise to a general conception. Having advanced thus far, when we now take in hand a third object we naturally ask ourselves the question, not merely whether this third object agrees with the first, but whether it agrees with it in the same circumstances in which the second did? in other words, whether it agrees with the general conception which has been obtained by abstraction from the first and second? Thus we see the tendency of general conceptions, as soon as formed, to substitute themselves as types, for whatever individual objects previously answered that purpose in our comparisons. We may, perhaps, find that no considerable number of other objects agree with this first general conception; and that we must drop the conception, and beginning again with a different individual case, proceed by fresh comparisons to a different general conception. Sometimes, again, we find that the same conception will serve, by merely leaving out some of its circumstances; and by this higher effort of abstraction, we obtain a still more general conception; as in the case formerly referred to, the scientific world rose from the conception of poles to the general conception of opposite properties in opposite directions; or as those South-Sea islanders, whose conception of a quadruped had been abstracted from hogs (the only animals of that description which they had seen), when they afterwards compared that conception with other quadrupeds, dropped some of the circumstances, and arrived at the more general conception which Europeans associate with the term.

These brief remarks contain, I believe, all that is well-grounded in the doctrine, that the conception by which the mind arranges and gives unity to phenomena must be furnished by the mind itself, and that we find the right conception by a tentative process, trying first one and then another until we hit the mark. The conception is not furnished *by* the mind until it has been furnished *to* the mind; and the facts which supply it are sometimes extraneous facts, but more often the very facts which we are attempting to arrange by it. It is quite true, however, that in endeavouring to arrange the facts, at whatever point we begin, we never advance three steps without forming a general conception, more or less distinct and precise; and that this general conception becomes the clue which we instantly endeavour to trace through the rest of the facts, or rather, becomes the standard with which we thenceforth compare them. If we are not satisfied with the agreements which we discover among the phenomena by comparing them with this type, or with some still more general conception which by an additional stage of abstraction we can form from the type; we change our path, and look out for other agreements: we recommence the comparison from a different starting-point, and so generate a different set of general conceptions. This is the tentative process which Dr. Whewell speaks of; and which has not unnaturally suggested the theory, that the conception is supplied by the mind itself: since the different conceptions which the mind successively tries, it either already possessed from its previous experience, or they were supplied to it in the first stage of the corresponding act of comparison; so that, in the subsequent part of the process, the conception manifested itself as something compared with the phenomena, not evolved from them.

§ 4. If this be a correct account of the instrumentality of general conceptions in the comparison which necessarily precedes Induction, we shall easily be able to translate into our own language what Dr. Whewell means by saying that conceptions, to be subservient to Induction, must be "clear" and "appropriate."

If the conception corresponds to a real agreement among the phenomena; if the comparison which we have made of a set of objects has led us to class them according to real resemblances and differences; the conception which does this cannot fail to be appropriate, for some purpose or other. The question of appropriateness is relative to the particular object we have in view. As soon as, by our comparison, we have ascertained some agreement, something which can be predicated in common of a number of objects; we have obtained a basis on which an inductive process is capable of being founded. But the agreements, or the ulterior consequences to which those agreements lead, may be of very different degrees of importance. If, for instance, we only compare animals according to their colour, and class those together which are coloured alike, we form the general conceptions of a white animal, a black animal, &c., which are conceptions legitimately formed; and if an induction were to be attempted concerning the causes of the colours of animals, this comparison would be the proper and necessary preparation for such an induction, but would not help us towards a knowledge of the laws of any other of the properties of animals : while if, with Cuvier, we compare and class them according to the structure of the skeleton, or, with Blainville, according to the nature of their outward integuments, the agreements and differences which are observable in these respects are not only of much greater importance in themselves, but are marks of agreements and differences in many other important particulars of the structure and mode of life of the animals. If, therefore, the study of their structure and habits be our object, the conceptions generated by these last comparisons are far more "appropriate" than those generated by the former. Nothing, other than this, can be meant by the appropriateness of a conception.

When Dr. Whewell says that the ancients, or the schoolmen, or any modern inquirers, missed discovering the real law of a phenomenon because they applied to it an inappropriate instead of an appropriate conception; he can only mean that in comparing various instances of the phenomenon, to ascertain in what those instances agreed, they missed the important points of agreement; and fastened upon such as were either imaginary, and not agreements at all, or if real agreements, were comparatively trifling, and had no connexion with the phenomenon, the law of which was sought.

Aristotle, philosophizing on the subject of motion, remarked that certain motions apparently take place spontaneously; bodies fall to the ground, flame ascends, bubbles of air rise in water, &c.: and these he called natural motions; while others not only never take place without external incitement, but even when such incitement is applied, tend spontaneously to cease; which, to distinguish them from the former, he called violent motions. Now, in comparing the so-called natural motions with one another, it appeared to Aristotle that they agreed in one circumstance, namely, that the body which moved (or seemed to move) spontaneously, was moving *towards its own place;* meaning thereby the place from whence it originally came, or the place where a great quantity of matter similar to itself was assembled. In the other class of motions, as when bodies are thrown up in the air, they are, on the contrary, moving *from* their own place. Now, this conception of a body moving towards its own place may justly be considered inappropriate; because, though it expresses a circumstance really found in some of the most familiar instances of motion apparently spontaneous, yet, first, there are many other cases of such motion, in which that circumstance is absent: the motion, for instance, of the earth and planets. Secondly, even when it is present, the motion, on closer examination, would often be seen not to be spontaneous: as, when air rises in water, it does not rise by its own nature, but is pushed up by the superior weight of the water which presses upon it. Finally, there are many cases in which the spontaneous motion takes place in the contrary direction to what the theory considers as the body's own place; for instance, when a fog rises from a lake, or when water dries up. The agreement, therefore, which Aristotle selected as his principle of classification, did not extend to all cases of the phenomenon he wanted to study, spontaneous motion; while it did include cases of the absence of the phenomenon, cases of motion not spontaneous. The conception was hence "inappropriate." We may add that, in the case in question, no conception would be appropriate; there is no agreement which runs through all the cases of spontaneous or apparently spontaneous motion and no others: they cannot be brought under one law: it is a case of Plurality of Causes.[*]

§ 5. So much for the first of Dr. Whewell's conditions, that conceptions must be appropriate. The second is, that they shall be "clear:" and let us consider what this implies. Unless the conception corresponds to a real agreement, it has a worse defect than that of not being clear; it is not applicable to the case at all. Among the phenomena, therefore, which we are attempting to connect by means of the conception, we must suppose that there really is an agreement, and that the conception is a conception of that agreement. In order, then, that it may be clear, the only requisite is, that we shall know exactly in what the agreement consists; that it shall have been carefully observed, and accurately remembered. We are said not to have a clear conception of the resemblance among a set of objects, when we have only a gene-

---

[*] Other examples of inappropriate conceptions are given by Dr. Whewell (*Phil. Ind. Sc.* ii. 185) as follows:—"Aristotle and his followers endeavoured in vain to account for the mechanical relation of forces in the lever, by applying the *inappropriate* geometrical conceptions of the properties of the circle: they failed in explaining the *form* of the luminous spot made by the sun shining through a hole, because they applied the *inappropriate* conception of a circular *quality* in the sun's light: they speculated to no purpose about the elementary composition of bodies, because they assumed the *inappropriate* conception of *likeness* between the elements and the compound, instead of the genuine notion of elements merely *determining* the qualities of the compound." But in these cases there is more than an inappropriate conception; there is a false conception; one which has no prototype in nature, nothing corresponding to it in facts. This is evident in the last two examples, and is equally true in the first; the "properties of the circle" which were referred to, being purely fantastical. There is, therefore, an error beyond the wrong choice of a principle of generalization; there is a false assumption of matters of fact. The attempt is made to resolve certain laws of nature into a more general law, that law not being one which, though real, is inappropriate, but one wholly imaginary.

ral feeling that they resemble, without having analysed their resemblance, or perceived in what points it consists, and fixed in our memory an exact recollection of those points. This want of clearness, or, as it may be otherwise called, this vagueness, in the general conception, may be owing either to our having no accurate knowledge of the objects themselves, or merely to our not having carefully compared them. Thus a person may have no clear idea of a ship because he has never seen one, or because he remembers but little, and that faintly, of what he has seen. Or he may have a perfect knowledge and remembrance of many ships of various kinds, frigates among the rest, but he may have no clear but only a confused idea of a frigate, because he has never been told, and has not compared them sufficiently to have remarked and remembered, in what particular points a frigate differs from some other kind of ship.

It is not, however, necessary, in order to have clear ideas, that we should know all the common properties of the things which we class together. That would be to have our conception of the class complete as well as clear. It is sufficient if we never class things together without knowing exactly why we do so,—without having ascertained exactly what agreements we are about to include in our conception; and if, after having thus fixed our conception, we never vary from it, never include in the class anything which has not those common properties, nor exclude from it anything which has. A clear conception means a determinate conception; one which does not fluctuate, which is not one thing to-day and another to-morrow, but remains fixed and invariable, except when, from the progress of our knowledge, or the correction of some error, we consciously add to it or alter it. A person of clear ideas, is a person who always knows in virtue of what properties his classes are constituted; what attributes are connoted by his general names.

The principal requisites, therefore, of clear conceptions, are habits of attentive observation, an extensive experience, and a memory which receives and retains an exact image of what is observed. And in proportion as any one has the habit of

observing minutely and comparing carefully a particular class
of phenomena, and an accurate memory for the results of the
observation and comparison, so will his conceptions of that
class of phenomena be clear; provided he has the indispensable habit, (naturally, however, resulting from those other
endowments,) of never using general names without a precise
connotation.

As the clearness of our conceptions chiefly depends on the
carefulness and *accuracy* of our observing and comparing faculties, so their appropriateness, or rather the chance we have of
hitting upon the appropriate conception in any case, mainly
depends on the *activity* of the same faculties. He who by
habit, grounded on sufficient natural aptitude, has acquired a
readiness in accurately observing and comparing phenomena,
will perceive so many more agreements and will perceive them
so much more rapidly than other people, that the chances are
much greater of his perceiving, in any instance, the agreement
on which the important consequences depend.

§ 6. It is of so much importance that the part of the
process of investigating truth, discussed in this chapter, should
be rightly understood, that I think it is desirable to restate the
results we have arrived at, in a somewhat different mode of
expression.

We cannot ascertain general truths, that is, truths applicable to classes, unless we have formed the classes in such a
manner that general truths can be affirmed of them. In the
formation of any class, there is involved a conception of it as a
class, that is, a conception of certain circumstances as being
those which characterize the class, and distinguish the objects
composing it from all other things. When we know exactly
what these circumstances are, we have a clear idea (or conception) of the class, and of the meaning of the general name
which designates it. The primary condition implied in having
this clear idea, is that the class be really a class; that it correspond to a real distinction; that the things it includes really
do agree with one another in certain particulars, and differ, in
those same particulars, from all other things. A person with-

out clear ideas, is one who habitually classes together, under
the same general names, things which have no common properties, or none which are not possessed also by other things;
or who, if the usage of other people prevents him from actually
misclassing things, is unable to state to himself the common
properties in virtue of which he classes them rightly.

But it is not the sole requisite of classification that the
classes should be real classes, framed by a legitimate mental
process. Some modes of classing things are more valuable
than others for human uses, whether of speculation or of
practice; and our classifications are not well made, unless the
things which they bring together not only agree with each
other in something which distinguishes them from all other
things, but agree with each other and differ from other
things in the very circumstances which are of primary importance for the purpose (theoretical or practical) which we
have in view, and which constitutes the problem before us.
In other words, our conceptions, though they may be clear,
are not *appropriate* for our purpose, unless the properties we
comprise in them are those which will help us towards what
we wish to understand—i. e., either those which go deepest
into the nature of the things, if our object be to understand
that, or those which are most closely connected with the particular property which we are endeavouring to investigate.

We cannot, therefore, frame good general conceptions
beforehand. That the conception we have obtained is the one
we want, can only be known when we have done the work for
the sake of which we wanted it; when we completely understand the general character of the phenomena, or the conditions
of the particular property with which we concern ourselves.
General conceptions formed without this thorough knowledge,
are Bacon's "notiones temerè à rebus abstractæ." Yet such
premature conceptions we must be continually making up, in
our progress to something better. They are an impediment
to the progress of knowledge, only when they are permanently
acquiesced in. When it has become our habit to group things
in wrong classes—in groups which either are not really classes,
having no distinctive points of agreement (absence of *clear*

ideas), or which are not classes of which anything important to our purpose can be predicated (absence of *appropriate* ideas); and when, in the belief that these badly made classes are those sanctioned by Nature, we refuse to exchange them for others, and cannot or will not make up our general conceptions from any other elements; in that case all the evils which Bacon ascribes to his "notiones temerè abstractæ" really occur. This was what the ancients did in physics, and what the world in general does in morals and politics to the present day.

It would thus, in my view of the matter, be an inaccurate mode of expression to say, that obtaining appropriate conceptions is a condition precedent to generalization. Throughout the whole process of comparing phenomena with one another for the purpose of generalization, the mind is trying to make up a conception; but the conception which it is trying to make up is that of the really important point of agreement in the phenomena. As we obtain more knowledge of the phenomena themselves, and of the conditions on which their important properties depend, our views on this subject naturally alter; and thus we advance from a less to a more "appropriate" general conception, in the progress of our investigations.

We ought not, at the same time, to forget that the really important agreement cannot always be discovered by mere comparison of the very phenomena in question, without the aid of a conception acquired elsewhere; as in the case, so often referred to, of the planetary orbits.

The search for the agreement of a set of phenomena is in truth very similar to the search for a lost or hidden object. At first we place ourselves in a sufficiently commanding position, and cast our eyes round us, and if we can see the object it is well; if not, we ask ourselves mentally what are the places in which it may be hid, in order that we may there search for it: and so on, until we imagine the place where it really is. And here too we require to have had a previous conception, or knowledge, of those different places. As in this familiar process, so in the philosophical operation which it illustrates, we first endeavour to find the lost object or recognise the common attribute, without conjecturally invoking the aid of any previously acquired conception, or in other words, of any hypothesis. Having failed in this, we call upon our imagination for some hypothesis of a possible place, or a possible point of resemblance, and then look, to see whether the facts agree with the conjecture.

For such cases something more is required than a mind accustomed to accurate observation and comparison. It must be a mind stored with general conceptions, previously acquired, of the sorts which bear affinity to the subject of the particular inquiry. And much will also depend on the natural strength and acquired culture of what has been termed the scientific imagination ; on the faculty possessed of mentally arranging known elements into new combinations, such as have not yet been observed in nature, though not contradictory to any known laws.

But the variety of intellectual habits, the purposes which they serve, and the modes in which they may be fostered and cultivated, are considerations belonging to the Art of Education : a subject far wider than Logic, and which this treatise does not profess to discuss. Here, therefore, the present chapter may properly close.

CHAPTER III.

§ 1. IT does not belong to the present undertaking to dwell on the importance of language as a medium of human intercourse, whether for purposes of sympathy or of information. Nor does our design admit of more than a passing allusion to that great property of names, on which their functions as an intellectual instrument are, in reality, ultimately dependent; their potency as a means of forming, and of riveting, associations among our other ideas: a subject on which an able thinker* has thus written :—

"Names are impressions of sense, and as such take the strongest hold on the mind, and of all other impressions can be most easily recalled and retained in view. They therefore serve to give a point of attachment to all the more volatile objects of thought and feeling. Impressions that when passed might be dissipated for ever, are, by their connexion with language, always within reach. Thoughts, of themselves, are perpetually slipping out of the field of immediate mental vision; but the name abides with us, and the utterance of it restores them in a moment. Words are the custodiers of every product of mind less impressive than themselves. All extensions of human knowledge, all new generalizations, are fixed and spread, even unintentionally, by the use of words. The child growing up learns, along with the vocables of his mother-tongue, that things which he would have believed to be different, are, in important points, the same. Without any formal instruction, the language in which we grow up teaches us all the common philosophy of the age. It directs us to observe and know things which we should have over-

---

* Professor Bain.

looked; it supplies us with classifications ready made, by which things are arranged (as far as the light of by-gone generations admits) with the objects to which they bear the greatest total resemblance. The number of general names in a language, and the degree of generality of those names, afford a test of the knowledge of the era, and of the intellectual insight which is the birthright of any one born into it."

It is not, however, of the functions of Names, considered generally, that we have here to treat, but only of the manner and degree in which they are directly instrumental to the investigation of truth; in other words, to the process of induction.

§ 2. Observation and Abstraction, the operations which formed the subject of the two foregoing chapters, are conditions indispensable to induction; there can be no induction where they are not. It has been imagined that Naming is also a condition equally indispensable. There are thinkers who have held that language is not solely, according to a phrase generally current, *an* instrument of thought, but *the* instrument: that names, or something equivalent to them, some species of artificial signs, are necessary to reasoning; that there could be no inference, and consequently no induction, without them. But if the nature of reasoning was correctly explained in the earlier part of the present work, this opinion must be held to be an exaggeration, though of an important truth. If reasoning be from particulars to particulars, and if it consist in recognising one fact as a mark of another, or a mark of a mark of another, nothing is required to render reasoning possible, except senses and association: senses to perceive that two facts are conjoined; association, as the law by which one of those two facts raises up the idea of the other.* For these mental phenomena, as well as for the

---

* This sentence having been erroneously understood as if I had meant to assert that belief is nothing but an irresistible association, I think it necessary to observe that I express no theory respecting the ultimate analysis either of reasoning or of belief, two of the most obscure points in analytical psychology.

belief or expectation which follows, and by which we recognise as having taken place, or as about to take place, that of which we have perceived a mark, there is evidently no need of language. And this inference of one particular fact from another is a case of induction. It is of this sort of induction that brutes are capable : it is in this shape that uncultivated minds make almost all their inductions, and that we all do so in the cases in which familiar experience forces our conclusions upon us without any active process of inquiry on our part, and in which the belief or expectation follows the suggestion of the evidence, with the promptitude and certainty of an instinct.*

§ 3. But though inference of an inductive character is possible without the use of signs, it could never, without them, be carried much beyond the very simple cases which we have just described, and which form, in all probability, the limit of the reasonings of those animals to whom conventional language is unknown. Without language, or something equivalent to it, there could only be as much reasoning from experience as can take place without the aid of general propositions. Now, though in strictness we may reason from past experience to a fresh individual case without the intermediate stage of a general proposition, yet without general propositions we should seldom remember what past experience we have had, and scarcely ever what conclusions that experience will warrant. The division of the inductive process into two parts, the first ascertaining what is a mark of the given fact, the second whether in the new case that mark exists,

---

I am speaking not of the powers themselves, but of the previous conditions necessary to enable those powers to exert themselves : of which conditions I am contending that language is not one, senses and association being sufficient without it.

* Mr. Bailey agrees with me in thinking that whenever "from something actually present to my senses conjoined with past experience, I feel satisfied that something has happened, or will happen, or is happening, beyond the sphere of my personal observation," I may with strict propriety be said to reason : and of course to reason inductively, for demonstrative reasoning is excluded by the circumstances of the case. (*The Theory of Reasoning*, 2nd ed. p. 27.)

is natural, and scientifically indispensable. It is, indeed, in a majority of cases, rendered necessary by mere distance of time. The experience by which we are to guide our judgments may be other people's experience, little of which can be communicated to us otherwise than by language: when it is our own, it is generally experience long past; unless, therefore, it were recorded by means of artificial signs, little of it (except in cases involving our intenser sensations or emotions, or the subjects of our daily and hourly contemplation) would be retained in the memory. It is hardly necessary to add, that when the inductive inference is of any but the most direct and obvious nature—when it requires several observations or experiments, in varying circumstances, and the comparison of one of these with another—it is impossible to proceed a step, without the artificial memory which words bestow. Without words, we should, if we had often seen A and B in immediate and obvious conjunction, expect B whenever we saw A; but to discover their conjunction when not obvious, or to determine whether it is really constant or only casual, and whether there is reason to expect it under any given change of circumstances, is a process far too complex to be performed without some contrivance to make our remembrance of our own mental operations accurate. Now, language is such a contrivance. When that instrument is called to our aid, the difficulty is reduced to that of making our remembrance of the meaning of words accurate. This being secured, whatever passes through our minds may be remembered accurately, by putting it carefully into words, and committing the words either to writing or to memory.

The function of Naming, and particularly of General Names, in Induction, may be recapitulated as follows. Every inductive inference which is good at all, is good for a whole class of cases: and, that the inference may have any better warrant of its correctness than the mere clinging together of two ideas, a process of experimentation and comparison is necessary; in which the whole class of cases must be brought to view, and some uniformity in the course of nature evolved and ascertained, since the existence of such an uniformity is required as a justification for drawing the inference in even a single case. This uniformity, therefore, may be ascertained once for all; and if, being ascertained, it can be remembered, it will serve as a formula for making, in particular cases, all such inferences as the previous experience will warrant. But we can only secure its being remembered, or give ourselves even a chance of carrying in our memory any considerable number of such uniformities, by registering them through the medium of permanent signs; which (being, from the nature of the case, signs not of an individual fact, but of an uniformity, that is, of an indefinite number of facts similar to one another) are general signs; universals; general names, and general propositions.

§ 4. And here I cannot omit to notice an oversight committed by some eminent thinkers; who have said that the cause of our using general names is the infinite multitude of individual objects, which, making it impossible to have a name for each, compels us to make one name serve for many. This is a very limited view of the function of general names. Even if there were a name for every individual object, we should require general names as much as we now do. Without them we could not express the result of a single comparison, nor record any one of the uniformities existing in nature; and should be hardly better off in respect to Induction than if we had no names at all. With none but names of individuals, (or in other words, proper names,) we might, by pronouncing the name, suggest the idea of the object, but we could not assert any proposition; except the unmeaning ones formed by predicating two proper names one of another. It is only by means of general names that we can convey any information, predicate any attribute, even of an individual, much more of a class. Rigorously speaking we could get on without any other general names than the abstract names of attributes; all our propositions might be of the form "such an individual object possesses such an attribute," or "such an attribute is always (or never) conjoined with such another attribute." In fact, however, mankind have always given general names to objects as well as attributes, and indeed before attributes: but the general names given to objects imply attributes, derive their whole meaning from attributes; and are chiefly useful as the language by means of which we predicate the attributes which they connote.

It remains to be considered what principles are to be adhered to in giving general names, so that these names, and the general propositions in which they fill a place, may conduce most to the purposes of Induction.

CHAPTER IV.

§ 1. In order that we may possess a language perfectly suitable for the investigation and expression of general truths, there are two principal, and several minor, requisites. The first is, that every general name should have a meaning, steadily fixed, and precisely determined. When, by the fulfilment of this condition, such names as we possess are fitted for the due performance of their functions, the next requisite, and the second in order of importance, is that we should possess a name wherever one is needed; wherever there is anything to be designated by it, which it is of importance to express.

The former of these requisites is that to which our attention will be exclusively directed in the present chapter.

§ 2. Every general name, then, must have a certain and knowable meaning. Now the meaning (as has so often been explained) of a general connotative name, resides in the connotation; in the attribute on account of which, and to express which, the name is given. Thus, the name animal being given to all things which possess the attributes of sensation and voluntary motion, the word connotes those attributes exclusively, and they constitute the whole of its meaning. If the name be abstract, its denotation is the same with the connotation of the corresponding concrete: it designates directly the attribute, which the concrete term implies. To give a precise meaning to general names is, then, to fix with steadiness the attribute or attributes connoted by each concrete general name, and denoted by the corresponding abstract. Since abstract names, in the order of their creation, do not precede but follow concrete ones, as is proved by the etymological fact that they are almost always derived from them ; we may consider their meaning as determined by, and dependent on, the meaning of their concrete : and thus the problem of giving a distinct meaning to general language, is all included in that of giving a precise connotation to all concrete general names.

This is not difficult in the case of new names; of the technical terms created by scientific inquirers for the purposes of science or art. But when a name is in common use, the difficulty is greater; the problem in this case not being that of choosing a convenient connotation for the name, but of ascertaining and fixing the connotation with which it is already used. That this can ever be a matter of doubt, is a sort of paradox. But the vulgar (including in that term all who have not accurate habits of thought) seldom know exactly what assertion they intend to make, what common property they mean to express, when they apply the same name to a number of different things. All which the name expresses with them, when they predicate it of an object, is a confused feeling of resemblance between that object and some of the other things which they have been accustomed to denote by the name. They have applied the name Stone to various objects previously seen; they see a new object, which appears to them somewhat like the former, and they call it a stone, without asking themselves in what respect it is like, or what mode or degree of resemblance the best authorities, or even they themselves, require as a warrant for using the name. This rough general impression of resemblance is, however, made up of particular circumstances of resemblance; and into these it is the business of the logician to analyse it; to ascertain what points of resemblance among the different things commonly called by the name, have produced in the common mind this vague feeling of likeness; have given to the things the similarity of aspect, which has made them a class, and has caused the same name to be bestowed upon them.

But though general names are imposed by the vulgar

without any more definite connotation than that of a vague resemblance; general propositions come in time to be made, in which predicates are applied to those names, that is, general assertions are made concerning the *whole* of the things which are denoted by the name. And since by each of these propositions some attribute, more or less precisely conceived, is of course predicated, the ideas of these various attributes thus become associated with the name, and in a sort of uncertain way it comes to connote them; there is a hesitation to apply the name in any new case in which any of the attributes familiarly predicated of the class do not exist. And thus, to common minds, the propositions which they are in the habit of hearing or uttering concerning a class, make up in a loose way a sort of connotation for the class-name. Let us take, for instance, the word Civilized. How few could be found, even among the most educated persons, who would undertake to say exactly what the term Civilized connotes. Yet there is a feeling in the minds of all who use it, that they are using it with a meaning; and this meaning is made up, in a confused manner, of everything which they have heard or read that civilized men, or civilized communities, are, or may be expected to be.

It is at this stage, probably, in the progress of a concrete name, that the corresponding abstract name generally comes into use. Under the notion that the concrete name must of course convey a meaning, or in other words, that there is some property common to all things which it denotes, people give a name to this common property; from the concrete Civilized, they form the abstract Civilization. But since most people have never compared the different things which are called by the concrete name, in such a manner as to ascertain what properties these things have in common, or whether they have any; each is thrown back upon the marks by which he himself has been accustomed to be guided in his application of the term: and these, being merely vague hearsays and current phrases, are not the same in any two persons, nor in the same person at different times. Hence the word (as Civilization, for example) which professes to be the designation of the unknown common property, conveys scarcely to any two minds the same idea. No two persons agree in the things they predicate of it; and when it is itself predicated of anything, no other person knows, nor does the speaker himself know with precision, what he means to assert. Many other words which could be named, as the word *honour*, or the word *gentleman*, exemplify this uncertainty still more strikingly.

It needs scarcely be observed, that general propositions of which no one can tell exactly what they assert, cannot possibly have been brought to the test of a correct induction. Whether a name is to be used as an instrument of thinking, or as a means of communicating the result of thought, it is imperative to determine exactly the attribute or attributes which it is to express : to give it, in short, a fixed and ascertained connotation.

§ 3. It would, however, be a complete misunderstanding of the proper office of a logician in dealing with terms already in use, if we were to think that because a name has not at present an ascertained connotation, it is competent to any one to give it such a connotation at his own choice. The meaning of a term actually in use is not an arbitrary quantity to be fixed, but an unknown quantity to be sought.

In the first place, it is obviously desirable to avail ourselves, as far as possible, of the associations already connected with the name; not enjoining the employment of it in a manner which conflicts with all previous habits, and especially not so as to require the rupture of those strongest of all associations between names, which are created by familiarity with propositions in which they are predicated of one another. A philosopher would have little chance of having his example followed, if he were to give such a meaning to his terms as should require us to call the North American Indians a civilized people, or the higher classes in France or England savages; or to say that civilized people live by hunting, and savages by agriculture. Were there no other reason, the extreme difficulty of effecting so complete a revolution in speech would be more than a sufficient one. The endeavour should be, that all generally received propositions into which the term enters, should be at least as true after its meaning is fixed, as they were before; and that the concrete name, therefore, should not receive such a connotation as shall prevent it from denoting things which, in common language, it is currently affirmed of. The fixed and precise connotation which it receives, should not be in deviation from, but in agreement (as far as it goes) with, the vague and fluctuating connotation which the term already had.

To fix the connotation of a concrete name, or the denotation of the corresponding abstract, is to define the name. When this can be done without rendering any received assertions inadmissible, the name can be defined in accordance with its received use, which is vulgarly called defining not the name but the thing. What is meant by the improper expression of defining a thing, (or rather a class of things—for nobody talks of defining an individual,) is to define the name, subject to the condition that it shall denote those things. This, of course, supposes a comparison of the things, feature by feature and property by property, to ascertain what attributes they agree in; and not unfrequently an operation strictly inductive, for the purpose of ascertaining some unobvious agreement, which is the cause of the obvious agreements.

For, in order to give a connotation to a name, consistently with its denoting certain objects, we have to make our selection from among the various attributes in which those objects agree. To ascertain in what they do agree is, therefore, the first logical operation requisite. When this has been done as far as is necessary or practicable, the question arises, which of these common attributes shall be selected to be associated with the name. For if the class which the name denotes be a Kind, the common properties are innumerable; and even if not, they are often extremely numerous. Our choice is first limited by the preference to be given to properties which are well known, and familiarly predicated of the class; but even these are often too numerous to be all included in the definition, and, besides, the properties most generally known may

not be those which serve best to mark out the class from all
others. We should therefore select from among the common
properties, (if among them any such are to be found,) those
on which it has been ascertained by experience, or proved by
deduction, that many others depend; or at least which are
sure marks of them, and from whence, therefore, many others
will follow by inference. We thus see that to frame a good
definition of a name already in use, is not a matter of choice
but of discussion, and discussion not merely respecting the
usage of language, but respecting the properties of things,
and even the origin of those properties. And hence every
enlargement of our knowledge of the objects to which the name
is applied, is liable to suggest an improvement in the definition. It is impossible to frame a perfect set of definitions on
any subject, until the theory of the subject is perfect: and as
science makes progress, its definitions are also progressive.

§ 4. The discussion of Definitions, in so far as it does
not turn on the use of words but on the properties of things,
Dr. Whewell calls the Explication of Conceptions. The act
of ascertaining, better than before, in what particulars any
phenomena which are classed together agree, he calls in his
technical phraseology, unfolding the general conception in
virtue of which they are so classed. Making allowance for
what appears to me the darkening and misleading tendency of
this mode of expression, several of his remarks are so much
to the purpose, that I shall take the liberty of transcribing
them.

He observes,* that many of the controversies which have
had an important share in the formation of the existing body
of science, have " assumed the form of a battle of Definitions.
For example, the inquiry concerning the laws of falling bodies,
led to the question whether the proper definition of a *uniform
force* is that it generates a velocity proportional to the *space*
from rest, or to the *time*. The controversy of the *vis viva* was
what was the proper definition of the *measure of force*. A

---

* *Novum Organum Renovatum*, pp. 35-37.

principal question in the classification of minerals is, what is the definition of a *mineral species.* Physiologists have endeavoured to throw light on their subject by defining *organisation,* or some similar term." Questions of the same nature are still open respecting the definitions of Specific Heat, Latent Heat, Chemical Combination, and Solution.

"It is very important for us to observe, that these controversies have never been questions of insulated and *arbitrary* definitions, as men seem often tempted to imagine them to have been. In all cases there is a tacit assumption of some proposition which is to be expressed by means of the definition, and which gives it its importance. The dispute concerning the definition thus acquires a real value, and becomes a question concerning true and false. Thus in the discussion of the question, What is a uniform force? it was taken for granted that gravity is a uniform force. In the debate of the *vis viva,* it was assumed that in the mutual action of bodies the whole effect of the force is unchanged. In the zoological definition of species, (that it consists of individuals which have, or may have, sprung from the same parents,) it is presumed that individuals so related resemble each other more than those which are excluded by such a definition; or, perhaps, that species so defined have permanent and definite differences. A definition of organization, or of some other term, which was not employed to express some principle, would be of no value.

"The establishment, therefore, of a right definition of a term, may be a useful step in the explication of our conceptions; but this will be the case then only when we have under our consideration some proposition in which the term is employed. For then the question really is, how the conception shall be understood and defined in order that the proposition may be true.

"To unfold our conceptions by means of definitions has never been serviceable to science, except when it has been associated with an immediate use of the definitions. The endeavour to define a Uniform Force was combined with the assertion that gravity is a uniform force: the attempt to

define Accelerating Force was immediately followed by the
doctrine that accelerating forces may be compounded : the
process of defining Momentum was connected with the principle that momenta gained and lost are equal : naturalists
would have given in vain the definition of Species which we
have quoted, if they had not also given the characters of
species so separated. . . . . . Definition may be the best
mode of explaining our conception, but that which alone
makes it worth while to explain it in any mode, is the opportunity of using it in the expression of truth.   When a definition is propounded to us as a useful step in knowledge, we
are always entitled to ask what principle it serves to enunciate."

In giving, then, an exact connotation to the phrase,
"an uniform force," the condition was understood, that the
phrase should continue to denote gravity. The discussion,
therefore, respecting the definition, resolved itself into this
question, What is there of an uniform nature in the motions
produced by gravity? By observations and comparisons, it
was found, that what was uniform in those motions was the
ratio of the velocity acquired to the time elapsed ; equal velocities being added in equal times. An uniform force, therefore, was defined, a force which adds equal velocities in equal
times. So, again, in defining momentum. It was already a
received doctrine, that when two objects impinge upon one
another, the momentum lost by the one is equal to that
gained by the other.   This proportion it was deemed necessary
to preserve, not from the motive (which operates in many
other cases) that it was firmly fixed in popular belief; for the
proposition in question had never been heard of by any but
the scientifically instructed.   But it was felt to contain a
truth : even a superficial observation of the phenomena left no
doubt that in the propagation of motion from one body to
another, there was something of which the one body gained
precisely what the other lost; and the word momentum had
been invented to express this unknown something. The
settlement, therefore, of the definition of momentum, involved the determination of the question, What is that of

which a body, when it sets another body in motion, loses exactly as much as it communicates? And when experiment had shown that this *something* was the product of the velocity of the body by its mass, or quantity of matter, this became the definition of momentum.

The following remarks,* therefore, are perfectly just: "The business of definition is part of the business of discovery. . . . . To define, so that our definition shall have any scientific value, requires no small portion of that sagacity by which truth is detected. . . . . When it has been clearly seen what ought to be our definition, it must be pretty well known what truth we have to state. The definition, as well as the discovery, supposes a decided step in our knowledge to have been made. The writers on Logic, in the middle ages, made Definition the last stage in the progress of knowledge; and in this arrangement at least, the history of science, and the philosophy derived from the history, confirm their speculative views." For in order to judge finally how the name which denotes a class may best be defined, we must know all the properties common to the class, and all the relations of causation or dependence among those properties.

If the properties which are fittest to be selected as marks of other common properties are also obvious and familiar, and especially if they bear a great part in producing that general air of resemblance which was the original inducement to the formation of the class, the definition will then be most felicitous. But it is often necessary to define the class by some property not familiarly known, provided that property be the best mark of those which are known. M. de Blainville, for instance, founded his definition of life on the process of decomposition and recomposition which incessantly takes place in every living body, so that the particles composing it are never for two instants the same. This is by no means one of the most obvious properties of living bodies; it might escape altogether the notice of an unscientific observer. Yet great authorities (independently of M. de Blainville, who is himself

* *Nov. Org. Renov.*, pp. 39, 40.

a first-rate authority) have thought that no other property so well answers the conditions required for the definition.

§ 5. Having laid. down the principles which ought for the most part to be observed in attempting to give a precise connotation to a term in use, I must now add, that it is not always practicable to adhere to those principles, and that even when practicable, it is occasionally not desirable.

Cases in which it is impossible to comply with all the conditions of a precise definition of a name in agreement with usage, occur very frequently. There is often no one connotation capable of being given to a word, so that it shall still denote everything it is accustomed to denote; or that all the propositions into which it is accustomed to enter, and which have any foundation in truth, shall remain true. Independently of accidental ambiguities, in which the different meanings have no connexion with one another; it continually happens that a word is used in two or more senses derived from each other, but yet radically distinct. So long as a term is vague; that is, so long as its connotation is not ascertained and permanently fixed, it is constantly liable to be applied by *extension* from one thing to another, until it reaches things which have little, or even no, resemblance to those which were first designated by it.

Suppose, says Dugald Stewart, in his *Philosophical Essays*,* " that the letters A, B, C, D, E, denote a series of objects; that A possesses some one quality in common with B; B a quality in common with C; C a quality in common with D; D a quality in common with E; while at the same time, no quality can be found which belongs in common to any *three* objects in the series. Is it not conceivable, that the affinity between A and B may produce a transference of the name of the first to the second; and that, in consequence of the other affinities which connect the remaining objects together, the same name may pass in succession from B to C; from C to D; and from D to E? In this manner, a common

---

* P. 217, 4to edition.

appellation will arise between A and E, although the two objects may, in their nature and properties, be so widely distant from each other, that no stretch of imagination can conceive how the thoughts were led from the former to the latter. The transitions, nevertheless, may have been all so easy and gradual, that, were they successfully detected by the fortunate ingenuity of a theorist, we should instantly recognise, not only the verisimilitude, but the truth of the conjecture: in the same way as we admit, with the confidence of intuitive conviction, the certainty of the well-known etymological process which connects the Latin preposition *e* or *ex* with the English substantive *stranger*, the moment that the intermediate links of the chain are submitted to our examination."[*]

The applications which a word acquires by this gradual extension of it from one set of objects to another, Stewart, adopting an expression from Mr. Payne Knight, calls its *transitive* applications; and after briefly illustrating such of them as are the result of local or casual associations, he proceeds as follows:[†]—

"But although by far the greater part of the transitive or derivative applications of words depend on casual and unaccountable caprices of the feelings or the fancy, there are certain cases in which they open a very interesting field of philosophical speculation. Such are those, in which an analogous transference of the corresponding term may be remarked universally, or very generally, in other languages; and in which, of course, the uniformity of the result must be ascribed to the essential principles of the human frame. Even in such cases, however, it will by no means be always found, on

---

[*] " E, ex, extra, extraneus, étranger, stranger."
Another etymological example sometimes cited is the derivation of the English *uncle* from the Latin *avus*. It is scarcely possible for two words to bear fewer outward marks of relationship, yet there is but one step between them; *avus, avunculus, uncle*.

So *pilgrim*, from *ager: per agrum, peragrinus, peregrinus, pellegrino, pilgrim*.
[†] P. 226-7.

examination, that the various applications of the same term
have arisen from any common quality or qualities in the objects
to which they relate. In the greater number of instances,
they may be traced to some natural and universal associations
of ideas, founded in the common faculties, common organs,
and common condition of the human race. . . . . . According
to the different degrees of intimacy and strength in the associations on which the *transitions* of language are founded, very
different effects may be expected to arise. Where the association is slight and casual, the several meanings will remain
distinct from each other, and will often, in process of time,
assume the appearance of capricious varieties in the use of the
same arbitrary sign. *Where the association is so natural and
habitual as to become virtually indissoluble, the transitive meanings will coalesce in one complex conception; and every new
transition will become a more comprehensive generalization of
the term in question."*

I solicit particular attention to the law of mind expressed
in the last sentence, and which is the source of the per-
plexity so often experienced in detecting these transitions of
meaning. Ignorance of that law is the shoal on which some
of the most powerful intellects which have adorned the
human race have been stranded. The inquiries of Plato
into the definitions of some of the most general terms of
moral speculation are characterized by Bacon as a far nearer
approach to a true inductive method than is elsewhere to
be found among the ancients, and are, indeed, almost
perfect examples of the preparatory process of comparison and
abstraction : but, from being unaware of the law just mentioned, he often wasted the powers of this great logical
instrument on inquiries in which it could realize no result,
since the phenomena, whose common properties he so elaborately endeavoured to detect, had not really any common
properties. Bacon himself fell into the same error in his
speculations on the nature of heat, in which he evidently
confounded under the name hot, classes of phenomena which
had no property in common. Stewart certainly overstates
the matter when he speaks of " a prejudice which has de-

scended to modern times from the scholastic ages, that when a word admits of a variety of significations, these different significations must all be species of the same genus, and must consequently include some essential idea common to every individual to which the generic term can be applied:"* for both Aristotle and his followers were well aware that there are such things as ambiguities of language, and delighted in distinguishing them. But they never suspected ambiguity in the cases where (as Stewart remarks) the association on which the transition of meaning was founded is so natural and habitual, that the two meanings blend together in the mind, and a real transition becomes an apparent generalization. Accordingly they wasted infinite pains in endeavouring to find a definition which would serve for several distinct meanings at once: as in an instance noticed by Stewart himself, that of "causation; the ambiguity of the word which, in the Greek language, corresponds to the English word *cause*, having suggested to them the vain attempt of tracing the common idea which, in the case of any *effect*, belongs to the *efficient*, to the *matter*, to the *form*, and to the *end*. The idle generalities" (he adds) "we meet with in other philosophers, about the ideas of the *good*, the *fit*, and the *becoming*, have taken their rise from the same undue influence of popular epithets on the speculations of the learned."†

Among the words which have undergone so many successive transitions of meaning that every trace of a property common to all the things they are applied to, or at least common and also peculiar to those things, has been lost, Stewart considers the word Beautiful to be one. And (without attempting to decide a question which in no respect belongs to logic) I cannot but feel, with him, considerable doubt, whether the word beautiful connotes the same property when we speak of a beautiful colour, a beautiful face, a beautiful scene, a beautiful character, and a beautiful poem. The word was doubtless extended from one of these objects

---

* *Essays*, p. 214.　　　　† Ibid. 215.

15—2

thought could not take place with anything like the rapidity which we know they possess. Very often, indeed, when we are employing a word in our mental operations, we are so far from waiting until the complex idea which corresponds to the meaning of the word is consciously brought before us in all its parts, that we run on to new trains of ideas by the other associations which the mere word excites, without having realized in our imagination any part whatever of the meaning: thus using the word, and even using it well and accurately, and carrying on important processes of reasoning by means of it, in an almost mechanical manner; so much so, that some metaphysicians, generalizing from an extreme case, have fancied that all reasoning is but the mechanical use of a set of terms according to a certain form. We may discuss and settle the most important interests of towns or nations, by the application of general theorems or practical maxims previously laid down, without having had consciously suggested to us, once in the whole process, the houses and green fields, the thronged market-places and domestic hearths, of which not only those towns and nations consist, but which the words town and nation confessedly mean.

Since, then, general names come in this manner to be used (and even to do a portion of their work well) without suggesting to the mind the whole of their meaning, and often with the suggestion of a very small, or no part at all of that meaning; we cannot wonder that words so used come in time to be no longer capable of suggesting any other of the ideas appropriated to them, than those with which the association is most immediate and strongest, or most kept up by the incidents of life: the remainder being lost altogether; unless the mind, by often consciously dwelling on them, keeps up the association. Words naturally retain much more of their meaning to persons of active imagination, who habitually represent to themselves things in the concrete, with the detail which belongs to them in the actual world. To minds of a different description, the only antidote to this corruption of language is predication. The habit of predicating of the name, all the various properties which it originally connoted, keeps up the association between the name and those properties.

But in order that it may do so, it is necessary that the predicates should themselves retain their association with the properties which they severally connote. For the propositions cannot keep the meaning of the words alive, if the meaning of the propositions themselves should die. And nothing is more common than for propositions to be mechanically repeated, mechanically retained in the memory, and their truth undoubtingly assented to and relied on, while yet they carry no meaning distinctly home to the mind; and while the matter of fact or law of nature which they originally expressed is as much lost sight of, and practically disregarded, as if it never had been heard of at all. In those subjects which are at the same time familiar and complicated, and especially in those which are so in as great a degree as moral and social subjects are, it is a matter of common remark how many important propositions are believed and repeated from habit, while no account could be given, and no sense is practically manifested, of the truths which they convey. Hence it is, that the traditional maxims of old experience, though seldom questioned, have often so little effect on the conduct of life; because their meaning is never, by most persons, really felt, until personal experience has brought it home. And thus also it is that so many doctrines of religion, ethics, and even politics, so full of meaning and reality to first converts, have manifested (after the association of that meaning with the verbal formulas has ceased to be kept up by the controversies which accompanied their first introduction) a tendency to degenerate rapidly into lifeless dogmas; which tendency, all the efforts of an education expressly and skilfully directed to keeping the meaning alive, are barely sufficient to counteract.

Considering, then, that the human mind, in different generations, occupies itself with different things, and in one age is led by the circumstances which surround it to fix more of its attention upon one of the properties of a thing, in another age upon another; it is natural and inevitable that in every age a certain portion of our recorded and traditional knowledge, not being continually suggested by the pursuits and inquiries with which mankind are at that time engrossed, should fall asleep, as it were, and fade from the memory. It would be in danger of being totally lost, if the propositions or formulas, the results of the previous experience, did not remain, as forms of words it may be, but of words that once really conveyed, and are still supposed to convey, a meaning: which meaning, though suspended, may be historically traced, and when suggested, may be recognised by minds of the necessary endowments as being still matter of fact, or truth.   While the formulas remain, the meaning may at any time revive; and as on the one hand the formulas progressively lose the meaning they were intended to convey, so, on the other, when this forgetfulness has reached its height and begun to produce obvious consequences, minds arise which from the contemplation of the formulas rediscover the truth, when truth it was, which was contained in them, and announce it again to mankind, not as a discovery, but as the meaning of that which they have been taught, and still profess to believe.

.   Thus there is a perpetual oscillation in spiritual truths, and in spiritual doctrines of any significance, even when not truths. Their meaning is almost always in a process either of being lost or of being recovered. Whoever has attended to the history of the more serious convictions of mankind—of the opinions by which the general conduct of their lives is, or as they conceive ought to be, more especially regulated—is aware that even when recognising verbally the same doctrines, they attach to them at different periods a greater or a less quantity, and even a different kind, of meaning. The words in their original acceptation connoted, and the propositions expressed, a complication of outward facts and inward feelings, to different portions of which the general mind is more particularly alive in different generations of mankind. To common minds, only that portion of the meaning is in each generation suggested, of which that generation possesses the counterpart in its own habitual expe-

rience. But the words and propositions lie ready to, suggest
to any mind duly prepared the remainder of the meaning.
Such individual minds are almost always to be found : and the
lost meaning, revived by them, again by degrees works its way
into the general mind.

The arrival of this salutary reaction may however be
materially retarded, by the shallow conceptions and incautious
proceedings of mere logicians. It sometimes happens that
towards the close of the downward period, when the words have
lost part of their significance, and have not yet begun to recover it, persons arise whose leading and favourite idea is the
importance of clear conceptions and precise thought, and the
necessity, therefore, of definite language. These persons, in
examining the old formulas, easily perceive that words are
used in them without a meaning; and if they are not the sort
of persons who are capable of rediscovering the lost signification, they naturally enough dismiss the formula, and define
the name without reference to it. In so doing they fasten
down the name to what it connotes in common use at the time
when it conveys the smallest quantity of meaning; and introduce the practice of employing it, consistently and uniformly,
according to that connotation. The word in this way acquires
an extent of denotation far beyond what it ́had before; it
becomes extended to many things to which it was previously,
in appearance capriciously, refused. Of the propositions in
which it was formerly used, those which were true in virtue of
the forgotten part of its meaning are now, by the clearer light
which the definition diffuses, seen not to be true according to
the definition; which, however, is the recognised and sufficiently correct expression of all that is perceived to be in the
mind of any one by whom the term is used at the present day.
The ancient formulas are consequently treated as prejudices;
and people are no longer taught as before, though not to
understand them, yet to believe that there is truth in them.
They no longer remain in the general mind surrounded by
respect, and ready at any time to suggest their original meaning. Whatever truths they contain are not only, in these
circumstances, rediscovered far more slowly, but, when redis-

covered, the prejudice with which novelties are regarded is now, in some degree at least, against them, instead of being on their side.

An example may make these remarks more intelligible. In all ages, except where moral speculation has been silenced by outward compulsion, or where the feelings which prompt to it still continue to be satisfied by the traditional doctrines of an established faith, one of the subjects which have most occupied the minds of thinking persons is the inquiry, What is virtue? or, What is a virtuous character? Among the different theories on the subject which have, at different times, grown up and obtained partial currency, every one of which reflected as in the clearest mirror, the express image of the age which gave it birth; there was one, according to which virtue consists in a correct calculation of our own personal interests, either in this world only, or also in another. To make this theory plausible, it was of course necessary that the only beneficial actions which people in general were accustomed to see, or were therefore accustomed to praise, should be such as were, or at least might without contradicting obvious facts be supposed to be, the result of a prudential regard to self-interest; so that the words really connoted no more, in common acceptation, than was set down in the definition.

Suppose, now, that the partisans of this theory had contrived to introduce a consistent and undeviating use of the term according to this definition. Suppose that they had seriously endeavoured, and had succeeded in the endeavour, to banish the word disinterestedness from the language; had obtained the disuse of all expressions attaching odium to selfishness or commendation to self-sacrifice, or which implied generosity or kindness to be anything but doing a benefit in order to receive a greater personal advantage in return. Need we say, that this abrogation of the old formulas for the sake of preserving clear ideas and consistency of thought, would have been a great evil? while the very inconsistency incurred by the coexistence of the formulas with philosophical opinions which seemed to condemn them as absurdities, operated as a stimulus to the re-examination of the subject; and thus the very doctrines originating in the oblivion into which a part of the truth had fallen, were rendered indirectly, but powerfully, instrumental to its revival.

The doctrine of the Coleridge school, that the language of any people among whom culture is of old date, is a sacred deposit, the property of all ages, and which no one age should consider itself empowered to alter—borders indeed, as thus expressed, on an extravagance; but it is grounded on a truth, frequently overlooked by that class of logicians who think more of having a clear than of having a comprehensive meaning; and who perceive that every age is adding to the truths which it has received from its predecessors, but fail to see that a counter process of losing truths already possessed, is also constantly going on, and requiring the most sedulous attention to counteract it. Language is the depository of the accumulated body of experience to which all former ages have contributed their part, and which is the inheritance of all yet to come. We have no right to prevent ourselves from transmitting to posterity a larger portion of this inheritance than we may ourselves have profited by. However much we may be able to improve on the conclusions of our forefathers, we ought to be careful not inadvertently to let any of their premises slip through our fingers. It may be good to alter the meaning of a word, but it is bad to let any part of the meaning drop. Whoever seeks to introduce a more correct use of a term with which important associations are connected, should be required to possess an accurate acquaintance with the history of the particular word, and of the opinions which in different stages of its progress it served to express. To be qualified to define the name, we must know all that has ever been known of the properties of the class of objects which are, or originally were, denoted by it. For if we give it a meaning according to which any proposition will be false which has ever been generally held to be true, it is incumbent on us to be sure that we know and have considered all which those, who believed the proposition, understood by it.

CHAPTER V.

§ 1. IT is not only in the mode which has now been pointed out, namely by gradual inattention to a portion of the ideas conveyed, that words in common use are liable to shift their connotation. The truth is, that the connotation of such words is perpetually varying; as might be expected from the manner in which words in common use acquire their connotation. A technical term, invented for purposes of art or science, has, from the first, the connotation given to it by its inventor; but a name which is in every one's mouth before any one thinks of defining it, derives its connotation only from the circumstances which are habitually brought to mind when it is pronounced. Among these circumstances, the properties common to the things denoted by the name, have naturally a principal place; and would have the sole place, if language were regulated by convention rather than by custom and accident. But besides these common properties, which if they exist are *certainly* present whenever the name is employed, any other circumstance may *casually* be found along with it, so frequently as to become associated with it in the same manner, and as strongly, as the common properties themselves. In proportion as this association forms itself, people give up using the name in cases in which those casual circumstances do not exist. They prefer using some other name, or the same name with some adjunct, rather than employ an expression which will call up an idea they do not want to excite. The circumstance originally casual, thus becomes regularly a part of the connotation of the word.

It is this continual incorporation of circumstances originally accidental, into the permanent signification of words, which is the cause that there are so few exact synonymes. It is this also which renders the dictionary meaning of a word, by universal remark so imperfect an exponent of its real meaning. The dictionary meaning is marked out in a broad, blunt way, and probably includes all that was originally necessary for the correct employment of the term; but in process of time so many collateral associations adhere to words, that whoever should attempt to use them with no other guide than the dictionary, would confound a thousand nice distinctions and subtle shades of meaning which dictionaries take no account of; as we notice in the use of a language in conversation or writing by a foreigner not thoroughly master of it. The history of a word, by showing the causes which determine its use, is in these cases a better guide to its employment than any definition; for definitions can only show its meaning at the particular time, or at most the series of its successive meanings, but its history·may show the law by which the succession was produced. The word *gentleman*, for instance, to the correct employment of which a dictionary would be no guide, originally meant simply a man born in a certain rank. From this it came by degrees to connote all such qualities or adventitious circumstances as were usually found to belong to persons of that rank. This consideration at once explains why in one of its vulgar acceptations it means any one who lives without labour, in another without manual labour, and in its more elevated signification it has in every age signified the conduct, character, habits, and outward appearance, in whomsoever found, which, according to the ideas of that age, belonged or were expected to belong to persons born and educated in a high social position.

It continually happens that of two words, whose dictionary meanings are either the same or very slightly different, one will be the proper word to use in one set of circumstances, another in another, without its being possible to show how the custom of so employing them originally grew up. The accident that one of the words was used and not the other on a particular occasion or in a particular social circle, will be sufficient to produce so strong an association between the word and some speciality of circumstances, that mankind abandon the use of it in any other case, and the speciality becomes part of its signification. The tide of custom first drifts the word on the shore of a particular meaning, then retires and leaves it there.

An instance in point is the remarkable change which, in the English language at least, has taken place in the signification of the word *loyalty*. That word originally meant in English, as it still means in the language from whence it came, fair, open dealing, and fidelity to engagements; in that sense the quality it expressed was part of the ideal chivalrous or knightly character. By what process, in England, the term became restricted to the single case of fidelity to the throne, I am not sufficiently versed in the history of courtly language to be able to pronounce. The interval between a *loyal chevalier* and a loyal subject is certainly great. I can only suppose that the word was, at some period, the favourite term at court to express fidelity to the oath of allegiance; until at length those who wished to speak of any other, and as it was probably deemed, inferior sort of fidelity, either did not venture to use so dignified a term, or found it convenient to employ some other in order to avoid being misunderstood.

§ 2. Cases are not unfrequent in which a circumstance, at first casually incorporated into the connotation of a word which originally had no reference to it, in time wholly supersedes the original meaning, and becomes not merely a part of the connotation, but the whole of it. This is exemplified in the word pagan, *paganus;* which originally, as its etymology imports, was equivalent to *villager;* the inhabitant of a *pagus*, or village. At a particular era in the extension of Christianity over the Roman empire, the adherents of the old religion, and the villagers or country people, were nearly the

same body of individuals, the inhabitants of the towns having
been earliest converted; as in our own day, and at all times,
the greater activity of social intercourse renders them the
earliest recipients of new opinions and modes, while old habits
and prejudices linger longest among the country people: not
to mention that the towns were more immediately under the
direct influence of the government, which at that time had
embraced Christianity. From this casual coincidence, the
word *paganus* carried with it, and began more and more
steadily to suggest, the idea of a worshipper of the ancient
divinities; until at length it suggested that idea so forcibly
that people who did not desire to suggest the idea avoided
using the word. But when *paganus* had come to connote
heathenism, the very unimportant circumstance, with reference
to that fact, of the place of residence, was soon disregarded in
the employment of the word. As there was seldom any occasion for making separate assertions respecting heathens who
lived in the country, there was no need for a separate word to
denote them; and pagan came not only to mean heathen, but
to mean that exclusively.

A case still more familiar to most readers is that of the
word *villain* or *villein*. This term, as everybody knows, had
in the middle ages a connotation as strictly defined as a word
could have, being the proper legal designation for those persons who were the subjects of the less onerous forms of feudal
bondage. The scorn of the semibarbarous military aristocracy
for these their abject dependants, rendered the act of likening
any person to this class of people a mark of the greatest con-
tumely: the same scorn led them to ascribe to the same people
all manner of hateful qualities, which doubtless also, in the
degrading situation in which they were held, were often not
unjustly imputed to them. These circumstances combined to
attach to the term villain, ideas of crime and guilt in so forcible
a manner, that the application of the epithet even to those to
whom it legally belonged became an affront, and was abstained
from whenever no affront was intended. From that time
guilt was part of the connotation; and soon became the whole

of it, since mankind were not prompted by any urgent motive to continue making a distinction in their language between bad men of servile station and bad men of any other rank in life.

These and similar instances in which the original signification of a term is totally lost—another and an entirely distinct meaning being first engrafted upon the former, and finally substituted for it—afford examples of the double movement which is always taking place in language : two counter-movements, one of Generalization, by which words are perpetually losing portions of their connotation, and becoming of less meaning and more general acceptation ; the other of Specialization, by which other, or even these same words, are continually taking on fresh connotation ; acquiring additional meaning, by being restricted in their employment to a part only of the occasions on which they might properly be used before.   This double movement is of sufficient importance in the ·natural history of language, (to which natural history the artificial modifications ought always to have some degree of reference,) to justify our dwelling a little longer on the nature of the twofold phenomenon, and the causes to which it owes its existence.

§ 3. To begin with the movement of generalization. It is unnecessary to dwell on the changes in the meaning of names which take place merely from their being used ignorantly, by persons who, not having properly mastered the received connotation of a word, apply it in a looser and wider sense than belongs to it.   This, however, is a real source of alterations in the language ; for when a word, from being often employed in cases where one of the qualities which it connotes does not exist, ceases to suggest that quality with certainty, then even those who are under no mistake as to the proper meaning of the word, prefer expressing that meaning in some other way, and leave the original word to its fate. The word 'Squire as standing for an owner of a landed estate ; Parson, as denoting not the rector of the parish, but clergymen in general ; Artist, to denote only a painter or sculptor ; are cases in point.* Independently, however, of the generalization of names through their ignorant misuse, there is a tendency in the same direction, consistently with a perfect knowledge of their meaning; arising from the fact, that the number of

---

* Such cases give a clear insight into the process of the degeneration of languages in periods of history when literary culture was suspended; and we are now in danger of experiencing a similar evil through the superficial extension of the same culture. So many persons without anything deserving the name of education have become writers by profession, that written language may almost be said to be principally wielded by persons ignorant of the proper use of the instrument, and who are spoiling it more and more for those who understand it. Vulgarisms, which creep in nobody knows how, are daily depriving the English language of valuable modes of expressing thought. To take a present instance: the verb *transpire* formerly conveyed very expressively its correct meaning, viz. to *become known* through unnoticed channels—to exhale, as it were, into publicity through invisible pores, like a vapour or gas disengaging itself. But of late a practice has commenced of employing this word, for the sake of finery, as a mere synonym of *to happen:* "the events which have *transpired* in the Crimea," meaning the incidents of the war. This vile specimen of bad English is already seen in the despatches of noblemen and viceroys: and the time is apparently not far distant when nobody will understand the word if used in its proper sense. It is a great error to think that these corruptions of language do no harm. Those who are struggling with the difficulty (and who know by experience how great it already is) of expressing oneself clearly with precision, find their resources continually narrowed by illiterate writers, who seize and twist from its purpose some form of speech which once served to convey briefly and compactly an unambiguous meaning. It would hardly be believed how often a writer is compelled to a circumlocution by the single vulgarism, introduced during the last few years, of using the word *alone* as an adverb, *only* not being fine enough for the rhetoric of ambitious ignorance. A man will say "to which I am not alone bound by honour but also by law," unaware that what he has unintentionally said is, that he is *not alone* bound, some other person being bound with him. Formerly if any one said, "I am not alone responsible for this," he was understood to mean, (what alone his words mean in correct English,) that he is not the sole person responsible; but if he now used such an expression, the reader would be confused between that and two other meanings; that he is not *only responsible* but something more; or that he is responsible *not only for this* but for something besides. The time is coming when Tennyson's Œnone could not say "I will not die alone," lest she should be supposed to mean that she would not only die but do something else.

The blunder of writing *predicate* for *predict* has become so widely diffused that it bids fair to render one of the most useful terms in the scientific vocabulary of Logic unintelligible. The mathematical and logical term "to eliminate" is undergoing a similar destruction. All who are acquainted either with the proper use of the word or with its etymology, know that to eliminate a thing is

things known to us, and of which we feel a desire to speak, multiply faster than the names for them.   Except on subjects for which there has been constructed a scientific terminology, with which unscientific persons do not meddle, great difficulty is generally found in bringing a new name into use; and independently of that difficulty, it is natural to prefer giving to a new object a name which at least expresses its resemblance to something already known, since by predicating of it a name entirely new we at first convey no information.   In this manner the name of a species often becomes the name of a genus; as *salt*, for example, or *oil;* the former of which words originally denoted only the muriate of soda, the latter, as its etymology indicates, only olive oil; but which now denote large and diversified classes of substances resembling these in some of their qualities, and connote only those common qualities, instead of the whole of the distinctive properties of olive oil and sea salt.   The words *glass* and *soap* are used by modern chemists in a similar manner, to denote genera of which the substances vulgarly so called are single species. And it often happens, as in those instances, that the term keeps its special signification in addition to its more general one, and becomes ambiguous, that is, two names instead of one.

These changes, by which words in ordinary use become more and more generalized, and less and less expressive, take place in a still greater degree with the words which express the complicated phenomena of mind and society.   Historians,

---

to thrust it out; but those who know nothing about it, except that it is a fine-looking phrase, use it in a sense precisely the reverse, to denote, not turning anything out, but bringing it in.   They talk of *eliminating* some truth, or other useful result, from a mass of details.   I suspect that this error must at first have arisen from some confusion between *to eliminate* and *to enucleate.*

Though no such evil consequences as take place in these instances, are likely to arise from the modern freak of writing *sanatory* instead of *sanitary,* it deserves notice as a charming specimen of pedantry engrafted upon ignorance. Those who thus undertake to correct the spelling of the classical English writers, are not aware that the meaning of *sanatory,* if there were such a word in the language, would have reference not to the preservation of health, but to the cure of disease.

travellers, and in general those who speak or write concerning
moral and social phenomena with which they are not familiarly
acquainted, are the great agents in this modification of language.   The vocabulary of all except unusually instructed as
well as thinking persons, is, on such subjects, eminently scanty.
They have a certain small set of words to which they are accustomed, and which they employ to express phenomena the most:
heterogeneous, because they have never sufficiently analysed:
the facts to which those words correspond in their own
country, to have attached perfectly definite ideas to the
words.   The first English conquerors of Bengal, for example,
carried with them the phrase *landed proprietor* into a country
where the rights of individuals over the soil were extremely
different in degree, and even in nature, from those recognised
in England.   Applying the term with all its English
associations in such a state of things; to one who had only a
limited right they gave an absolute right, from another
because he had not an absolute right they took away all right,
drove whole classes of people to ruin and despair, filled the
country with banditti, created, with the best intentions, a disorganiza-
secure, and produced, with the best intentions, a disorganiza-
tion of society which had not been produced in that country
by the most ruthless of its barbarian invaders.   Yet the
usage of persons capable of so gross a misapprehension, determines the meaning of language; and the words they thus
misuse grow in generality, until the instructed are obliged to
acquiesce; and to employ those words (first freeing them from
vagueness by giving them a definite connotation) as generic
terms, subdividing the genera into species.

§ 4.   While the more rapid growth of ideas than of names
thus creates a perpetual necessity for making the same names
serve, even if imperfectly, on a greater number of occasions; a
counter-operation is going on, by which names become on
the contrary restricted to fewer occasions, by taking on, as it
were, additional connotation, from circumstances not originally included in the meaning, but which have become connected with it in the mind by some accidental cause.   We

have seen above, in the words *pagan* and *villain*, remarkable examples of the specialization of the meaning of words from casual associations, as well as of the generalization of it in a new direction, which often follows.

Similar specializations are of frequent occurrence in the history even of scientific nomenclature. " It is by no means uncommon," says Dr. Paris, in his *Pharmacologia*,* " to find a word which is used to express general characters subsequently become the name of a specific substance in which such characters are predominant; and we shall find that some important anomalies in nomenclature may be thus explained. The term Aρσενίκον, from which the word Arsenic is derived, was an ancient epithet applied to those natural substances which possessed strong and acrimonious properties, and as the poisonous quality of arsenic was found to be remarkably powerful, the term was especially applied to Orpiment, the form in which this metal most usually occurred. So the term *Verbena* (quasi *Herbena*) originally denoted all those herbs that were held sacred on account of their being employed in the rites of sacrifice, as we learn from the poets; but as *one* herb was usually adopted upon these occasions, the word Verbena came to denote that particular herb *only*, and it is transmitted to us to this day under the same title, viz. Verbena or Vervain, and indeed until lately it enjoyed the medical reputation which its sacred origin conferred upon it, for it was worn suspended around the neck as an amulet. *Vitriol*, in the original application of the word, denoted *any* crystalline body with a certain degree of transparency (*vitrum*); it is hardly necessary to observe that the term is now appropriated to a particular species: in the same manner, Bark, which is a general term, is applied to express *one* genus, and by way of eminence, it has the article *The* prefixed, as *The* bark: the same observation will apply to the word Opium, which, in its primitive sense, signifies *any* juice (ὀπὸς, *Succus*), while it now only denotes *one* species, viz. that of the poppy. So, again, *Elaterium* was used by Hippocrates to signify

---

* *Historical Introduction*, vol. i. pp. 66—8.

various internal applications, especially purgatives, of a violent and drastic nature (from the word ἐλαύνω, *agito, moveo, stimulo*), but by succeeding authors it was exclusively applied to denote the active matter which subsides from the juice of the wild cucumber.   The word *Fecula*, again, originally meant to imply *any* substance which was derived by spontaneous subsidence from a liquid (from *fæx*, the grounds or settlement of *any* liquor) ; afterwards it was applied to Starch, which is deposited in this manner by agitating the flour of wheat in water ; and lastly, it has been applied to a peculiar vegetable principle, which, like starch, is insoluble in cold, but completely soluble in boiling water, with which it forms a gelatinous solution.   This indefinite meaning of the word *fecula* has created numerous mistakes in pharmaceutic chemistry ; Elaterium, for instance, is said to be *fecula,* and, in the original sense of the word, it is properly so called, inasmuch as it is procured from a vegetable juice by spontaneous subsidence, but in the limited and modern acceptation of the term, it conveys an erroneous idea ; for instead of the active principle of the juice residing in *fecula,* it is a peculiar proximate principle, *sui generis, to* which I have ventured to bestow the name of *Elatin.*   For the same reason, much doubt and obscurity involve the meaning of the word *Extract,* because it is applied *generally* to any substance obtained by the evaporation of a vegetable solution, and *specifically* to a peculiar proximate principle, possessed of certain characters, by which it is distinguished from every other elementary body."

A generic term is always liable to become thus limited to a single species, or even individual, if people have occasion to think and speak of that individual or species much oftener than of anything else which is contained in the genus.   Thus by cattle, a stage-coachman will understand horses ; beasts, in the language of agriculturists, stands for oxen ; and birds, with some sportsmen, for partridges only.   The law of language which operates in these trivial instances, is the very same in conformity to which the terms Θεός, Deus, and God, were adopted from Polytheism by Christianity, to express the single object of its own adoration.   Almost all the terminology of the Christian Church is made up of words originally used in a much more general acceptation: *Ecclesia*, Assembly; *Bishop*, Episcopus, Overseer; *Priest*, Presbyter, Elder; *Deacon*, Diaconus, Administrator; *Sacrament*, a vow of allegiance; *Evangelium*, good tidings; and some words, as *Minister*, are still used both in the general and in the limited sense. It would be interesting to trace the progress by which *author* came, in its most familiar sense, to signify a writer, and ποίητης, or maker, a poet.

Of the incorporation into the meaning of a term, of circumstances accidentally connected with it at some particular period, as in the case of Pagan, instances might easily be multiplied. Physician (φυσίκος, or naturalist) became, in England, synonymous with a healer of diseases, because until a comparatively late period medical practitioners were the only naturalists. *Clerc*, or clericus, a scholar, came to signify an ecclesiastic, because the clergy were for many centuries the only scholars.

Of all ideas, however, the most liable to cling by association to anything with which they have ever been connected by proximity, are those of our pleasures and pains, or of the things which we habitually contemplate as sources of our pleasures or pains. The additional connotation, therefore, which a word soonest and most readily takes on, is that of agreeableness or painfulness, in their various kinds and degrees: of being a good or bad thing; desirable or to be avoided; an object of hatred, of dread, contempt, admiration, hope, or love. Accordingly there is hardly a single name, expressive of any moral or social fact calculated to call forth strong affections either of a favourable or of a hostile nature, which does not carry with it decidedly and irresistibly a connotation of those strong affections, or, at the least, of approbation or censure; insomuch that to employ those names in conjunction with others by which the contrary sentiments were expressed, would produce the effect of a paradox, or even a contradiction in terms. The baneful influence of a connotation thus acquired, on the prevailing habits of thought, especially in morals and politics, has been well pointed out on many occasions by Bentham. It gives rise to the fallacy of "question-begging names." The very property which we are inquiring whether a thing possesses or not, has become so associated with the name of the thing as to be part of its meaning, insomuch that by merely uttering the name we assume the point which was to be made out : one of the most frequent sources of apparently self-evident propositions.

Without any further multiplication of examples to illustrate the changes which usage is continually making in the signification of terms, I shall add, as a practical rule, that the logician, not being able to prevent such transformations, should submit to them with a good grace when they are irrevocably effected, and if a definition is necessary, define the word according to its new meaning; retaining the former as a second signification, if it is needed, and if there is any chance of being able to preserve it either in the language of philosophy or in common use. Logicians cannot *make* the meaning of any but scientific terms : that of all other words is made by the collective human race. But logicians can ascertain clearly what it is which, working obscurely, has guided the general mind to a particular employment of a name; and when they have found this, they can clothe it in such distinct and permanent terms, that mankind shall see the meaning which before they only felt, and shall not suffer it to be afterwards forgotten or misapprehended.

CHAPTER VI.

THE PRINCIPLES OF A PHILOSOPHICAL LANGUAGE FURTHER CONSIDERED.

§ 1. WE have, thus far, considered only one of the requisites of a language adapted for the investigation of truth; that its terms shall each of them convey a determinate and unmistakeable meaning. There are, however, as we have already remarked, other requisites; some of them important only in the second degree, but one which is fundamental, and barely yields in point of importance, if it yields at all, to the quality which we have already discussed at so much length. That the language may be fitted for its purposes, not only should every word perfectly express its meaning, but there should be no important meaning without its word. Whatever we have occasion to think of often, and for scientific purposes, ought to have a name appropriated to it.

This requisite of philosophical language may be considered under three different heads; that number of separate conditions being involved in it.

§ 2. First: there ought to be all such names, as are needful for making such a record of individual observations that the words of the record shall exactly show what fact it is which has been observed. In other words, there should be an accurate Descriptive Terminology.

The only things which we can observe directly being our own sensations, or other feelings, a complete descriptive language would be one in which there should be a name for every variety of elementary sensation or feeling. Combinations of sensations or feelings may always be described, if we have a name for each of the elementary feelings which compose them; but brevity of description, and clearness (which often depends very much on brevity,) are greatly promoted by giving distinctive names not to the elements alone, but also to all combinations which are of frequent recurrence. On this occasion I cannot do better than quote from Dr. Whewell* some of the excellent remarks which he has made on this important branch of our subject.

"The meaning of [descriptive] technical terms can be fixed in the first instance only by convention, and can be made intelligible only by presenting to the senses that which the terms are to signify. The knowledge of a colour by its name can only be taught through the eye. No description can convey to a hearer what we mean by *apple-green* or *French-grey*. It might, perhaps, be supposed that, in the first example, the term *apple*, referring to so familiar an object, sufficiently suggests the colour intended. But it may easily be seen that this is not true; for apples are of many different hues of green, and it is only by a conventional selection that we can appropriate the term to one special shade. When this appropriation is once made, the term refers to the sensation, and not to the parts of the term; for these enter into the compound merely as a help to the memory, whether the suggestion be a natural connexion as in 'apple-green,' or a casual one as in 'French-grey.' In order to derive due advantage from technical terms of this kind, they must be associated *immediately* with the perception to which they belong; and not connected with it through the vague usages of common language. The memory must retain the sensation; and the technical word must be understood as directly as the most familiar word, and more distinctly. When we find such terms as *tin-white* or *pinchbeck-brown*, the metallic colour so denoted ought to start up in our memory without delay or search.

"This, which it is most important to recollect with respect to the simpler properties of bodies, as colour and form, is no less true with respect to more compound notions. In all cases the term is fixed to a peculiar meaning by con-

* *History of Scientific Ideas,* ii. 110, 111.

vention; and the student, in order to use the word, must be completely familiar with the convention, so that he has no need to frame conjectures from the word itself. Such conjectures would always be insecure, and often erroneous. Thus the term *papilionaceous* applied to a flower is employed to indicate, not only a resemblance to a butterfly, but a resemblance arising from five petals of a certain peculiar shape and arrangement; and even if the resemblance were much stronger than it is in such cases, yet, if it were produced in a different way, as for example, by one petal, or two only, instead of a 'standard,' two 'wings,' and a 'keel' consisting of two parts more or less united into one, we should be no longer justified in speaking of it as a 'papilionaceous' flower."

When, however, the thing named is, as in this last case, a combination of simple sensations, it is not necessary, in order to learn the meaning of the word, that the student should refer back to the sensations themselves; it may be communicated to him through the medium of other words; the terms, in short, may be defined. But the names of elementary sensations, or elementary feelings of any sort, cannot be defined; nor is there any mode of making their signification known but by making the learner experience the sensation, or referring him, through some known mark, to his remembrance of having experienced it before. Hence it is only the impressions on the outward senses, or those inward feelings which are connected in a very obvious and uniform manner with outward objects, that are really susceptible of an exact descriptive language. The countless variety of sensations which arise, for instance, from disease, or from peculiar physiological states, it would be in vain to attempt to name; for as no one can judge whether the sensation I have is the same with his, the name cannot have, to us two, real community of meaning. The same may be said, to a considerable extent, of purely mental feelings. But in some of the sciences which are conversant with external objects, it is scarcely possible to surpass the perfection to which this quality of a philosophical language has been carried.

"The formation* of an exact and extensive descriptive language for botany has been executed with a degree of skill and felicity, which, before it was attained, could hardly have been dreamt of as attainable. Every part of a plant has been named; and the form of every part, even the most minute, has had a large assemblage of descriptive terms appropriated to it, by means of which the botanist can convey and receive knowledge of form and structure, as exactly as if each minute part were presented to him vastly magnified. This acquisition was part of the Linnæan reform. . . . . . . 'Tournefort,' says Decandolle, 'appears to have been the first who really perceived the utility of fixing the sense of terms in such a way as always to employ the same word in the same sense, and always to express the same idea by the same words; but it was Linnæus who really created and fixed this botanical language, and this is his fairest claim to glory, for by this fixation of language he has shed clearness and precision over all parts of the science.'

"It is not necessary here to give any detailed account of the terms of botany. The fundamental ones have been gradually introduced, as the parts of plants were more carefully and minutely examined. Thus the flower was necessarily distinguished into the *calyx,* the *corolla,* the *stamens,* and the *pistils;* the sections of the corolla were termed *petals* by Columna; those of the calyx were called *sepals* by Necker. Sometimes terms of greater generality were devised; as *perianth,* to include the calyx and corolla, whether one or both of these were present; *pericarp,* for the part enclosing the grain, of whatever kind it be, fruit, nut, pod, &c. And it may easily be imagined, that descriptive terms may, by definition and combination, become very numerous and distinct. Thus leaves may be called *pinnatifid, pinnatipartite, pinnatisect, pinnatilobate, palmatifid, palmatipartite,* &c., and each of these words designates different combinations of the modes and extent of the divisions of the leaf with the divisions of its outline.

---

* *Hist. Sc. Id.* ii. 111—113.

In some cases, arbitrary numerical relations are introduced
into the definition : thus, a leaf is called *bilobate*, when it is
divided into two parts by a notch ; but if the notch go to the
middle of its length, it is *bifid;* if it go near the base of the
leaf, it is *bipartite ;* if to the base, it is *bisect.* Thus, too, a
pod of a cruciferous plant is a *siliqua*, if it is four times as long
as it is broad, but if it be shorter than this it is a *silicula*.
Such terms being established, the form of the very complex
leaf or frond of a fern (Hymenophyllum Wilsoni) is exactly
conveyed. by the following phrase :—' fronds rigid pinnate,
pinnæ recurved subunilateral, pinnatifid, the segments linear
undivided or bifid spinuloso-serrate.'

" Other characters, as well as form, are conveyed with the
like precision : Colour by means of a classified scale of colours.
. . . . . . This was done with most precision by Werner,
and his scale of colours is still the most usual standard of
naturalists. Werner also introduced a more exact terminology
with regard to other characters which are important in mineralogy, as lustre, hardness. But Mohs improved upon this
step by giving a numerical scale of hardness, in which talc is
1, gypsum 2, calc spar 3, and so on. . . . . Some properties,
as specific gravity, by their definition give at once a numerical
measure ; and others, as crystalline form, require a very considerable array of mathematical calculation and reasoning, to
point out their relations and gradations."

§ 3. Thus far of Descriptive Terminology, or of the
language requisite for placing on record our observation
of individual instances. But when we proceed from this
to Induction, or rather to that comparison of observed
instances which is the preparatory step towards it, we stand
in need of an additional and a different sort of general
names.

Whenever, for purposes of Induction, we find it necessary
to introduce (in Dr. Whewell's phraseology) some new general
conception ; that is, whenever the comparison of a set of
phenomena leads to the recognition in them of some common
circumstance, which, our attention not having been directed

to it on any former occasion, is to us a new phenomenon;
it is of importance that this new conception, or this new
result of abstraction, should have a name appropriated to it;
especially if the circumstance it involves be one which leads
to many consequences, or which is likely to be found also in
other classes of phenomena.   No doubt, in most cases of the
kind, the meaning might be conveyed by joining together
several words already in use.   But when a thing has to be
often spoken of, there are more reasons than the saving of
time and space, for speaking of it in the most concise manner
possible.   What darkness would be spread over geometrical
demonstrations, if wherever the word *circle* is used, the
definition of a circle were inserted instead of it.   In mathematics and its applications, where the nature of the processes
demands that the attention should be strongly concentrated,
but does not require that it should be widely diffused, the
importance of concentration also in the expressions has always
been duly felt; and a mathematician no sooner finds that
he shall often have occasion to speak of the same two
things together, than he at once creates a term to express
them whenever combined: just as, in his algebraical opera-
tions, he substitutes for $(a^m + b^n) \dfrac{p}{q}$, or for $\dfrac{a}{b} + \dfrac{b}{c} + \dfrac{c}{d} +$ &c., the
single letter P, Q, or S; not solely to shorten his symbolical
expressions, but to simplify the purely intellectual part of his
operations, by enabling the mind to give its exclusive attention to the relation between the quantity S and the other
quantities which enter into the equation, without being distracted by thinking unnecessarily of the parts of which S
is itself composed.

But there is another reason, in addition to that of promoting perspicuity, for giving a brief and compact name to
each of the more considerable results of abstraction which
are obtained in the course of our intellectual phenomena.
By naming them, we fix our attention upon them; we keep
them more constantly before the mind.   The names are
remembered, and being remembered, suggest their definition;
while if instead of specific and characteristic names, the

meaning had been expressed by putting together a number of other names, that particular combination of words already in common use for other purposes would have had nothing to make itself remembered by. If we want to render a particular combination of ideas permanent in the mind, there is nothing which clenches it like a name specially devoted to express it. If mathematicians had been obliged to speak of " that to which a quantity, in increasing or diminishing, is always approaching nearer, so that the difference becomes less than any assignable quantity, but to which it never becomes exactly equal," instead of expressing all this by the simple phrase, " the limit of a quantity," we should probably have long remained without most of the important truths which have been discovered by means of the relation between quantities of various kinds and their limits. If instead of speaking of *momentum*, it had been necessary to say, " the product of the number of units of velocity in the velocity by the number of units of mass in the mass," many of the dynamical truths now apprehended by means of this complex idea would probably have escaped notice, for want of recalling the idea itself with sufficient readiness and familiarity. And on subjects less remote from the topics of popular discussion, whoever wishes to draw attention to some new or unfamiliar distinction among things, will find no way so sure as to invent or select suitable names for the express purpose of marking it.

A volume devoted to explaining what the writer means by civilization, does not raise so vivid a conception of it as the single expression, that Civilization is a different thing from Cultivation; the compactness of that brief designation for the contrasted quality being an equivalent for a long discussion. So, if we would impress forcibly upon the understanding and memory the distinction between the two different conceptions of a representative government, we cannot more effectually do so than by saying that Delegation is not Representation. Hardly any original thoughts on mental or social subjects ever make their way among mankind, or assume their proper importance in the minds even of their inventors, until aptly-selected words or phrases have, as it were, nailed them down and held them fast.

§ 4. Of the three essential parts of a philosophical language, we have now mentioned two : a terminology suited for describing with precision the individual facts observed ; and a name for every common property of any importance or interest, which we detect by comparing those facts : including (as the concretes corresponding to those abstract terms) names for the classes which we artificially construct in virtue of those properties, or as many of them, at least, as we have frequent occasion to predicate anything of.

But there is a sort of classes, for the recognition of which no such elaborate process is necessary ; because each of them is marked out from all others not by some one property, the detection of which may depend on a difficult act of abstraction, but by its properties generally. I mean, the Kinds of things, in the sense which, in this treatise, has been specially attached to that term. By a Kind, it will be remembered, we mean one of those classes which are distinguished from all others not by one or a few definite properties, but by an unknown multitude of them : the combination of properties on which the class is grounded, being a mere index to an indefinite number of other distinctive attributes. The class horse is a Kind, because the things which agree in possessing the characters by which we recognise a horse, agree in a great number of other properties, as we know, and, it cannot be doubted, in many more than we know. Animal, again, is a Kind, because no definition that could be given of the name animal could either exhaust the properties common to all animals, or supply premises from which the remainder of those properties could be inferred. But a combination of properties which does not give evidence of the existence of any other independent peculiarities, does not constitute a Kind. White horse, therefore, is not a Kind ; because horses which agree in whiteness, do not agree in anything else, except the qualities common to all horses, and whatever may be the causes or effects of that particular colour.

On the principle that there should be a name for every-thing which we have frequent occasion to make assertions about, there ought evidently to be a name for every Kind; for as it is the very meaning of a Kind that the individuals com-posing it have an indefinite multitude of properties in common, it follows that, if not with our present knowledge, yet with that which we may hereafter acquire, the Kind is a subject to which there will have to be applied many predicates. The third component element of a philosophical language, there-fore, is that there shall be a name for every Kind. In other words, there must not only be a terminology, but also a nomenclature.

The words Nomenclature and Terminology are employed by most authors almost indiscriminately; Dr. Whewell being, as far as I am aware, the first writer who has regularly assigned to the two words different meanings. The distinction how-ever which he has drawn between them being real and important, his example is likely to be followed; and (as is apt to be the case when such innovations in language are felici-tously made) a vague sense of the distinction is found to have influenced the employment of the terms in common practice, before the expediency had been pointed out of discriminating them philosophically. Every one would say that the reform effected by Lavoisier and Guyton-Morveau in the language of chemistry consisted in the introduction of a new nomencla-ture, not of a new terminology. Linear, lanceolate, oval, or oblong, serrated, dentate, or crenate leaves, are expressions forming part of the terminology of botany, while the names " Viola odorata," and " Ulex Europæus," belong to its nomen-clature.

A nomenclature may be defined, the collection of the names of all the Kinds with which any branch of knowledge is con-versant; or more properly, of all the lowest Kinds, or *infimæ species*—those which may be subdivided indeed, but not into Kinds, and which generally accord with what in natural history are termed simply species. Science possesses two splendid examples of a systematic nomenclature; that of plants and animals, constructed by Linnæus and his successors, and that of chemistry, which we owe to the illustrious group of chemists who flourished in France towards the close of the eighteenth century. In these two departments, not only has every known species, or lowest Kind, a name assigned to it, but when new lowest Kinds are discovered, names are at once given to them on an uniform principle. In other sciences the nomenclature is not at present constructed on any system, either because the species to be named are not numerous enough to require one, (as in geometry for example,) or because no one has yet suggested a suitable principle for such a system, as in mineralogy; in which the want of a scientifically constructed nomenclature is now the principal cause which retards the progress of the science.

§ 5. A word which carries on its face that it belongs to a nomenclature, seems at first sight to differ from other concrete general names in this—that its meaning does not reside in its connotation, in the attributes implied in it, but in its denotation, that is, in the particular group of things which it is appointed to designate; and cannot, therefore, be unfolded by means of a definition, but must be made known in another way. This opinion, however, appears to me erroneous. Words belonging to a nomenclature differ, I conceive, from other words mainly in this, that besides the ordinary connotation, they have a peculiar one of their own: besides connoting certain attributes, they also connote that those attributes are distinctive of a Kind. The term "peroxide of iron," for example, belonging by its form to the systematic nomenclature of chemistry, bears on its face that it is the name of a peculiar Kind of substance. It moreover connotes, like the name of any other class, some portion of the properties common to the class; in this instance the property of being a compound of iron and the largest dose of oxygen with which iron will combine. These two things, the fact of being such a compound, and the fact of being a Kind, constitute the connotation of the name peroxide of iron. When we say of the substance before us, that it is the peroxide of iron, we thereby assert, first, that it is a compound of iron and a maximum of oxygen, and next, that the substance so composed is a peculiar Kind of substance.

Now, this second part of the connotation of any word belonging to a nomenclature is as essential a portion of its meaning as the first part, while the definition only declares the first: and hence the appearance that the signification of such terms cannot be conveyed by a definition : which appearance, however, is fallacious. The name Viola odorata denotes a Kind, of which a certain number of characters, sufficient to distinguish it, are enunciated in botanical works. This enumeration of characters is surely, as in other cases, a definition of the name. No, say some, it is not a definition, for the name Viola odorata does not mean those characters; it means that particular group of plants, and the characters are selected from among a much greater number, merely as marks by which to recognise the group. But to this I reply, that the name does not mean that group, for it would be applied to that group no longer than while the group is believed to be an *infima species;* if it were to be discovered that several distinct Kinds have been confounded under this one name, no one would any longer apply the name Viola odorata to the whole of the group, but would apply it, if retained at all, to one only of the Kinds contained therein. What is imperative, therefore, is not that the name shall denote one particular collection of objects, but that it shall denote a Kind, and a lowest Kind. The form of the name declares that, happen what will, it is to denote an *infima species ;* and that, therefore, the properties which it connotes, and which are expressed in the definition, are to be connoted by it no longer than while we continue to believe that those properties, when found together, indicate a Kind, and that the whole of them are found in no more than one Kind.

With the addition of this peculiar connotation, implied in the form of every word which belongs to a systematic nomenclature ; the set of characters which is employed to discriminate each Kind from all other Kinds (and which is a real definition) constitutes as completely as in any other case the whole meaning of the term. It is no objection to say that (as is often the case in natural history) the set of characters may be changed, and another substituted as being better suited for the purpose of distinction, while the word, still continuing to denote the same group of things, is not considered to have changed its meaning. For this is no more than may happen in the case of any other general name: we may, in reforming its connotation, leave its denotation untouched; and it is generally desirable to do so. The connotation, however, is not the less for this the real meaning, for we at once apply the name wherever the characters set down in the definition are found; and that which exclusively guides us in applying the term, must constitute its signification. If we find, contrary to our previous belief, that the characters are not peculiar to one species, we cease to use the term coextensively with the characters; but then it is because the other portion of the connotation fails; the condition that the class must be a Kind. The connotation, therefore, is still the meaning; the set of descriptive characters is a true definition; and the meaning is unfolded, not indeed (as in other cases) by the definition alone, but by the definition and the form of the word taken together.

§ 6. We have now analysed what is implied in the two principal requisites of a philosophical language; first, precision, or definiteness, and secondly, completeness. Any further remarks on the mode of constructing a nomenclature must be deferred until we treat of Classification; the mode of naming the Kinds of things being necessarily subordinate to the mode of arranging those Kinds into larger classes. With respect to the minor requisites of terminology, some of them are well stated and illustrated in the "Aphorisms concerning the Language of Science," included in Dr. Whewell's *Philosophy of the Inductive Sciences.* These, as being of secondary importance in the peculiar point of view of Logic, I shall not further refer to, but shall confine my observations to one more quality, which, next to the two already treated of, appears to be the most valuable which the language of science can possess. Of this quality a general notion may be conveyed by the following aphorism :

Whenever the nature of the subject permits our reasoning processes to be, without danger, carried on mechanically, the language should be constructed on as mechanical principles as possible; while in the contrary case, it should be so constructed that there shall be the greatest possible obstacles to a merely mechanical use of it.

I am aware that this maxim requires much explanation, which I shall at once proceed to give. And first, as to what is meant by using a language mechanically. The complete or extreme case of the mechanical use of language, is when it is used without any consciousness of a meaning, and with only the consciousness of using certain visible or audible marks in conformity to technical rules previously laid down. This extreme case is nowhere realized except in the figures of arithmetic and the symbols of algebra, a language unique in its kind, and approaching as nearly to perfection, for the purposes to which it is destined, as can, perhaps, be said of any creation of the human mind. Its perfection consists in the completeness of its adaptation to a purely mechanical use. The symbols are mere counters, without even the semblance of a meaning apart from the convention which is renewed each time they are employed, and which is altered at each renewal, the same symbol a or x being used on different occasions to represent things which (except that, like all things, they are susceptible of being numbered) have no property in common. There is nothing, therefore, to distract the mind from the set of mechanical operations which are to be performed upon the symbols, such as squaring both sides of the equation, multiplying or dividing them by the same or by equivalent symbols, and so forth. Each of these operations, it is true, corresponds to a syllogism; represents one step of a ratiocination relating not to the symbols, but to the things signified by them. But as it has been found practicable to frame a technical form, by conforming to which we can make sure of finding the conclusion of the ratiocination, our end can be completely attained without our ever thinking of anything but the symbols. Being thus intended to work merely as mechanism, they have the qualities which mechanism ought to have. They are of the least possible bulk, so that they take up scarcely any room, and waste no time in their manipulation; they are compact, and fit so closely together that the eye can take in the whole at once of almost every operation which they are employed to perform.

These admirable properties of the symbolical language of mathematics have made so strong an impression on the minds of many thinkers, as to have led them to consider the symbolical language in question as the ideal type of philosophical language generally; to think that names in general, or (as they are fond of calling them) signs, are fitted for the purposes of thought in proportion as they can be made to approximate to the compactness, the entire unmeaningness, and the capability of being used as counters without a thought of what they represent, which are characteristic of the a and b, the x and y, of algebra. This notion has led to sanguine views of the acceleration of the progress of science by means which, I conceive, cannot possibly conduce to that end, and forms part of that exaggerated estimate of the influence of signs, which has contributed in no small degree to prevent the real laws of our intellectual operations from being rightly understood.

In the first place, a set of signs by which we reason without consciousness of their meaning, can be serviceable, at most, only in our deductive operations. In our direct inductions we cannot for a moment dispense with a distinct mental image of the phenomena, since the whole operation turns on a perception of the particulars in which those phenomena agree and differ. But, further, this reasoning by counters is only suitable to a very limited portion even of our deductive processes. In our reasonings respecting numbers, the only general principles which we ever have occasion to introduce, are these, Things which are equal to the same thing are equal to one another, and The sums or differences of equal things are equal, with their various corollaries. Not only can no hesitation ever arise respecting the applicability of these principles, since they are true of all magnitudes whatever; but every possible application of which they are susceptible, may

be reduced to a technical rule; and such, in fact, the rules of
the calculus are. But if the symbols represent any other
things than mere numbers, let us say even straight or curve
lines, we have then to apply theorems of geometry not true of
all lines without exception, and to select those which are true
of the lines we are reasoning about. And how can we do this
unless we keep completely in mind what particular lines these
are? Since additional geometrical truths may be introduced
into the ratiocination in any stage of its progress, we cannot
suffer ourselves, during even the smallest part of it, to use the
names mechanically (as we use algebraical symbols) without
an image annexed to them. It is only after ascertaining that
the solution of a question concerning lines can be made to
depend on a previous question concerning numbers, or in
other words after the question has been (to speak technically)
reduced to an equation, that the unmeaning signs become
available, and that the nature of the facts themselves to which
the investigation relates can be dismissed from the mind. Up
to the establishment of the equation, the language in which
mathematicians carry on their reasoning does not differ in
character from that employed by close reasoners on any other
kind of subject.

I do not deny that every correct ratiocination, when
thrown into the syllogistic shape, is conclusive from the mere
form of the expression, provided none of the terms used be
ambiguous; and this is one of the circumstances which have
led some writers to think that if all names were so judiciously
constructed and so carefully defined as not to admit of any
ambiguity, the improvement thus made in language would
not only give to the conclusions of every deductive science
the same certainty with those of mathematics, but would
reduce all reasonings to the application of a technical form,
and enable their conclusiveness to be rationally assented to
after a merely mechanical process, as is undoubtedly the case
in algebra. But, if we except geometry, the conclusions of
which are already as certain and exact as they can be made,
there is no science but that of number, in which the practical

validity of a reasoning can be apparent to any person who has looked only at the form of the process. Whoever has assented to what was said in the last Book concerning the case of the Composition of Causes, and the still stronger case of the entire supersession of one set of laws by another, is aware that geometry and algebra are the only sciences of which the propositions are categorically true: the general propositions of all other sciences are true only hypothetically, supposing that no counteracting cause happens to interfere. A conclusion, therefore, however correctly deduced, in point of form, from admitted laws of nature, will have no other than an hypothetical certainty. At every step we must assure ourselves that no other law of nature has superseded, or intermingled its operation with, those which are the premises of the reasoning; and how can this be done by merely looking at the words? We must not only be constantly thinking of the phenomena themselves, but we must be constantly studying them; making ourselves acquainted with the peculiarities of every case to which we attempt to apply our general principles.

The algebraic notation, considered as a philosophical language, is perfect in its adaptation to the subjects for which it is commonly employed, namely those of which the investigations have already been reduced to the ascertainment of a relation between numbers. But, admirable as it is for its own purpose, the properties by which it is rendered such are so far from constituting it the ideal model of philosophical language in general, that the more nearly the language of any other branch of science approaches to it, the less fit that language is for its own proper functions. On all other subjects, instead of contrivances to prevent our attention from being distracted by thinking of the meaning of our signs, we ought to wish for contrivances to make it impossible that we should ever lose sight of that meaning even for an instant.

With this view, as much meaning as possible should be thrown into the formation of the word itself; the aids of derivation and analogy being made available to keep alive a consciousness of all that is signified by it. In this respect those languages have an immense advantage which form their compounds and derivatives from native roots, like the German, and not from those of a foreign or dead language, as is so much the case with English, French, and Italian: and the best are those which form them according to fixed analogies, corresponding to the relations between the ideas to be expressed. All languages do this more or less, but especially, among modern European languages, the German; while even that is inferior to the Greek, in which the relation between the meaning of a derivative word and that of its primitive is in general clearly marked by its mode of formation; except in the case of words compounded with prepositions, which are often, in both those languages, extremely anomalous.

But all that can be done, by the mode of constructing words, to prevent them from degenerating into sounds passing through the mind without any distinct apprehension of what they signify, is far too little for the necessity of the case. Words, however well constructed originally, are always tending, like coins, to have their inscription worn off by passing from hand to hand; and the only possible mode of reviving it is to be ever stamping it afresh, by living in the habitual contemplation of the phenomena themselves, and not resting in our familiarity with the words that express them. If any one, having possessed himself of the laws of phenomena as recorded in words, whether delivered to him originally by others, or even found out by himself, is content from thenceforth to live among these formulæ, to think exclusively of them, and of applying them to cases as they arise, without keeping up his acquaintance with the realities from which these laws were collected—not only will he continually fail in his practical efforts, because he will apply his formulæ without duly considering whether, in this case and in that, other laws of nature do not modify or supersede them; but the formulæ themselves will progressively lose their meaning to him, and he will cease at last even to be capable of recognising with certainty whether a case falls within the contemplation of his formula or not.   It is, in short, as necessary, on all subjects not mathematical, that the things on which we reason should be conceived by us in the concrete, and "clothed in circumstances," as it is in algebra that we should keep all individualizing peculiarities sedulously out of view.

With this remark we close our observations on the Philosophy of Language.

CHAPTER VII.

OF CLASSIFICATION, AS SUBSIDIARY TO INDUCTION.

§ 1. THERE is, as has been frequently remarked in this work, a classification of things, which is inseparable from the fact of giving them general names. Every name which connotes an attribute, divides, by that very fact, all things whatever into two classes, those which have the attribute and those which have it not; those of which the name can be predicated, and those of which it cannot. And the division thus made is not merely a division of such things as actually exist, or are known to exist, but of all such as may hereafter be discovered, and even of all which can be imagined.

On this kind of Classification we have nothing to add to what has previously been said. The Classification which requires to be discussed as a separate act of the mind, is altogether different. In the one, the arrangement of objects in groups, and distribution of them into compartments, is a mere incidental effect consequent on the use of names given for another purpose, namely that of simply expressing some of their qualities. In the other the arrangement and distribution are the main object, and the naming is secondary to, and purposely conforms itself to, instead of governing, that more important operation.

Classification, thus regarded, is a contrivance for the best possible ordering of the ideas of objects in our minds; for causing the ideas to accompany or succeed one another in such a way as shall give us the greatest command over our knowledge already acquired, and lead most directly to the acquisition of more. The general problem of Classification, in reference to these purposes, may be stated as follows: To provide that things shall be thought of in such groups, and those groups in such an order, as will best conduce to the remembrance and to the ascertainment of their laws.

Classification thus considered, differs from classification in the wider sense, in having reference to real objects exclusively, and not to all that are imaginable: its object being the due co-ordination in our minds of those things only, with the properties of which we have actually occasion to make ourselves acquainted. But, on the other hand, it embraces *all* really existing objects. We cannot constitute any one class properly, except in reference to a general division of the whole of nature; we cannot determine the group in which any one object can most conveniently be placed, without taking into consideration all the varieties of existing objects, all at least which have any degree of affinity with it. No one family of plants or animals could have been rationally constituted, except as part of a systematic arrangement of all plants or animals; nor could such a general arrangement have been properly made, without first determining the exact place of plants and animals in a general division of nature.

§ 2. There is no property of objects which may not be taken, if we please, as the foundation for a classification or mental grouping of those objects; and in our first attempts we are likely to select for that purpose properties which are simple, easily conceived, and perceptible on a first view, without any previous process of thought. Thus Tournefort's arrangement of plants was founded on the shape and divisions of the corolla; and that which is commonly called the Linnæan (though Linnæus also suggested another and more scientific arrangement) was grounded chiefly on the number of the stamens and pistils.

But these classifications, which are at first recommended by the facility they afford of ascertaining to what class any individual belongs, are seldom much adapted to the ends of that Classification which is the subject of our present remarks. The Linnæan arrangement answers the purpose of making us think together of all those kinds of plants which possess the same number of stamens and pistils; but to think of them

in that manner is of little use, since we seldom have anything
to affirm in common of the plants which have a given number
of stamens and pistils. If plants of the class Pentandria,
order Monogynia, agreed in any other properties, the habit of
thinking and speaking of the plants under a common designation would conduce to our remembering those common properties so far as they were ascertained, and would dispose us
to be on the look-out for such of them as were not yet known.
But since this is not the case, the only purpose of thought
which the Linnæan classification serves is that of causing us
to remember, better than we should otherwise have done, the
exact number of stamens and pistils of every species of plants.
Now, as this property is of little importance or interest, the
remembering it with any particular accuracy is of no moment.
And, inasmuch as, by habitually thinking of plants in those
groups, we are prevented from habitually thinking of them in
groups which have a greater number of properties in common,
the effect of such a classification, when systematically adhered
to, upon our habits of thought, must be regarded as mischievous.

The ends of scientific classification are best answered, when
the objects are formed into groups respecting which a greater
number of general propositions can be made, and those propositions more important, than could be made respecting any
other groups into which the same things could be distributed.
The properties, therefore, according to which objects are
classified, should, if possible, be those which are causes of
many other properties: or at any rate, which are sure marks
of them. Causes are preferable, both as being the surest and
most direct of marks, and as being themselves the properties
on which it is of most use that our attention should be
strongly fixed. But the property which is the cause of the
chief peculiarities of a class, is unfortunately seldom fitted to
serve also as the diagnostic of the class. Instead of the cause, we
must generally select some of its more prominent effects, which
may serve as marks of the other effects and of the cause.

A classification thus formed is properly scientific or philosophical, and is commonly called a Natural, in contradistinc-

tion to a Technical or Artificial, classification or arrangement. The phrase Natural Classification seems most peculiarly appropriate to such arrangements as correspond, in the groups which they form, to the spontaneous tendencies of the mind, by placing together the objects most similar in their general aspect: in opposition to those technical systems which, arranging things according to their agreement in some circumstance arbitrarily selected, often throw into the same group objects which in the general aggregate of their properties present no resemblance, and into different and remote groups, others which have the closest similarity. It is one of the most valid recommendations of any classification to the character of a scientific one, that it shall be a natural classification in this sense also; for the test of its scientific character is the number and importance of the properties which can be asserted in common of all objects included in a group; and properties on which the general aspect of the things depends, are, if only on that ground, important, as well as, in most cases, numerous. But, though a strong recommendation, this circumstance is not a *sine quâ non;* since the most obvious properties of things may be of trifling importance compared with others that are not obvious. I have seen it mentioned as a great absurdity in the Linnæan classification, that it places (which by the way it does not) the violet by the side of the oak: it certainly dissevers natural affinities, and brings together things quite as unlike as the oak and the violet are. But the difference, apparently so wide, which renders the juxtaposition of those two vegetables so suitable an illustration of a bad arrangement, depends, to the common eye, mainly on mere size and texture; now if we made it our study to adopt the classification which would involve the least peril of similar *rapprochements,* we should return to the obsolete division into trees, shrubs, and herbs, which though of primary importance with regard to mere general aspect, yet (compared even with so petty and unobvious a distinction as that into dicotyledons and monocotyledons) answers to so few differences in the other properties of plants, that a classification founded on it (independently of the indistinctness of the lines of demarcation) would be as completely artificial and technical as the Linnæan.

Our natural groups, therefore, must often be founded not on the obvious, but on the unobvious properties of things, when these are of greater importance.   But in such cases it is essential that there should be some other property or set of properties, more readily recognisable by the observer, which coexist with, and may be received as marks of, the properties which are the real groundwork of the classification. A natural arrangement, for example, of animals, must be founded in the main on their internal structure, but (as has been justly remarked) it would be absurd that we should not be able to determine the genus and species of an animal without first killing it.   On this ground, the preference, among zoological classifications, is probably due to that of M. de Blainville, founded on the differences in the external integuments; differences which correspond, much more accurately than might be supposed, to the really important varieties, both in the other parts of the structure, and in the habits and history of the animals.

This shows, more strongly than ever, how extensive a knowledge of the properties of objects is necessary for making a good classification of them.   And as it is one of the uses of such a classification that by drawing attention to the properties on which it is founded, and which if the classification be good are marks of many others, it facilitates the discovery of those others; we see in what manner our knowledge of things, and our classification of them, tend mutually and indefinitely to the improvement of each other.

We said just now that the classification of objects should follow those of their properties which indicate not only the most numerous, but also the most important peculiarities. What is here meant by importance?   It has reference to the particular end in view; and the same objects, therefore, may admit with propriety of several different classifications.   Each science or art forms its classification of things according to the properties which fall within its special cognizance, or of which it must take account in order to accomplish its peculiar practical end. A farmer does not divide plants, like a botanist, into dicotyledonous and monocotyledonous, but into useful plants and weeds. A geologist divides fossils, not like a zoologist, into families corresponding to those of living species, but into fossils of the secondary and of the tertiary periods, above the coal and below the coal, &c. Whales are or are not fish, according to the purpose for which we are considering them. "If we are speaking of the internal structure and physiology of the animal, we must not call them fish; for in these respects they deviate widely from fishes : they have warm blood, and produce and suckle their young as land quadrupeds do. But this would not prevent our speaking of the *whale fishery*, and calling such animals *fish* on all occasions connected with this employment; for the relations thus rising depend upon the animal's living in the water, and being caught in a manner similar to other fishes. A plea that human laws which mention fish do not apply to whales, would be rejected at once by an intelligent judge."*

These different classifications are all good, for the purposes of their own particular departments of knowledge or practice. But when we are studying objects not for any special practical end, but for the sake of extending our knowledge of the whole of their properties and relations, we must consider as the most important attributes, those which contribute most, either by themselves or by their effects, to render the things like one another, and unlike other things ; which give to the class composed of them the most marked individuality ; which fill, as it were, the largest space in their existence, and would most impress the attention of a spectator who knew all their properties but was not specially interested in any. Classes formed on this principle may be called, in a more emphatic manner than any others, natural groups.

§ 3. On the subject of these groups Dr. Whewell lays

---

* *Nov. Org. Renov.* pp. 286, 287.

down a theory, grounded on an important truth, which he has, in some respects, expressed and illustrated very felicitously; but also, as it appears to me, with some admixture of error. It will be advantageous, for both these reasons, to extract the statement of his doctrine in the very words he has used.

"Natural groups," according to this theory,[*] are "given by Type, not by Definition." And this consideration accounts for that "indefiniteness and indecision which we frequently find in the descriptions of such groups, and which must appear so strange and inconsistent to any one who does not suppose these descriptions to assume any deeper ground of connexion than an arbitrary choice of the botanist. Thus in the family of the rose-tree, we are told that the *ovules* are *very rarely* erect, the *stigmata usually* simple. Of what use, it might be asked, can such loose accounts be? To which the answer is, that they are not inserted in order to distinguish the species, but in order to describe the family, and the total relations of the ovules and the stigmata of the family are better known by this general statement. A similar observation may be made with regard to the Anomalies of each group, which occur so commonly, that Mr. Lindley, in his *Introduction to the Natural System of Botany*, makes the 'Anomalies' an article in each family. Thus, part of the character of the Rosaceæ is, that they have alternate *stipulate* leaves, and that the *albumen* is *obliterated;* but yet in *Lowea*, one of the genera of this family, the stipulæ are *absent;* and the albumen is *present* in another, *Neillia*. This implies, as we have already seen, that the artificial character (or *diagnosis*, as Mr. Lindley calls it,) is imperfect. It is, though very nearly, yet not exactly, commensurate with the natural group: and hence in certain cases this character is made to yield to the general weight of natural affinities.

"These views,—of classes determined by characters which cannot be expressed in words,—of propositions which state, not what happens in all cases, but only usually,—of particulars

---

* *Hist. Sc. Id.* ii. 120—122.

which are included in a class, though they transgress the definition of it, may probably surprise the reader. They are so contrary to many of the received opinions respecting the use of definitions, and the nature of scientific propositions, that they will probably appear to many persons highly illogical and unphilosophical. But a disposition to such a judgment arises in a great measure from this, that the mathematical and mathematico-physical sciences have, in a great degree, determined men's views of the general nature and form of scientific truth; while Natural History has not yet had time or opportunity to exert its due influence upon the current habits of philosophizing. The apparent indefiniteness and inconsistency of the classifications and definitions of Natural History belongs, in a far higher degree, to all other except mathematical speculations; and the modes in which approximations to exact distinctions and general truths have been made in Natural History, may be worthy our attention, even for the light they throw upon the best modes of pursuing truth of all kinds.

"Though in a Natural group of objects a definition can no longer be of any use as a regulative principle, classes are not therefore left quite loose, without any certain standard or guide. The class is steadily fixed, though not precisely limited; it is given, though not circumscribed; it is determined, not by a boundary line without, but by a central point within; not by what it strictly excludes, but by what it eminently includes; by an example, not by a precept; in short, instead of a Definition we have a Type for our director.

"A Type is an example of any class, for instance a species of a genus, which is considered as eminently possessing the character of the class. All the species which have a greater affinity with this type-species than with any others, form the genus, and are arranged about it, deviating from it in various directions and different degrees. Thus a genus may consist of several species which approach very near the type, and of which the claim to a place with it is obvious; while there may be other species which straggle further from this central knot, and which yet are clearly more connected with it than with any other. And even if there should be some species of which the place is dubious, and which appear to be equally bound to two generic types, it is easily seen that this would not destroy the reality of the generic groups, any more than the scattered trees of the intervening plain prevent our speaking intelligibly of the distinct forests of two separate hills.

" The type-species of every genus, the type-genus of every family, is, then, one which possesses all the characters and properties of the genus in a marked and prominent manner. The type of the Rose family has alternate stipulate leaves, wants the albumen, has the ovules not erect, has the stigmata simple, and besides these features, which distinguish it from the exceptions or varieties of its class, it has the features which make it prominent in its class. It is one of those which possess clearly several leading attributes; and thus, though we cannot say of any one genus that it *must* be the type of the family, or of any one species that it *must* be the type of the genus, we are still not wholly to seek; the type must be connected by many affinities with most of the others of its group; it must be near the centre of the crowd, and not one of the stragglers."

In this passage (the latter part of which especially I cannot help noticing as an admirable example of philosophic style) Dr. Whewell has stated very clearly and forcibly, but (I think) without making all necessary distinctions, one of the principles of a Natural Classification. What this principle is, what are its limits, and in what manner he seems to me to have overstepped them, will appear when we have laid down another rule of Natural Arrangement, which appears to me still more fundamental.

§ 4. The reader is by this time familiar with the general truth (which I restate so often on account of the great confusion in which it is commonly involved), that there are in nature distinctions of Kind; distinctions not consisting in a given number of definite properties, *plus* the effects which follow from those properties, but running through the whole nature, through the attributes generally, of the things so distinguished. Our knowledge of the properties of a Kind is never complete. We are always discovering, and expecting to discover, new ones. Where the distinction between two classes of things is not one of Kind, we expect to find their properties alike, except where there is some reason for their being different. On the contrary, when the distinction is in Kind, we expect to find the properties different unless there be some cause for their being the same. All knowledge of a Kind must be obtained by observation and experiment upon the Kind itself; no inference respecting its properties from the properties of things not connected with it by Kind, goes for more than the sort of presumption usually characterized as an analogy, and generally in one of its fainter degrees.

Since the common properties of a true Kind, and consequently the general assertions which can be made respecting it, or which are certain to be made hereafter as our knowledge extends, are indefinite and inexhaustible; and since the very first principle of natural classification is that of forming the classes so that the objects composing each may have the greatest number of properties in common; this principle prescribes that every such classification shall recognise and adopt into itself all distinctions of Kind, which exist among the objects it professes to classify. To pass over any distinctions of Kind, and substitute definite distinctions, which, however considerable they may be, do not point to ulterior unknown differences, would be to replace classes with more by classes with fewer attributes in common; and would be subversive of the Natural Method of Classification.

Accordingly all natural arrangements, whether the reality of the distinction of Kinds was felt or not by their framers, have been led, by the mere pursuit of their own proper end, to conform themselves to the distinctions of Kind, so far as these had been ascertained at the time. The Species of

Plants are not only real Kinds, but are probably,* all of them, real lowest Kinds, Infimæ Species; which if we were to subdivide, as of course it is open to us to do, into subclasses, the subdivision would necessarily be founded on *definite* distinctions, not pointing (apart from what may be known of their causes or effects) to any difference beyond themselves.

In so far as a natural classification is grounded on real Kinds, its groups are certainly not conventional; it is perfectly true that they do not depend upon an arbitrary choice of the naturalist. But it does not follow, nor, I conceive, is it true, that these classes are determined by a type, and not by characters. To determine them by a type would be as sure a way of missing the Kind, as if we were to select a set of characters arbitrarily. They are determined by characters, but these are not arbitrary. The problem is, to find a few definite characters which point to the multitude of indefinite ones. Kinds are Classes between which there is an impassable barrier; and what we have to seek is, marks whereby we may determine on which side of the barrier an object takes its place. The characters which will best do this should be chosen: if they are also important in themselves, so much the better. When we have selected the characters, we parcel out the objects according to those characters, and not, I conceive, according to resemblance to a type. We do not compose the species Ranunculus acris, of all plants which bear a satisfactory degree of resemblance to a model-buttercup, but of those which possess certain characters selected as marks by which we might recognise the possibility of a common

---

* I say probably, not certainly, because this is not the consideration by which a botanist determines what shall or shall not be admitted as a species. In natural history those objects belong to the same species, which are, or consistently with experience might have been, produced from the same stock. But this distinction, in most, and probably in all cases, happily accords with the other. It seems to be a law of physiology, that animals and plants do really, in the philosophical as well as the popular sense, propagate their kind; transmitting to their descendants all the distinctions of Kind (down to the most special or lowest Kind) which they themselves possess.

parentage; and the enumeration of those characters is the definition of the species.

The question next arises, whether, as all Kinds must have a place among the classes, so all the classes in a natural arrangement must be Kinds? And to this I answer, certainly not. The distinctions of Kinds are not numerous enough to make up the whole of a classification. Very few of the genera of plants, or even of the families, can be pronounced with certainty to be Kinds. The great distinctions of Vascular and Cellular, Dicotyledonous or Exogenous and Monocotyledonous or Endogenous plants, are perhaps differences of Kind; the lines of demarcation which divide those classes seem (though even on this I would not pronounce positively) to go through the whole nature of the plants. But the different species of a genus, or genera of a family, usually have in common only a limited number of characters. A Rose does not seem to differ from a Rubus, or the Umbelliferæ from the Ranunculaceæ, in much else than the characters botanically assigned to those genera or those families. Unenumerated differences certainly do exist in some cases; there are families of plants which have peculiarities of chemical composition, or yield products having peculiar effects on the animal economy. The Cruciferæ and Fungi contain an unusual proportion of nitrogen; the Labiatæ are the chief sources of essential oils, the Solaneæ are very commonly narcotic, &c. In these and similar cases there are possibly distinctions of Kind; but it is by no means indispensable that there should be. Genera and Families may be eminently natural, though marked out from one another by properties limited in number; provided those properties are important, and the objects contained in each genus or family resemble each other more than they resemble anything which is excluded from the genus or family.

After the recognition and definition, then, of the *infimæ species*, the next step is to arrange those *infimæ species* into larger groups: making these groups correspond to Kinds wherever it is possible, but in most cases without any such guidance. And in doing this it is true that we are naturally

and properly guided, in most cases at least, by resemblance
to a type. We form our groups round certain selected Kinds,
each of which serves as a sort of exemplar of its group. But
though the groups are suggested by types, I cannot think that
a group when formed is *determined* by the type; that in
deciding whether a species belongs to the group, a reference is
made to the type, and not to the characters; that the characters "cannot be expressed in words." This assertion is
inconsistent with Dr. Whewell's own statement of the fundamental principle of classification, namely, that "general assertions shall be possible." If the class did not possess any
characters in common, what general assertions would be
possible respecting it? Except that they all resemble each
other more than they resemble anything else, nothing whatever could be predicated of the class.

The truth is, on the contrary, that every genus or family
is framed with distinct reference to certain characters, and is
composed, first and principally, of species which agree in possessing all those characters. To these are added, as a sort of
appendix, such other species, generally in small number, as
possess *nearly* all the properties selected; wanting some of
them one property, some another, and which, while they agree
with the rest *almost* as much as these agree with one another,
do not resemble in an equal degree any other group. Our
conception of the class continues to be grounded on the characters; and the class might be defined, those things which
either possess that set of characters, *or* resemble the things
that do so, more than they resemble anything else.

And this resemblance itself is not, like resemblance between simple sensations, an ultimate fact, unsusceptible of
analysis. Even the inferior degree of resemblance is created
by the possession of common characters. Whatever resembles
the genus Rose more than it resembles any other genus, does
so because it possesses a greater number of the characters of
that genus, than of the characters of any other genus. Nor
can there be any real difficulty in representing, by an enumeration of characters, the nature and degree of the resemblance which is strictly sufficient to include any object in

o all
hich
.ions.
e not
some
here-
.cters
ccep-
ided,
sibly
want
t be
etter
its
lass,
hose

any
nsti-
But
are
oup,
d in
ence
h is
the
siug
und,
iner,
ard,
lass,
any

individual or species belongs to the class or not. And this,
as it seems to me, is the amount of truth contained in the
doctrine of Types.

We shall see presently that where the classification is made
for the express purpose of a special inductive inquiry, it is not
optional, but necessary for fulfilling the conditions of a correct

and properly guided, in most cases at least, by resemblance
to a type. We form our groups round certain selected Kinds,
each of which serves as a sort of exemplar of its group. But
though the groups are suggested by types, I cannot think that
a group when formed is *determined* by the type; that in
deciding whether a species belongs to the group, a reference is
made to the type, and not to the characters; that the characters "cannot be expressed in words." This assertion is
inconsistent with Dr. Whewell's own statement of the fundamental principle of classification, namely, that "general assertions shall be possible." If the class did not possess any
characters in common, what general assertions would be
possible respecting it? Except that they all resemble each
other more than they resemble anything else, nothing whatever could be predicated of the class.

The truth is, on the contrary, that every genus or family
is framed with distinct reference to certain characters, and is
composed, first and principally, of species which agree in possessing all those characters. To these are added, as a sort of
appendix, such other species, generally in small number, as
possess *nearly* all the properties selected; wanting some of
them one property, some another, and which, while they agree
with the rest *almost* as much as these agree with one another,
do not resemble in an equal degree any other group. Our
conception of the class continues to be grounded on the characters; and the class might be defined, those things which
either possess that set of characters, *or* resemble the things
that do so, more than they resemble anything else.

And this resemblance itself is not, like resemblance between simple sensations, an ultimate fact, unsusceptible of
analysis. Even the inferior degree of resemblance is created
by the possession of common characters. Whatever resembles
the genus Rose more than it resembles any other genus, does
so because it possesses a greater number of the characters of
that genus, than of the characters of any other genus. Nor
can there be any real difficulty in representing, by an enumeration of characters, the nature and degree of the resemblance which is strictly sufficient to include any object in

the class. There are always some properties common to all things which are included. Others there often are, to which some things, which are nevertheless included, are exceptions. But the objects which are exceptions to one character are not exceptions to another: the resemblance which fails in some particulars must be made up for in others. The class, therefore, is constituted by the possession of *all* the characters which are universal, and *most* of those which admit of exceptions. If a plant had the ovules erect, the stigmata divided, possessed the albumen, and was without stipules, it possibly would not be classed among the Rosaceæ. But it may want any one, or more than one of these characters, and not be excluded. The ends of a scientific classification are better answered by including it. Since it agrees so nearly, in its known properties, with the sum of the characters of the class, it is likely to resemble that class more than any other in those of its properties which are still undiscovered.

Not only, therefore, are natural groups, no less than any artificial classes, determined by characters; they are constituted in contemplation of, and by reason of, characters. But it is in contemplation not of those characters only which are rigorously common to all the objects included in the group, but of the entire body of characters, all of which are found in most of those objects, and most of them in all. And hence our conception of the class, the image in our minds which is representative of it, is that of a specimen complete in all the characters; most naturally a specimen which, by possessing them all in the greatest degree in which they are ever found, is the best fitted to exhibit clearly, and in a marked manner, what they are. It is by a mental reference to this standard, not instead of, but in illustration of, the definition of the class, that we usually and advantageously determine whether any individual or species belongs to the class or not. And this, as it seems to me, is the amount of truth contained in the doctrine of Types.

We shall see presently that where the classification is made for the express purpose of a special inductive inquiry, it is not optional, but necessary for fulfilling the conditions of a correct

Inductive Method, that we should establish a type-species or genus, namely, the one which exhibits in the most eminent degree the particular phenomenon under investigation. But of this hereafter. It remains, for completing the theory of natural groups, that a few words should be said on the principles of the nomenclature adapted to them.

§ 5. A Nomenclature in science, is, as we have said, a system of the names of Kinds. These names, like other class-names, are defined by the enumeration of the characters distinctive of the class. The only merit which a set of names can have beyond this, is to convey, by the mode of their construction, as much information as possible: so that a person who knows the thing, may receive all the assistance which the name can give in remembering what he knows, while he who knows it not, may receive as much knowledge respecting it as the case admits of, by merely being told its name.

There are two modes of giving to the name of a Kind this sort of significance. The best, but which unfortunately is seldom practicable, is when the word can be made to indicate, by its formation, the very properties which it is designed to connote. The name of a Kind does not, of course, connote all the properties of the Kind, since these are inexhaustible, but such of them as are sufficient to distinguish it; such as are sure marks of all the rest. Now, it is very rarely that one property, or even any two or three properties, can answer this purpose. To distinguish the common daisy from all other species of plants would require the specification of many characters. And a name cannot, without being too cumbrous for use, give indication, by its etymology or mode of construction, of more than a very small number of these. The possibility, therefore, of an ideally perfect Nomenclature, is probably confined to the one case in which we are happily in possession of something nearly approaching to it; the Nomenclature of elementary Chemistry. The substances, whether simple or compound, with which chemistry is conversant, are Kinds, and, as such, the properties which distinguish each of them from the rest are innumerable; but in the case of compound substances (the simple ones are not numerous enough to require a systematic nomenclature), there is one property, the chemical composition, which is of itself sufficient to distinguish the Kind; and is (with certain reservations not yet thoroughly understood) a sure mark of all the other properties of the compound. All that was needful, therefore, was to make the name of every compound express, on the first hearing, its chemical composition; that is, to form the name of the compound, in some uniform manner, from the names of the simple substances which enter into it as elements. This was done, most skilfully and successfully, by the French chemists. The only thing left unexpressed by them was the exact proportion in which the elements were combined; and even this, since the establishment of the atomic theory, it has been found possible to express by a simple adaptation of their phraseology.

But where the characters which must be taken into consideration in order sufficiently to designate the Kind, are too numerous to be all signified in the derivation of the name, and where no one of them is of such preponderant importance as to justify its being singled out to be so indicated, we may avail ourselves of a subsidiary resource. Though we cannot indicate the distinctive properties of the Kind, we may indicate its nearest natural affinities, by incorporating into its name the name of the proximate natural group of which it is one of the species. On this principle is founded the admirable binary nomenclature of botany and zoology. In this nomenclature the name of every species consists of the name of the genus, or natural group next above it, with a word added to distinguish the particular species. The last portion of the compound name is sometimes taken from some *one* of the peculiarities in which that species differs from others of the genus; as Clematis *integrifolia*, Potentilla *alba*, Viola *palustris*, Artemisia *vulgaris;* sometimes from a circumstance of an historical nature, as Narcissus *poeticus*, Potentilla *tormentilla* (indicating that the plant was formerly known by the latter name), Exacum *Candollii* (from the fact that De Candolle was its first discoverer); and sometimes the word is purely conventional, as Thlaspi *bursa-pastoris*, Ranunculus *thora;* it is

of little consequence which; since the second, or as it is usually
called, the specific name, could at most express, independently
of convention, no more than a very small portion of the connotation of the term.　But by adding to this the name of the
superior genus, we may make the best amends we can for the
impossibility of so contriving the name as to express all the
distinctive characters of the Kind.　We make it, at all events,
express as many of those characters as are common to the
proximate natural group in which the Kind is included.　If
even those common characters are so numerous or so little
familiar as to require a further extension of the same resource,
we might, instead of a binary, adopt a ternary nomenclature,
employing not only the name of the genus, but that of the
next natural group in order of generality above the genus,
commonly called the Family.　This was done in the mineralo-
gical nomenclature proposed by Professor Mohs.　"The names
framed by him were not composed of two, but of three elements, designating respectively the Species, the Genus, and
the Order; thus he has such species as *Rhombohedral Lime
Haloide, Octohedral Fluor Haloide, Prismatic Hal Baryte.*"[*]
The binary construction, however, has been found sufficient in
botany and zoology, the only sciences in which this general
principle has hitherto been successfully adopted in the construction of a nomenclature.

Besides the advantage which this principle of nomenclature
possesses, in giving to the names of species the greatest quantity
of independent significance which the circumstances of the case
admit of, it answers the further end of immensely economizing
the use of names, and preventing an otherwise intolerable
burden on the memory.　When the names of species become
extremely numerous, some artifice (as Dr. Whewell[†] observes)
becomes absolutely necessary to make it possible to recollect
or apply them.　"The known species of plants, for example,
were ten thousand in the time of Linnæus, and are now probably sixty thousand.　It would be useless to endeavour to
frame and employ separate names for each of these species.

---

[*] *Nov. Org. Renov.* p. 274.　　　　[†] *Hist. Sc. Id.* i. 138.

The division of the objects into a subordinated system of classification enables us to introduce a Nomenclature which does not require this enormous number of names. Each of the genera has its name, and the species are marked by the addition of some epithet to the name of the genus. In this manner about seventeen hundred generic names, with a moderate number of specific names, were found by Linnæus sufficient to designate with precision all the species of vegetables known at .his time." And though the number of generic names has since greatly increased, it has not increased in anything like the proportion of the multiplication of known species.

CHAPTER VIII.

OF CLASSIFICATION BY SERIES.

§ 1. THUS far, we have considered the principles of scientific classification so far only as relates to the formation of natural groups; and at this point most of those who have attempted a theory of natural arrangement, including, among the rest, Dr. Whewell, have stopped. There remains, however, another, and a not less important portion of the theory, which has not yet, as far as I am aware, been systematically treated of by any writer except M. Comte. This is, the arrangement of the natural groups into a natural series.*

The end of Classification, as an instrument for the investigation of nature, is (as before stated) to make us think of those objects together, which have the greatest number of important common properties; and which therefore we have oftenest occasion, in the course of our inductions, for taking into joint consideration. Our ideas of objects are thus brought into the order most conducive to the successful prosecution of inductive inquiries generally. But when the purpose is to facilitate some particular inductive inquiry, more is required. To be instrumental to that purpose, the classification must bring those objects together, the simultaneous contemplation

---

* Dr. Whewell, in his reply (*Philosophy of Discovery*, p. 270) says that he "stopped short of, or rather passed by, the doctrine of a series of organised beings," because he "thought it bad and narrow philosophy." If he did, it was evidently without understanding this form of the doctrine; for he proceeds to quote a passage from his "History," in which the doctrine he condemns is designated as that of "a mere linear progression in nature, which would place each genus in contact only with the preceding and succeeding ones." Now the series treated of in the text agrees with this linear progression in nothing whatever but in being a progression.

It would surely be possible to arrange all *places* (for example) in the order of their distance from the North Pole, though there would be not merely a plurality, but a whole circle of places at every single gradation in the scale.

of which is likely to throw most light upon the particular subject. That subject being the laws of some phenomenon or some set of connected phenomena; the very phenomenon or set of phenomena in question must be chosen as the ground-work of the classification.

The requisites of a classification intended to facilitate the study of a particular phenomenon, are, first, to bring into one class all Kinds of things which exhibit that phenomenon, in whatever variety of forms or degrees; and secondly, to arrange those Kinds in a series according to the degree in which they exhibit it, beginning with those which exhibit most of it, and terminating with those which exhibit least. The principal example, as yet, of such a classification, is afforded by comparative anatomy and physiology, from which, therefore, our illustrations shall be taken.

§ 2. The object being supposed to be, the investigation of the laws of animal life; the first step, after forming the most distinct conception of the phenomenon itself, possible in the existing state of our knowledge, is to erect into one great class (that of animals) all the known Kinds of beings where that phenomenon presents itself; in however various combinations with other properties, and in however different degrees. As some of these Kinds manifest the general phenomenon of animal life in a very high degree, and others in an insignificant degree, barely sufficient for recognition; we must, in the next place, arrange the various Kinds in a series, following one another according to the degrees in which they severally exhibit the phenomenon; beginning therefore with man, and ending with the most imperfect kinds of zoophytes.

This is merely saying that we should put the instances, from which the law is to be inductively collected, into the order which is implied in one of the four Methods of Experimental Inquiry discussed in the preceding Book; the fourth Method, that of Concomitant Variations. As formerly remarked, this is often the only method to which recourse can be had, with assurance of a true conclusion, in cases in which we have but limited means of effecting, by artificial experiments, a separation of circumstances usually conjoined. The principle of the method is, that facts which increase or diminish together, and disappear together, are either cause and effect, or effects of a common cause. When it has been ascertained that this relation really subsists between the variations, a connexion between the facts themselves may be confidently laid down, either as a law of nature or only as an empirical law, according to circumstances.

That the application of this Method must be preceded by the formation of such a series as we have described, is too obvious to need being pointed out; and the mere arrangement of a set of objects in a series, according to the degrees in which they exhibit some fact of which we are seeking the law, is too naturally suggested by the necessities of our inductive operations, to require any lengthened illustration here. But there are cases in which the arrangement required for the special purpose, becomes the determining principle of the classification of the same objects for general purposes. This will naturally and properly happen, when those laws of the objects which are sought in the special inquiry enact so principal a part in the general character and history of those objects—exercise so much influence in determining all the phenomena of which they are either the agents or the theatre —that all other differences existing among the objects are fittingly regarded as mere modifications of the one phenomenon sought; effects determined by the co-operation of some incidental circumstance with the laws of that phenomenon. Thus in the case of animated beings, the differences between one class of animals and another may reasonably be considered as mere modifications of the general phenomenon, animal life; modifications arising. either from the different degrees in which that phenomenon is manifested in different animals, or from the intermixture of the effects of incidental causes peculiar to the nature of each, with the effects produced by the general laws of life; those laws still exercising a predominant influence over the result. Such being the case, no other inductive inquiry respecting animals can be successfully carried on, except in subordination to the great inquiry into the universal laws of animal life. And the classification of animals best suited to that one purpose, is the most suitable to all the other purposes of zoological science.

§ 3. To establish a classification of this sort, or even to apprehend it when established, requires the power of recognising the essential similarity of a phenomenon, in its minuter degrees and obscurer forms, with what is called the *same* phenomenon in the greatest perfection of its development; that is, of identifying with each other all phenomena which differ only in degree, and in properties which we suppose to be caused by difference of degree. In order to recognise this identity, or in other words, this exact similarity of quality, the assumption of a type-species is indispensable. We must consider as the type of the class, that among the Kinds included in it, which exhibits the properties constitutive of the class, in the highest degree; conceiving the other varieties as instances of degeneracy, as it were, from that type; deviations from it by inferior intensity of the characteristic property or properties. For every phenomenon is best studied (*cæteris paribus*) where it exists in the greatest intensity. It is there that the effects which either depend on it, or depend on the same causes with it, will also exist in the greatest degree. It is there, consequently, and only there, that those effects of it, or joint effects with it, can become fully known to us, so that we may learn to recognise their smaller degrees, or even their mere rudiments, in cases in which the direct study would have been difficult or even impossible. Not to mention that the phenomenon in its higher degrees may be attended by effects or collateral circumstances which in its smaller degrees do not occur at all, requiring for their production in any sensible amount a greater degree of intensity of the cause than is there met with. In man, for example, (the species in which both the phenomenon of animal and that of organic life exist in the highest degree,) many subordinate phenomena develop themselves in the course of his animated existence, which the inferior varieties of animals do not show. The knowledge of these properties may nevertheless be of great avail towards

the discovery of the conditions and laws of the general
phenomenon of life, which is common to man with those
inferior animals.   And they are, even, rightly considered as
properties of animated nature itself; because they may evidently
be affiliated to the general laws of animated nature; because
we may fairly presume that some rudiments or feeble degrees
of those properties would be recognised in all animals by more
perfect organs, or even by more perfect instruments, than ours;
and because those may be correctly termed properties of a class,
which a thing exhibits exactly in proportion as it belongs to
the class, that is, in proportion as it possesses the main
attributes constitutive of the class.

§ 4.   It remains to consider how the internal distribution of the series may most properly take place: in what
manner it should be divided into Orders, Families, and
Genera.

The main principle of division must of course be natural
affinity; the classes formed must be natural groups : and the
formation of these has already been sufficiently treated of.
But the principles of natural grouping must be applied in
subordination to the principle of a natural series.   The groups
must not be so constituted as to place in the same group things
which ought to occupy different points of the general scale.
The precaution necessary to be observed for this purpose is,
that the *primary* divisions must be grounded not on all
distinctions indiscriminately, but on those which correspond
to variations in the degree of the main phenomenon.   The
series of Animated Nature should be broken into parts at the
points where the variation in the degree of intensity of the
main phenomenon (as marked by its principal characters,
Sensation, Thought, Voluntary Motion, &c.) begins to be
attended by conspicuous changes in the miscellaneous properties of the animal.   Such well-marked changes take place,
for example, where the class Mammalia ends ; at the points
where Fishes are separated from Insects, Insects from Mol-
lusca, &c.   When so formed, the primary natural groups will
compose the series by mere juxtaposition, without redistri-

bution; each of them corresponding to a definite portion of the scale. In like manner each family should, if possible, be so subdivided, that one portion of it shall stand higher and the other lower, though of course contiguous, in the general scale; and only when this is impossible is it allowable to ground the remaining subdivisions on characters having no determinable connexion with the main phenomenon.

Where the principal phenomenon so far transcends in importance all other properties on which a classification could be grounded, as it does in the case of animated existence, any considerable deviation from the rule last laid down is in general sufficiently guarded against by the first principle of a natural arrangement, that of forming the groups according to the most important characters. All attempts at a scientific classification of animals, since first their anatomy and physiology were successfully studied, have been framed with a certain degree of instinctive reference to a natural series, and have accorded in many more points than they have differed, with the classification which would most naturally have been grounded on such a series. But the accordance has not always been complete; and it still is often a matter of discussion, which of several classifications best accords with the true scale of intensity of the main phenomenon. Cuvier, for example, has been justly criticized for having formed his natural groups with an undue degree of reference to the mode of alimentation, a circumstance directly connected only with organic life, and not leading to the arrangement most appropriate for the purposes of an investigation of the laws of animal life, since both carnivorous and herbivorous or frugivorous animals are found at almost every degree in the scale of animal perfection. Blainville's classification has been considered by high authorities to be free from this defect; as representing correctly, by the mere order of the principal groups, the successive degeneracy of animal nature from its highest to its most imperfect exemplification.

§ 5. A classification of any large portion of the field of nature in conformity to the foregoing principles, has hitherto been found practicable only in one great instance, that of animals. In the case even of vegetables, the natural arrangement has not been carried beyond the formation of natural groups. Naturalists have found, and probably will continue to find it impossible to form those groups into any series, the terms of which correspond to real gradations in the phenomenon of vegetative or organic life. Such a difference of degree may be traced between the class of Vascular Plants and that of Cellular, which includes lichens, algæ, and other substances whose organization is simpler and more rudimentary than that of the higher order of vegetables, and which therefore approach nearer to mere inorganic nature. But when we rise much above this point, we do not find any sufficient difference in the degree in which different plants possess the properties of organization and life. The dicotyledons are of more complex structure, and somewhat more perfect organization, than the monocotyledons: and some dicotyledonous families, such as the Compositæ, are rather more complex in their organization than the rest. But the differences are not of a marked character, and do not promise to throw any particular light upon the conditions and laws of vegetable life and development. If they did, the classification of vegetables would have to be made, like that of animals, with reference to the scale or series indicated.

Although the scientific arrangements of organic nature afford as yet the only complete example of the true principles of rational classification, whether as to the formation of groups or of series, those principles are applicable to all cases in which mankind are called upon to bring the various parts of any extensive subject into mental co-ordination. They are as much to the point when objects are to be classed for purposes of art or business, as for those of science. The proper arrangement, for example, of a code of laws, depends on the same scientific conditions as the classifications in natural history; nor could there be a better preparatory discipline for that important function, than the study of the principles of a natural arrangement, not only in the abstract, but in their actual application to the class of phenomena for which they were first elaborated, and which are still the best school for learning their use. Of this the great authority on codification, Bentham, was perfectly aware: and his early *Fragment on Government*, the admirable introduction to a series of writings unequalled in their department, contains clear and just views (as far as they go) on the meaning of a natural arrangement, such as could scarcely have occurred to any one who lived anterior to the age of Linnæus and Bernard de Jussieu.

BOOK V.

ON FALLACIES.

"Errare non modo affirmando et negando, sed etiam sentiendo, et in tacitâ hominum cogitatione contingit."—Hobbes, *Computatio sive Logica*, ch. v.

"Il leur semble qu'il n'y a qu'à douter par fantaisie, et qu'il n'y a qu'à dire en général que notre nature est infirme ; que notre esprit est plein d'aveuglement ; qu'il faut avoir un grand soin de se défaire de ses préjugés, et autres choses semblables. Ils pensent que cela suffit pour ne plus se laisser séduire à ses sens, et pour ne plus se tromper du tout. Il ne suffit pas de dire que l'esprit est foible, il faut lui faire sentir ses foiblesses. Ce n'est pas assez de dire qu'il est sujet à l'erreur, il faut lui découvrir en quoi consistent ses erreurs."—Malebranche, *Recherche de la Vérité.*

CHAPTER I.

§ 1. It is a maxim of the schoolmen, that " contra-
riorum eadem est scientia:" we never really know what a thing
is, unless we are also able to give a sufficient account of its
opposite. Conformably to this maxim, one considerable section,
in most treatises on Logic, is devoted to the subject of Fallacies ; and the practice is too well worthy of observance, to allow
of our departing from it. The philosophy of reasoning, to be
complete, ought to comprise the theory of bad as well as of
good reasoning.

We have endeavoured to ascertain the principles by which
the sufficiency of any proof can be tested, and by which the
nature and amount of evidence needful to prove any given
conclusion can be determined beforehand. If these principles
were adhered to, then although the number and value of the
truths ascertained would be limited by the opportunities, or
by the industry, ingenuity, and patience, of the individual
inquirer, at least error would not be embraced instead of truth.
But the general consent of mankind, founded on their
experience, vouches for their being far indeed from even this
negative kind of perfection in the employment of their
reasoning powers.

In the conduct of life—in the practical business of mankind
—wrong inferences, incorrect interpretations of experience,
unless after much culture of the thinking faculty, are absolutely inevitable: and with most people, after the highest
degree of culture they ever attain, such erroneous inferences,
producing corresponding errors in conduct, are lamentably
frequent. Even in the speculations to which eminent intellects have systematically devoted themselves, and in reference
to which the collective mind of the scientific world is always

at hand to aid the efforts and correct the aberrations of individuals, it is only from the more perfect sciences, from those
of which the subject-matter is the least complicated, that
opinions not resting on a correct induction have at length,
generally speaking, been expelled. In the departments of
inquiry relating to the more complex phenomena of nature,
and especially those of which the subject is man, whether as
a moral and intellectual, a social, or even as a physical being ;
the diversity of opinions still prevalent among instructed
persons, and the equal confidence with which those of the
most contrary ways of thinking cling to their respective tenets,
are proof not only that right modes of philosophizing are not
yet generally adopted on those subjects, but that wrong ones
are : that inquirers have not only in general missed the
truth, but have often embraced error ; that even the most
cultivated portion of our species have not yet learned to
abstain from drawing conclusions which the evidence does not
warrant.

The only complete safeguard against reasoning ill, is the
habit of reasoning well ; familiarity with the principles of correct
reasoning, and practice in applying those principles. It is,
however, not unimportant to consider what are the most
common modes of bad reasoning ; by what appearances the
mind is most likely to be seduced from the observance of true
principles of induction ; what, in short, are the most common
and most dangerous varieties of Apparent Evidence, whereby
persons are misled into opinions for which there does not exist
evidence really conclusive.

A catalogue of the varieties of apparent evidence which are
not real evidence, is an enumeration of Fallacies. Without
such an enumeration, therefore, the present work would be
wanting in an essential point. And while writers who included
in their theory of reasoning nothing more than ratiocination,
have, in consistency with this limitation, confined their
remarks to the fallacies which have their seat in that portion
of the process of investigation ; we, who profess to treat of the
whole process, must add to our directions for performing it
rightly, warnings against performing it wrongly in any of its

parts: whether the ratiocinative or the experimental portion of it be in fault, or the fault lie in dispensing with ratiocination and induction altogether.

§ 2. In considering the sources of unfounded inference, it is unnecessary to reckon the errors which arise, not from a wrong method, nor even from ignorance of the right one, but from a casual lapse, through hurry or inattention, in the application of the true principles of induction. Such errors, like the accidental mistakes in casting up a sum, do not call for philosophical analysis or classification; theoretical considerations can throw no light upon the means of avoiding them. In the present treatise our attention is required, not to mere inexpertness in performing the operation in the right way, (the only remedies for which are increased attention and more sedulous practice,) but to the modes of performing it in a way fundamentally wrong; the conditions under which the human mind persuades itself that it has sufficient grounds for a conclusion which it has not arrived at by any of the legitimate methods of induction—which it has not, even carelessly or overhastily, endeavoured to test by those legitimate methods.

§ 3. There is another branch of what may be called the Philosophy of Error, which must be mentioned here, though only to be excluded from our subject. The sources of erroneous opinions are twofold, moral and intellectual. Of these, the moral do not fall within the compass of this work. They may be classed under two general heads; Indifference to the attainment of truth, and Bias: of which last the most common case is that in which we are biassed by our wishes; but the liability is almost as great to the undue adoption of a conclusion which is disagreeable to us, as of one which is agreeable, if it be of a nature to bring into action any of the stronger passions. Persons of timid character are the more predisposed to believe any statement, the more it is calculated to alarm them. Indeed it is a psychological law, deducible from the most general laws of the mental constitution of man, that any strong passion renders us credulous as to the existence of objects suitable to excite it.

But the moral causes of opinions, though with most persons the most powerful of all, are but remote causes: they do not act directly, but by means of the intellectual causes; to which they bear the same relation that the circumstances called, in the theory of medicine, *predisposing* causes, bear to *exciting* causes. Indifference to truth cannot, in and by itself, produce erroneous belief; it operates by preventing the mind from collecting the proper evidences, or from applying to them the test of a legitimate and rigid induction; by which omission it is exposed unprotected to the influence of any species of apparent evidence which offers itself spontaneously, or which is elicited by that smaller quantity of trouble which the mind may be willing to take. As little is Bias a direct source of wrong conclusions. We cannot believe a proposition only by wishing, or only by dreading, to believe it. The most violent inclination to find a set of propositions true, will not enable the weakest of mankind to believe them without a vestige of intellectual grounds—without any, even apparent, evidence. It acts indirectly, by placing the intellectual grounds of belief in an incomplete or distorted shape before his eyes. It makes him shrink from the irksome labour of a rigorous induction, when he has a misgiving that its result may be disagreeable; and in such examination as he does institute, it makes him exert that which *is* in a certain measure voluntary, his attention, unfairly, giving a larger share of it to the evidence which seems favourable to the desired conclusion, a smaller to that which seems unfavourable. It operates, too, by making him look out eagerly for reasons, or apparent reasons, to support opinions which are conformable, or resist those which are repugnant, to his interests or feelings; and when the interests or feelings are common to great numbers of persons, reasons are accepted and pass current, which would not for a moment be listened to in that character, if the conclusion had nothing more powerful than its reasons to speak in its behalf. The natural or acquired partialities of mankind are continually throwing up philosophical theories, the sole recommendation of which consists in the premises they afford for proving cherished doctrines, or justifying favourite feelings : and when any one of these theories has been so thoroughly discredited as no longer to serve the purpose, another is always ready to take its place. This propensity, when exercised in favour of any widely-spread persuasion or sentiment, is often decorated with complimentary epithets ; and the contrary habit of keeping the judgment in complete subordination to evidence, is stigmatized by various hard names, as scepticism, immorality, coldness, hard-heartedness, and similar expressions according to the nature of the case. But though the opinions of the generality of mankind, when not dependent on mere habit and inculcation, have their root much more in the inclinations than in the intellect, it is a necessary condition to the triumph of the moral bias that it should first pervert the understanding. Every erroneous inference, though originating in moral causes, involves the intellectual operation of admitting insufficient evidence as sufficient; and whoever was on his guard against all kinds of inconclusive evidence which can be mistaken for conclusive, would be in no danger of being led into error even by the strongest bias. There are minds so strongly fortified on the intellectual side, that they could not blind themselves to the light of truth, however really desirous of doing so ; they could not, with all the inclination in the world, pass off upon themselves bad arguments for good ones. If the sophistry of the intellect could be rendered impossible, that of the feelings, having no instrument to work with, would be powerless. A comprehensive classification of all those things which, not being evidence, are liable to appear such to the understanding, will, therefore, of itself include all errors of judgment arising from moral causes, to the exclusion only of errors of practice committed against better knowledge.

To examine, then, the various kinds of apparent evidence which are not evidence at all, and of apparently conclusive evidence which do not really amount to conclusiveness, is the object of that part of our inquiry into which we are about to enter.

The subject is not beyond the compass of classification and comprehensive survey. The things, indeed, which are not evidence of any given conclusion, are manifestly endless, and this negative property, having no dependence on any positive ones, cannot be made the groundwork of a real classification. But the things which, not being evidence, are susceptible of being mistaken for it, are capable of a classification having reference to the positive property which they possess, of appearing to be evidence. We may arrange them, at our choice, on either of two principles; according to the cause which makes them appear to be evidence, not being so; or according to the particular kind of evidence which they simulate. The Classification of Fallacies which will be attempted in the ensuing chapter, is founded on these considerations jointly.

CHAPTER II.

CLASSIFICATION OF FALLACIES.

§ 1. IN attempting to establish certain general distinctions which shall mark out from one another the various kinds of Fallacious Evidence, we propose to ourselves an altogether different aim from that of several eminent thinkers, who have given, under the name of Political or other Fallacies, a mere enumeration of a certain number of erroneous opinions; false general propositions which happen to be often met with; *loci communes* of bad arguments on some particular subject. Logic is not concerned with the false opinions which people happen to entertain, but with the manner in which they come to entertain them. The question is not, what facts have at any time been erroneously supposed to be proof of certain other facts, but what property in the facts it was which led any one to this mistaken supposition.

When a fact is supposed, though incorrectly, to be evidentiary of, or a mark of, some other fact, there must be a cause of the error; the supposed evidentiary fact must be connected in some particular manner with the fact of which it is deemed evidentiary,—must stand in some particular relation to it, without which relation it would not be regarded in that light. The relation may either be one resulting from the simple contemplation of the two facts side by side with one another, or it may depend on some process of mind, by which a previous association has been established between them. Some peculiarity of relation, however, there must be; the fact which can, even by the wildest aberration, be supposed to prove another fact, must stand in some special position with regard to it; and if we could ascertain and define that special position, we should perceive the origin of the error.

We cannot regard one fact as evidentiary of another, unless we believe that the two are always, or in the majority of cases, conjoined. If we believe A to be evidentiary of B, if when we see A we are inclined to infer B from it, the reason is because we believe that wherever A is, B also either always or for the most part exists, either as an antecedent, a consequent, or a concomitant. If when we see A we are inclined not to expect B—if we believe A to be evidentiary of the absence of B—it is because we believe that where A is, B either is never, or at least seldom, found. Erroneous conclusions, in short, no less than correct conclusions, have an invariable relation to a general formula, either expressed or tacitly implied. When we infer some fact from some other fact which does not really prove it, we either have admitted, or, if we maintained consistency, ought to admit, some groundless general proposition respecting the conjunction of the two phenomena.

For every property, therefore, in facts, or in our mode of considering facts, which leads us to believe that they are habitually conjoined when they are not, or that they are not when in reality they are, there is a corresponding kind of Fallacy; and an enumeration of fallacies would consist in a specification of those properties in facts, and those peculiarities in our mode of considering them, which give rise to this erroneous opinion.

§ 2. To begin, then; the supposed connexion, or repugnance, between the two facts, may either be a conclusion from evidence (that is, from some other proposition or propositions) or may be admitted without any such ground; admitted, as the phrase is, on its own evidence; embraced as self-evident, as an axiomatic truth. This gives rise to the first great distinction, that between Fallacies of Inference, and Fallacies of Simple Inspection. In the latter division must be included not only all cases in which a proposition is believed and held for true, literally without any extrinsic evidence, either of specific experience or general reasoning; but those more frequent cases in which simple inspection

creates a *presumption* in favour of a proposition; not sufficient for belief, but sufficient to cause the strict principles of
a regular induction to be dispensed with, and creating a pre-
disposition to believe it on evidence which would be seen to be
insufficient if no such presumption existed. This class, comprehending the whole of what may be termed Natural
Prejudices, and which I shall call indiscriminately Fallacies of
Simple Inspection or Fallacies *à priori*, shall be placed at the
head of our list.

Fallacies of Inference, or erroneous conclusions from
supposed evidence, must be subdivided according to the nature
of the apparent evidence from which the conclusions are
drawn; or (what is the same thing) according to the particular kind of sound argument which the fallacy in question
simulates. But there is a distinction to be first drawn, which
does not answer to any of the divisions of sound arguments,
but arises out of the nature of bad ones. We may know
exactly what our evidence is, and yet draw a false conclusion
from it; we may conceive precisely what our premises are,
what alleged matters of fact, or general principles, are the
foundation of our inference; and yet, because the premises
are false, or because we have inferred from them what they
will not support, our conclusion may be erroneous. But a
case, perhaps even more frequent, is that in which the error
arises from not conceiving our premises with due clearness,
that is, (as shown in the preceding Book,*) with due fixity:
forming one conception of our evidence when we collect or
receive it, and another when we make use of it; or unadvisedly,
and in general unconsciously, substituting, as we proceed, different premises in the place of those with which we set out, or
a different conclusion for that which we undertook to prove.
This gives existence to a class of fallacies which may be justly
termed (in a phrase borrowed from Bentham) Fallacies of
Confusion; comprehending, among others, all those which have
their source in language, whether arising from the vagueness or
ambiguity of our terms, or from casual associations with them.

---

* Supra, p. 204.

When the fallacy is not one of Confusion, that is, when the proposition believed, and the evidence on which it is believed, are steadily apprehended and unambiguously expressed, there remain to be made two cross divisions. The Apparent Evidence may be either particular facts, or foregone generalizations; that is, the process may simulate either simple Induction, or Deduction; and again, the evidence, whether consisting of supposed facts or of general propositions, may be false in itself, or, being true, may fail to bear out the conclusion attempted to be founded on it. This gives us first, Fallacies of Induction and Fallacies of Deduction, and then a subdivision of each of these, according as the supposed evidence is false, or true but inconclusive.

Fallacies of Induction, where the facts on which the induction proceeds are erroneous, may be termed Fallacies of Observation. The term is not strictly accurate, or rather, not accurately coextensive with the class of fallacies which I propose to designate by it. Induction is not always grounded on facts immediately observed, but sometimes on facts inferred: and when these last are erroneous, the error may not be, in the literal sense of the term, an instance of bad observation, but of bad inference. It will be convenient, however, to make only one class of all the inductions of which the error lies in not sufficiently ascertaining the facts on which the theory is grounded; whether the cause of failure be mal-observation, or simple non-observation, and whether the mal-observation be direct, or by means of intermediate marks which do not prove what they are supposed to prove. And in the absence of any comprehensive term to denote the ascertainment, by whatever means, of the facts on which an induction is grounded, I will venture to retain for this class of fallacies, under the explanation now given, the title of Fallacies of Observation.

The other class of inductive fallacies, in which the facts are correct, but the conclusion not warranted by them, are properly denominated Fallacies of Generalization: and these, again, fall into various subordinate classes or natural groups, some of which will be enumerated in their proper place.

When we now turn to Fallacies of Deduction, namely those modes of incorrect argumentation in which the premises, or some of them, are general propositions, and the argument a ratiocination ; we may of course subdivide these also into two species similar to the two preceding, namely, those which proceed on false premises, and those of which the premises, though true, do not support the conclusion. But of these species, the first must necessarily fall under some one of the heads already enumerated. For the error must be either in those premises which are general propositions, or in those which assert individual facts. In the former case it is an Inductive Fallacy, of one or the other class ; in the latter it is a Fallacy of Observation : unless, in either case, the erroneous premise has been assumed on simple inspection, in which case the fallacy is *à priori*. Or finally, the premises, of whichever kind they are, may never have been conceived in so distinct a manner as to produce any clear consciousness by what means they were arrived at ; as in the case of what is called reasoning in a circle : and then the fallacy is one of Confusion.

There remain, therefore, as the only class of fallacies having properly their seat in deduction, those in which the premises of the ratiocination do not bear out its conclusion ; the various cases, in short, of vicious argumentation, provided against by the rules of the syllogism. We shall call these, Fallacies of Ratiocination.

We have thus five distinguishable classes of fallacy, which may be expressed in the following synoptic table :—

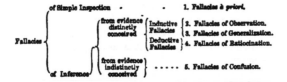

§ 3. We must not, however, expect to find that men's actual errors always, or even commonly, fall so unmistakeably

under some one of these classes, as to be incapable of being
referred to any other. Erroneous arguments do not admit of
such a sharply cut division as valid arguments do. An argument fully stated, with all its steps distinctly set out, in
language not susceptible of misunderstanding, must, if it be
erroneous, be so in some one of these five modes unequivocally :
or indeed of the first four, since the fifth, on such a supposition, would vanish. But it is not in the nature of bad reasoning to express itself thus unambiguously. When a sophist,
whether he is imposing on himself or attempting to impose
on others, can be constrained to throw his sophistry into so
distinct a form, it needs, in a large proportion of cases, no
further exposure.

In all arguments, everywhere but in the schools, some of
the links are suppressed ; à fortiori when the arguer either
intends to deceive, or is a lame and inexpert thinker, little
accustomed to bring his reasoning processes to any test : and
it is in those steps of the reasoning which are made in this
tacit and half-conscious, or even wholly unconscious manner,
that the error oftenest lurks. In order to detect the fallacy,
the proposition thus silently assumed must be supplied ; but
the reasoner, most likely, has never really asked himself what
he was assuming : his confuter, unless permitted to extort it
from him by the Socratic mode of interrogation, must himself
judge what the suppressed premise ought to be in order to
support the conclusion. And hence, in the words of Archbishop Whately, "it must be often a matter of doubt, or
rather, of arbitrary choice, not only to which genus each kind
of fallacy should be referred, but even to which kind to refer
any one individual fallacy ; for since, in any course of argument, one premise is usually suppressed, it frequently happens
in the case of a fallacy, that the hearers are left to the alternative of supplying either a premise which is not true, or else, one
which does not prove the conclusion : e. g. if a man expatiates
on the distress of the country, and thence argues that the
government is tyrannical, we must suppose him to assume
either that 'every distressed country is under a tyranny,' which
is a manifest falsehood, or merely that ' every country under a

tyranny is distressed,' which, however true, proves nothing, the middle term being undistributed." The former would be ranked, in our distribution, among fallacies of generalization, the latter among those of ratiocination. "Which are we to suppose the speaker meant us to understand? Surely" (if he understood himself) "just whichever each of his hearers might happen to prefer: some might assent to the false premise; others allow the unsound syllogism."

Almost all fallacies, therefore, might in strictness be brought under our fifth class, Fallacies of Confusion. A fallacy can seldom be absolutely referred to any of the other classes; we can only say, that if all the links were filled up which should be capable of being supplied in a valid argument, it would either stand thus (forming a fallacy of one class), or thus (a fallacy of another); or at furthest we may say, that the conclusion is most *likely* to have originated in a fallacy of such and such a class. Thus in the illustration just quoted, the error committed may be traced with most probability to a fallacy of generalization; that of mistaking an uncertain mark, or piece of evidence, for a certain one; concluding from an effect to some one of its possible causes, when there are others which would have been equally capable of producing it.

Yet, though the five classes run into each other, and a particular error often seems to be arbitrarily assigned to one of them rather than to any of the rest, there is considerable use in so distinguishing them. We shall find it convenient to set apart, as Fallacies of Confusion, those of which confusion is the most obvious characteristic; in which no other cause can be assigned for the mistake committed, than neglect or inability to state the question properly, and to apprehend the evidence with definiteness and precision. In the remaining four classes I shall place not only the cases in which the evidence is clearly seen to be what it is, and yet a wrong conclusion drawn from it, but also those in which, although there be confusion, the confusion is not the sole cause of the error, but there is some shadow of a ground for it in the nature of the evidence itself. And in distributing these cases of partial

confusion among the four classes, I shall, when there can
be any hesitation as to the precise seat of the fallacy, suppose it to be in that part of the process in which, from
the nature of the case, and the tendencies of the human mind,
an error would in the particular circumstances be the most
probable.

After these observations we shall proceed, without further
preamble, to consider the five classes in their order.

CHAPTER III.

§ 1. THE tribe of errors of which we are to treat in the first instance, are those in which no actual inference takes place at all: the proposition (it cannot in such cases be called a conclusion) being embraced, not as proved, but as requiring no proof; as a self-evident truth; or else as having such intrinsic verisimilitude, that external evidence not in itself amounting to proof, is sufficient in aid of the antecedent presumption.

An attempt to treat this subject comprehensively would be a transgression of the bounds prescribed to this work, since it would necessitate the inquiry which, more than any other, is the grand question of what is called metaphysics, viz. What are the propositions which may reasonably be received without proof? That there must be some such propositions all are agreed, since there cannot be an infinite series of proof, a chain suspended from nothing. But to determine what these propositions are, is the *opus magnum* of the more recondite mental philosophy. Two principal divisions of opinion on the subject have divided the schools of philosophy from its first dawn. The one recognises no ultimate premises but the facts of our subjective consciousness; our sensations, emotions, intellectual states of mind, and volitions. These, and whatever by strict rules of induction can be derived from these, it is possible, according to this theory, for us to know; of all else we must remain in ignorance. The opposite school hold that there are other existences, suggested indeed to our minds by these subjective phenomena, but not inferrible from them, by any process either of deduction or of induction; which, however, we must, by the constitution of our mental nature recognise as realities; and realities, too, of a higher order than the phenomena of our consciousness, being the efficient causes and necessary substrata of all Phenomena. Among these entities they reckon Substances, whether matter or spirit; from the dust under our feet to the soul, and from that to Deity. All these, according to them, are preternatural or supernatural beings, having no likeness in experience, though experience is entirely a manifestation of their agency. Their existence, together with more or less of the laws to which they conform in their operations, are, on this theory, apprehended and recognised as real by the mind itself intuitively: experience (whether in the form of sensation or of mental feeling) having no other part in the matter than as affording facts which are consistent with these necessary postulates of reason, and which are explained and accounted for by them.

As it is foreign to the purpose of the present treatise to decide between these conflicting theories, we are precluded from inquiring into the existence, or defining the extent and limits, of knowledge *à priori*, and from characterizing the kind of correct assumption which the fallacy of incorrect assumption, now under consideration, simulates. Yet since it is allowed on both sides that such assumptions are often made improperly, we may find it practicable, without entering into the ultimate metaphysical grounds of the discussion, to state some speculative propositions, and suggest some practical cautions, respecting the forms in which such unwarranted assumptions are most likely to be made.

§ 2. In the cases in which, according to the thinkers of the ontological school, the mind apprehends, by intuition, things, and the laws of things, not cognizable by our sensitive faculty; those intuitive, or supposed intuitive, perceptions are undistinguishable from what the opposite school are accustomed to call ideas of the mind. When they themselves say that they perceive the things by an immediate act of a faculty given for that purpose by their Creator, it would be said of them by their opponents that they find an idea or conception

in their own minds, and from the idea or conception, infer the
existence of a corresponding objective reality. Nor would
this be an unfair statement, but a mere version into other
words of the account given by many of themselves; and one
to which the more clear-sighted of them might, and generally
do, without hesitation, subscribe. Since, therefore, in the
cases which lay the strongest claims to be examples of knowledge *à priori*, the mind proceeds from the idea of a thing to
the reality of the thing itself, we cannot be surprised by finding that illicit assumptions *à priori* consist in doing the same
thing erroneously: in mistaking subjective facts for objective,
laws of the percipient mind for laws of the perceived object,
properties of the ideas or conceptions for properties of the
things conceived.

Accordingly, a large proportion of the erroneous thinking
which exists in the world proceeds on a tacit assumption, that
the same order must obtain among the objects in nature
which obtains among our ideas of them. That if we always
think of two things together, the two things must always
exist together. That if one thing makes us think of another
as preceding or following it, that other must precede it or
follow it in actual fact. And conversely, that when we cannot
conceive two things together they cannot exist together, and
that their combination may, without further evidence, be rejected from the list of possible occurrences.

Few persons, I am inclined to think, have reflected on the
great extent to which this fallacy has prevailed, and prevails,
in the actual beliefs and actions of mankind. For a first illustration of it, we may refer to a large class of popular superstitions. If any one will examine in what circumstances most
of those things agree, which in different ages and by different
portions of the human race have been considered as omens or
prognostics of some interesting event, whether calamitous or
fortunate; they will be found very generally characterized by
this peculiarity, that they cause the mind to *think* of that, of
which they are therefore supposed to forebode the actual
occurrence. "Talk of the devil, and he will appear," has
passed into a proverb. Talk of the devil, that is, raise the

idea, and the reality will follow. In times when the appearance of that personage in a visible form was thought to be no unfrequent occurrence, it has doubtless often happened to persons of vivid imagination and susceptible nerves, that talking of the devil has caused them to fancy they saw him; as, even in our more incredulous days, listening to ghost stories predisposes us to see ghosts; and thus, as a prop to the *à priori* fallacy, there might come to be added an auxiliary fallacy of mal-observation, with one of false generalization grounded on it. Fallacies of different orders often herd or cluster together in this fashion, one smoothing the way for another. But the origin of the superstition is evidently that which we have assigned. In like manner it has been universally considered unlucky to speak of misfortune. The day on which any calamity happened has been considered an unfortunate day, and there has been a feeling everywhere, and in some nations a religious obligation, against transacting any important business on that day. For on such a day our thoughts are likely to be of misfortune. For a similar reason, any untoward occurrence in commencing an undertaking has been considered ominous of failure; and often, doubtless, has really contributed to it, by putting the persons engaged in the enterprise more or less out of spirits: but the belief has equally prevailed where the disagreeable circumstance was, independently of superstition, too insignificant to depress the spirits by any influence of its own. All know the story of Cæsar's accidentally stumbling in the act of landing on the African coast; and the presence of mind with which he converted the direful presage into a favourable one by exclaiming, "Africa, I embrace thee." Such omens, it is true, were often conceived as warnings of the future, given by a friendly or a hostile deity; but this very superstition grew out of a pre-existing tendency; the god was supposed to send, as an indication of what was to come, something which people were already disposed to consider in that light. So in the case of lucky or unlucky names. Herodotus tells us how the Greeks, on the way to Mycale, were encouraged in their enterprise by the arrival of a deputation from Samos, one of the members of which was named Hegesistratus, the leader of armies.

Cases may be pointed out in which something which could have no real effect but to make persons *think* of misfortune, was regarded not merely as a prognostic, but as something approaching to an actual cause of it. The εὐφήμει of the Greeks, and *favete linguis*, or *bona verba quæso*, of the Romans, evince the care with which they endeavoured to repress the utterance of any word expressive or suggestive of ill fortune; not from notions of delicate politeness, to which their general mode of conduct and feeling had very little reference, but from *bonâ fide* alarm lest the event so suggested to the imagination should in fact occur. Some vestige of a similar superstition has been known to exist among uneducated persons even in our own day: it is thought an unchristian thing to talk of, or suppose, the death of any person while he is alive. It is known how careful the Romans were to avoid, by an indirect mode of speech, the utterance of any word directly expressive of death or other calamity: how instead of *mortuus est* they said *vixit*; and "be the event fortunate or *otherwise*" instead of *adverse*. The name Maleventum, of which Salmasius so sagaciously detected the Thessalian origin (Μαλόεις, Μαλόεντος), they changed into the highly propitious denomination, Beneventum; Egesta into Segesta; and Epidamnus, a name so interesting in its associations to the reader of Thucydides, they exchanged for Dyrrhachium, to escape the perils of a word suggestive of *damnum* or detriment.

"If an hare cross the highway," says Sir Thomas Browne,[*] "there are few above threescore that are not perplexed thereat; which notwithstanding is but an augurial terror, according to that received expression, *Inauspicatum dat iter oblatus lepus*. And the ground of the conceit was probably no greater than this, that a fearful animal passing by us portended unto us something to be feared; as upon the like consideration the meeting of a fox presaged some future imposture." Such

---

[*] *Vulgar Errors*, book v. chap. 21.

superstitions as these last must be the result of study; they are too recondite for natural or spontaneous growth. But when the attempt was once made to construct a science of predictions, any association, though ever so faint or remote, by which an object could be connected in however far-fetched a manner with ideas either of prosperity or of danger and misfortune, was enough to determine its being classed among good or evil omens.

An example of rather a different kind from any of these, but falling under the same principle, is the famous attempt on which so much labour and ingenuity were expended by the alchemists, to make gold potable. The motive to this was a conceit that potable gold could be no other than the universal medicine: and why gold? Because it was so precious. It must have all marvellous properties as a physical substance, because the mind was already accustomed to marvel at it.

From a similar feeling, "every substance," says Dr. Paris,* "whose origin is involved in mystery, has at different times been eagerly applied to the purposes of medicine. Not long since, one of those showers which are now known to consist of the excrements of insects, fell in the north of Italy; the inhabitants regarded it as manna, or some supernatural panacea, and they swallowed it with such avidity, that it was only by extreme address that a small quantity was obtained for a chemical examination." The superstition, in this instance, though doubtless partly of a religious character, probably in part also arose from the prejudice that a wonderful thing must of course have wonderful properties.

§ 3. The instances of *à priori* fallacy which we have hitherto cited belong to the class of vulgar errors, and do not now, nor in any but a rude age ever could, impose upon minds of any considerable attainments. But those to which we are about to proceed, have been, and still are, all but universally prevalent among thinkers. The same disposition

---

* *Pharmacologia*, Historical Introduction, p. 16.

to give objectivity to a law of the mind—to suppose that what is true of our ideas of things must be true of the things themselves—exhibits itself in many of the most accredited modes of philosophical investigation, both on physical and on metaphysical subjects.   In one of its most undisguised manifestations, it embodies itself in two maxims, which lay claim to axiomatic truth : Things which we cannot think of together, cannot coexist ; and Things which we cannot help thinking of together, must coexist.   I am not sure that the maxims were ever expressed in these precise words, but the history both of philosophy and of popular opinions abounds with exemplifications of both forms of the doctrine.

To begin with the latter of them: Things which we cannot think of except together, must exist together.   This is assumed in the generally received and accredited mode of reasoning which concludes that A must accompany B in point of fact, because "it is involved in the idea."   Such thinkers do not reflect that the idea, being a result of abstraction, ought to conform to the facts, and cannot make the facts conform to it.   The argument is at most admissible as an appeal to authority ; a surmise, that what is now part of the idea, must, before it became so, have been found by previous inquirers in the facts.   Nevertheless, the philosopher who more than all others made professions of rejecting authority, Descartes, constructed his system on this very basis.   His favourite device for arriving at truth, even in regard to outward things, was by looking into his own mind for it.   "Credidi me," says his celebrated maxim, " pro regulâ generali sumere posse, omne id quod valdè dilucidè et distinctè concipiebam, verum esse ;" whatever can be very clearly conceived, must certainly exist ; that is, as he afterwards explains it, if the idea includes existence.   And on this ground he infers that geometrical figures really exist, because they can be distinctly conceived.   Whenever existence is " involved in an idea," a thing conformable to the idea must really exist ; which is as much as to say, whatever the idea contains must have its equivalent in the thing ; and what we

are not able to leave out of the idea cannot be absent from
the reality.* This assumption pervades the philosophy not
only of Descartes, but of all the thinkers who received their
impulse mainly from him, in particular the two most remarkable among them, Spinoza and Leibnitz, from whom the
modern German metaphysical philosophy is essentially an
emanation. I am indeed disposed to think that the fallacy
now under consideration has been the cause of two-thirds of
the bad philosophy, and especially of the bad metaphysics,
which the human mind has never ceased to produce. Our
general ideas contain nothing but what has been put into
them, either by our passive experience, or by our active habits
of thought; and the metaphysicians in all ages, who have
attempted to construct the laws of the universe by reasoning
from our supposed necessities of thought, have always proceeded, and only could proceed, by laboriously finding in their
own minds what they themselves had formerly put there, and
evolving from their ideas of things what they had first
involved in those ideas. In this way all deeply-rooted opinions
and feelings are enabled to create apparent demonstrations of
their truth and reasonableness, as it were out of their own
substance.

The other form of the fallacy; Things which we cannot
think of together cannot exist together,—including as one of
its branches, that what we cannot think of as existing cannot
exist at all,—may thus be briefly expressed: Whatever is
inconceivable must be false.

Against this prevalent doctrine I have sufficiently argued
in a former Book,† and nothing is required in this place but

---

* The author of one of the Bridgewater Treatises has fallen, as it seems to
me, into a similar fallacy when, after arguing in rather a curious way to prove
that matter may exist without any of the known properties of matter, and may
therefore be changeable, he concludes that it cannot be eternal, because " eternal
(passive) existence necessarily involves incapability of change." I believe it
would be difficult to point out any other connexion between the facts of eternity
and unchangeableness, than a strong association between the two ideas. Most
of the *à priori* arguments, both religious and anti-religious, on the origin of
things, are fallacies drawn from the same source.

† Supra, book ii. chap. v. § 6, and ch. vii. § 1, 2, 3. See also *Examination
of Sir William Hamilton's Philosophy*, chap. vi. and elsewhere.

examples. It was long held that Antipodes were impossible because of the difficulty which was found in conceiving persons with their heads in the same direction as our feet. And it was one of the received arguments against the Copernican system, that we cannot conceive so great a void space as that system supposes to exist in the celestial·regions. When men's imaginations had always been used to conceive the stars as firmly set in solid spheres, they naturally found much difficulty in imagining them in so different, and, as it doubtless appeared to them, so precarious a situation. But they had no right to mistake the limitation (whether natural, or, as it in fact proved, only artificial) of their own faculties, for an inherent limitation of the possible modes of existence in the universe.

It may be said in objection, that the error in these cases was in the minor premise, not the major; an error of fact, not of principle; that it did not consist in supposing that what is inconceivable cannot be true, but in supposing antipodes to be inconceivable, when present experience proves that they can be conceived. Even if this objection were allowed, and the proposition that what is inconceivable cannot be true were suffered to remain unquestioned as a speculative truth, it would be a truth on which no practical consequence could ever be founded, since, on this showing, it is impossible to affirm of any proposition, not being a contradiction in terms, that it is inconceivable. Antipodes were really, not fictitiously, inconceivable to our ancestors: they are indeed conceivable to us; and as the limits of our power of conception have been so largely extended, by the extension of our experience and the more varied exercise of our imagination, so may posterity find many combinations perfectly conceivable to them which are inconceivable to us. But, as beings of limited experience, we must always and necessarily have limited conceptive powers; while it does not by any means follow that the same limitation obtains in the possibilities of nature, nor even in her actual manifestations.

Rather more than a century and a half ago it was a scientific maxim, disputed by no one, and which no one deemed to

require any proof, that "a thing cannot act where it is not."
With this weapon the Cartesians waged a formidable war
against the theory of gravitation, which, according to them,
involving so obvious an absurdity, must be rejected *in limine*:
the sun could not possibly act upon the earth, not being there.
It was not surprising that the adherents of the old systems
of astronomy should urge this objection against the new;
but the false assumption imposed equally on Newton himself,
who in order to turn the edge of the objection, imagined a
subtle ether which filled up the space between the sun and the
earth, and by its intermediate agency was the proximate cause
of the phenomena of gravitation.   "It is inconceivable," said
Newton, in one of his letters to Dr. Bentley,* "that inanimate
brute matter should, without the mediation of something else,
which is not material, operate upon and affect other matter
without mutual contact. . . . .   That gravity should be
innate, inherent, and essential to matter, so that one body may
act on another, at a distance, through a vacuum, without the
mediation of anything else, by and through which their action
and force may be conveyed from one to another, is to me so
great an absurdity, that I believe no man, who in philosophical
matters has a competent faculty of thinking, can ever fall into
it."   This passage should be hung up in the cabinet of every
cultivator of science who is ever tempted to pronounce a fact
impossible because it appears to him inconceivable.   In our
own day one would be more tempted, though with equal in-
justice, to reverse the concluding observation, and consider
the seeing any absurdity at all in a thing so simple and
natural, to be what really marks the absence of "a competent
faculty of thinking."   No one now feels any difficulty in conceiving gravity to be, as much as any other property is,
" inherent, and essential to matter," nor finds the comprehension of it facilitated in the smallest degree by the supposition
of an ether (though some recent inquirers do give this as an
explanation of it); nor thinks it at all incredible that the celestial bodies can and do act where they, in actual bodily presence,

---

* I quote this passage from Playfair's celebrated *Dissertations on the Pro-*
gress of Mathematical and Physical Science.

are not. To us it is not more wonderful that bodies should act upon one another "without mutual contact," than that they should do so when in contact; we are familiar with both these facts, and we find them equally inexplicable, but equally easy to believe. To Newton, the one, because his imagination was familiar with it, appeared natural and a matter of course, while the other, for the contrary reason, seemed too absurd to be credited.

It is strange that any one, after such a warning, should rely implicitly on the evidence à priori of such propositions as these, that matter cannot think; that space, or extension, is infinite; that nothing can be made out of nothing (ex nihilo nihil fit). Whether these propositions are true or not this is not the place to determine, nor even whether the questions are soluble by the human faculties. But such doctrines are no more self-evident truths, than the ancient maxim that a thing cannot act where it is not, which probably is not now believed by any educated person in Europe.* Matter cannot think; why? because we cannot conceive thought to be annexed to any arrangement of material particles. Space is infinite, because having never known any part of it which had not other parts beyond it, we cannot conceive an absolute termination. Ex nihilo nihil fit, because having never known any physical product without a pre-existing physical material, we cannot, or think we cannot, imagine a creation out of nothing. But these things may in themselves be as conceivable as gravitation without an intervening medium, which Newton thought too great an absurdity for any person of a competent faculty of philosophical thinking to admit: and even supposing them not conceivable, this, for aught we know, may be merely one of the limitations of our very limited minds, and not in nature at all.

No writer has more directly identified himself with the fallacy now under consideration, or has embodied it in more distinct terms, than Leibnitz. In his view, unless a thing was

---

* This statement I must now correct, as too unqualified. The maxim in question was maintained with full conviction by no less an authority than Sir William Hamilton. See my Examination, chap. xxiv.

not merely conceivable, but even explainable, it could not exist in nature. All *natural* phenomena, according to him, must be susceptible of being accounted for *à priori*. The only facts of which no explanation could be given but the will of God, were miracles properly so called. " Je reconnais," says he,* " qu'il n'est pas permis de nier ce qu'on n'entend pas; mais j'ajoute qu'on a droit de nier (au moins dans l'ordre naturel) ce qui absolument n'est point intelligible ni explicable. Je soutiens aussi . . . . qu'enfin la conception des créatures n'est pas la mesure du pouvoir de Dieu, mais que leur conceptivité, ou force de concevoir, est la mesure du pouvoir de la nature, tout ce qui est conforme à l'ordre naturel pouvant être conçu ou entendu par quelque créature."

Not content with assuming that nothing can be true which we are unable to conceive, scientific inquirers have frequently given a still further extension to the doctrine, and held that, even of things not altogether inconceivable, that which we can conceive with the greatest ease is likeliest to be true. It was long an admitted axiom, and is not yet entirely discredited, that " nature always acts by the simplest means," *i.e.* by those which are most easily conceivable.† A large proportion of all the errors ever committed in the investigation of the laws of nature, have arisen from the assumption that the most familiar explanation or hypothesis must be the truest. One of the most instructive facts in scientific history is the pertinacity with which the human mind clung to the belief that the heavenly bodies must move in circles, or be carried round by the revolution of spheres; merely because those were in themselves the simplest suppositions : though, to make them accord with the facts which were ever contradicting them more and more, it became necessary to add sphere to sphere and circle to circle, until the original simplicity was converted into almost inextricable complication.

---

* *Nouveaux Essais sur l'Entendement Humain—Avant-propos.* (Œuvres, Paris ed. 1842, vol. i. p. 19.)

† This doctrine also was accepted as true, and conclusions were grounded on it, by Sir William Hamilton. See *Examination*, chap. xxiv.

§ 4. We pass to another *à priori* fallacy or natural prejudice, allied to the former, and originating as that does, in the tendency to presume an exact correspondence between the laws of the mind and those of things external to it. The fallacy may be enunciated in this general form — Whatever can be thought of apart exists apart: and its most remarkable manifestation consists in the personification of abstractions. Mankind in all ages have had a strong propensity to conclude that wherever there is a name, there must be a distinguishable separate entity corresponding to the name; and every complex idea which the mind has formed for itself by 'operating upon its conceptions of individual things, was considered to have an outward objective reality answering to it. Fate, Chance, Nature, Time, Space, were real beings, nay, even gods. If the analysis of qualities in the earlier part of this work be correct, names of qualities and names of substances stand for the very same sets of facts or phenomena; *whiteness* and *a white thing* are only different phrases, required by convenience for speaking of the same external fact under different relations. Not such, however, was the notion which this verbal distinction suggested of old, either to the vulgar or to the scientific. Whiteness was an entity, inhering or sticking in the white substance: and so of all other qualities. So far was this carried, that even concrete general terms were supposed to be, not names of indefinite numbers of individual substances, but names of a peculiar kind of entities termed Universal Substances. Because we can think and speak of man in general, that is, of all persons in so far as possessing the common attributes of the species, without fastening our thoughts permanently on some one individual person; therefore man in general was supposed to be, not an aggregate of individual persons, but an abstract or universal man, distinct from these.

It may be imagined what havoc metaphysicians trained in these habits made with philosophy, when they came to the largest generalizations of all. *Substantiæ Secundæ* of any kind were bad enough, but such Substantiæ Secundæ as τὸ ὄν, for example, and τὸ ἕν, standing for peculiar entities supposed to be inherent in all things which *exist*, or which are said to be *one*, were enough to put an end to all intelligible discussion; especially since, with a just perception that the truths which philosophy pursues are *general* truths, it was soon laid down that these general substances were the only subjects of science, being immutable, while individual substances cognizable by the senses, being in a perpetual flux, could not be the subject of real knowledge. This misapprehension of the import of general language constitutes Mysticism, a word so much oftener written and spoken than understood. Whether in the Vedas, in the Platonists, or in the Hegelians, mysticism is neither more nor less than ascribing objective existence to the subjective creations of our own faculties, to ideas or feelings of the mind; and believing that by watching and contemplating these ideas of its own making, it can read in them what takes place in the world without.

§ 5. Proceeding with the enumeration of *à priori* fallacies, and endeavouring to arrange them with as much reference as possible to their natural affinities, we come to another, which is also nearly allied to the fallacy preceding the last, standing in the same relation to one variety of it as the fallacy last mentioned does to the other. This, too, represents nature as under incapacities corresponding to those of our intellect; but instead of only asserting that nature cannot do a thing because we cannot conceive it done, goes the still greater length of averring that nature does a particular thing, on the sole ground that we can see no reason why she should not. Absurd as this seems when so plainly stated, it is a received principle among scientific authorities for demonstrating *à priori* the laws of physical phenomena. A phenomenon must follow a certain law, because we see no reason why it should deviate from that law in one way rather than in another. This is called the Principle of the Sufficient Reason;* and by means of it philosophers often flatter themselves that they are able

---

* Not that of Leibnitz, but the principle commonly appealed to under that name by mathematicians.

to establish, without any appeal to experience, the most general truths of experimental physics.

Take, for example, two of the most elementary of all laws, the law of inertia and the first law of motion. A body at rest cannot, it is affirmed, begin to move unless acted upon by some external force: because, if it did, it must either move up or down, forward or backward, and so forth; but *if no outward force acts upon it, there can be no reason* for its moving up rather than down, or down rather than up, &c., *ergo,* it will not move at all.

This reasoning I conceive to be entirely fallacious, as indeed Dr. Brown, in his treatise on Cause and Effect, has shown with great acuteness and justness of thought. We have before remarked, that almost every fallacy may be referred to different genera by different modes of filling up the suppressed steps; and this particular one may, at our option, be brought under *petitio principii.* It supposes that nothing can be a "sufficient reason" for a body's moving in one particular direction, except some external force. But this is the very thing to be proved. Why not some *internal* force? Why not the law of the thing's own nature? Since these philosophers think it necessary to prove the law of inertia, they of course do not suppose *it* to be self-evident; they must, therefore, be of opinion that, previously to all proof, the supposition of a body's moving by internal impulse is an admissible hypothesis; but if so, why is not the hypothesis also admissible, that the internal impulse acts naturally in some one particular direction, not in another? If spontaneous motion might have been the law of matter, why not spontaneous motion towards the sun, towards the earth, or towards the zenith? Why not, as the ancients supposed, towards a particular place in the universe, appropriated to each particular kind of substance? Surely it is not allowable to say that spontaneity of motion is credible in itself, but not credible if supposed to take place in any determinate direction.

Indeed, if any one chose to assert that all bodies when uncontrolled set out in a direct line towards the north pole, he might equally prove his point by the principle of the

Sufficient Reason. By what right is it assumed that a state
of rest is the particular state which cannot be deviated from
without special cause? Why not a state of motion, and of
some particular sort of motion? Why may we not say that
the natural state of a horse left to himself is to amble, because
otherwise he must either trot, gallop, or stand still, and
because we know no reason why he should do one of these
rather than another? If this is to be called an unfair use of
the "sufficient reason," and the other a fair one, there must
be a tacit assumption that a state of rest is more natural to a
horse than a state of ambling. If this means that it is the
state which the animal will assume when left to himself, that
is the very point to be proved; and if it does not mean this,
it can only mean that a state of rest is the simplest state, and
therefore the most likely to prevail in nature, which is one of
the fallacies or natural prejudices we have already examined.

So again of the First Law of Motion; that a body once
moving will, if left to itself, continue to move uniformly in a
straight line. An attempt is made to prove this law by saying,
that if not, the body must deviate either to the right or to the
left, and that there is no reason why it should do one more than
the other. But who could know, antecedently to experience,
whether there was a reason or not? Might it not be the
nature of bodies, or of some particular bodies, to deviate
towards the right? or if the supposition is preferred, towards
the east, or south? It was long thought that bodies, terrestrial ones at least, had a natural tendency to deflect downwards; and there is no shadow of anything objectionable in
the supposition, except that it is not true. The pretended
proof of the law of motion is even more manifestly untenable
than that of the law of inertia, for it is flagrantly inconsistent; it assumes that the continuance of motion in the
direction first taken is more natural than deviation either to
the right or to the left, but denies that one of these can
possibly be more natural than the other. All these fancies
of the possibility of knowing what is natural or not natural
by any other means than experience, are, in truth, entirely
futile. The real and only proof of the laws of motion, or of

any other law of the universe, is experience; it is simply that no other suppositions explain or are consistent with the facts of universal nature.

Geometers have, in all ages, been open to the imputation of endeavouring to prove the most general facts of the outward world by sophistical reasoning, in order to avoid appeals to the senses. Archimedes, says Professor Playfair,* established some of the elementary propositions of statics by a process in which he "borrows no principle from experiment, but establishes his conclusion entirely by reasoning *à priori*. He assumes, indeed, that equal bodies, at the ends of the equal arms of a lever, will balance one another; and also that a cylinder or parallelopiped of homogeneous matter, will be balanced about its centre of magnitude. These, however, are not inferences from experience; they are, properly speaking, conclusions deduced from the principle of the Sufficient Reason." And to this day there are few geometers who would not think it far more scientific to establish these or any other premises in this way, than to rest their evidence on that familiar experience which in the case in question might have been so safely appealed to.

§ 6. Another natural prejudice, of most extensive prevalence, and which had a great share in producing the errors fallen into by the ancients in their physical inquiries, was this: That the differences in nature must correspond to our received distinctions; that effects which we are accustomed, in popular language, to call by different names, and arrange in different classes, must be of different natures, and have different causes. This prejudice, so evidently of the same origin with those already treated of, marks more especially the earliest stage of science, when it has not yet broken loose from the trammels of every-day phraseology. The extraordinary prevalence of the fallacy among the Greek philosophers may be accounted for by their generally knowing no other language than their own; from which it was a consequence

---

* *Dissertation,* ut supra, p. 27.

that their ideas followed the accidental or arbitrary combinations of that language, more completely than can happen among the moderns to any but illiterate persons. They had great difficulty in distinguishing between things which their language confounded, or in putting mentally together things which it distinguished; and could hardly combine the objects in nature, into any classes but those which were made for them by the popular phrases of their own country: or at least could not help fancying those classes to be natural, and all others arbitrary and artificial. Accordingly, scientific investigation among the Greek schools of speculation and their followers in the middle ages, was little more than a mere sifting and analysing of the notions attached to common language. They thought that by determining the meaning of words, they could become acquainted with facts. "They took for granted," says Dr. Whewell,* "that philosophy must result from the relations of those notions which are involved in the common use of language, and they proceeded to seek it by studying such notions." In his next chapter, Dr. Whewell has so well illustrated and exemplified this error, that I shall take the liberty of quoting him at some length.

"The propensity to seek for principles in the common usages of language may be discerned at a very early period. Thus we have an example of it in a saying which is reported of Thales, the founder of Greek philosophy. When he was asked, 'What is the *greatest* thing?' he replied '*Place;* for all other things are *in* the world, but the world is *in* it.' In Aristotle we have the consummation of this mode of speculation. The usual point from which he starts in his inquiries is, that *we say* thus or thus in common language. Thus, when he has to discuss the question whether there be, in any part of the universe, a void, or space in which there is nothing, he inquires first in how many senses we say that one thing is *in* another. He enumerates many of these; we say the part is in the whole, as the finger is *in* the hand; again we say, the species is in the genus, as man is included *in* animal; again,

* *Hist. Ind. Sc.* Book i. chap. i.

the government of Greece is *in* the king; and various other senses are described and exemplified, but of all these *the most proper* is when we say a thing is *in* a vessel, and generally *in place*. He next examines what *place* is, and comes to this conclusion, that 'if about a body there be another body including it, it is in place, and if not, not.' A body moves when it changes its place; but he adds, that if water be in a vessel, the vessel being at rest, the parts of the water may still move, for they are included by each other; so that while the whole does not change its place, the parts may change their place in a circular order. Proceeding then to the question of a *void*, he as usual examines the different senses in which the term is used, and adopts as the most proper, *place without matter*: with no useful result.

"Again, in a question concerning mechanical action, he says, 'When a man moves a stone by pushing it with a stick, *we say* both that the man moves the stone, and that the stick moves the stone, but the latter *more properly*.'

"Again, we find the Greek philosophers applying themselves to extract their dogmas from the most general and abstract notions which they could detect; for example, from the conception of the Universe as One or as Many things. They tried to determine how far we may, or must, combine with these conceptions that of a whole, of parts, of number, of limits, of place, of beginning or end, of full or void, of rest, or motion, of cause and effect, and the like. The analysis of such conceptions with such a view, occupies, for instance, almost the whole of Aristotle's Treatise on the Heavens."

The following paragraph merits particular attention:—
"Another mode of reasoning, very widely applied in these attempts, was the *doctrine of contrarieties*, in which it was assumed, that adjectives or substances which are in common language, or in some abstract mode of conception, opposed to each other, must point at some fundamental antithesis in nature, which it is important to study. Thus Aristotle says that the Pythagoreans, from the contrasts which number suggests, collected ten principles—Limited and Unlimited, Odd and Even, One and Many, Right and Left, Male and

Female, Rest and Motion, Straight and Curved, Light and Darkness, Good and Evil, Square and Oblong . . . Aristotle himself deduced the doctrine of four elements and other dogmas by oppositions of the same kind."

Of the manner in which, from premises obtained in this way, the ancients attempted to deduce laws of nature, an example is given in the same work a few pages further on. "Aristotle decides that there is no void, on such arguments as this. In a void there could be no difference of up and down; for as in nothing there are no differences, so there are none in a privation or negation; but a void is merely a privation or negation of matter; therefore, in a void, bodies could not move up and down, which it is in their nature to do. It is easily seen" (Dr. Whewell very justly adds) "that such a mode of reasoning elevates the familiar forms of language, and the intellectual connexions of terms, to a supremacy over facts; making truth depend upon whether terms are or are not privative, and whether we say that bodies fall *naturally*."

The propensity to assume that the same relations obtain between objects themselves, which obtain between our ideas of them, is here seen in the extreme stage of its development. For the mode of philosophizing, exemplified in the foregoing instances, assumes no less than that the proper way of arriving at knowledge of nature, is to study nature itself subjectively; to apply our observation and analysis not to the facts, but to the common notions entertained of the facts.

Many other equally striking examples may be given of the tendency to assume that things which for the convenience of common life are placed in different classes, must differ in every respect. Of this nature was the universal and deeply-rooted prejudice of antiquity and the middle ages, that celestial and terrestrial phenomena must be essentially different, and could in no manner or degree depend on the same laws. Of the same kind, also, was the prejudice against which Bacon contended, that nothing produced by nature could be successfully imitated by man: " Calorem solis et ignis toto genere differre;

ne scilicet homines putent se per opera ignis, aliquid simile iis
quæ in Natura fiunt, educere et formare posse :" and again,
" Compositionem tantum opus Hominis, Mistionem vero opus
solius Naturæ esse : ne scilicet homines sperent aliquam ex arte
Corporum naturalium generationem aut transformationem." *
The grand distinction in the ancient scientific speculations,
between natural and violent motions, though not without a
plausible foundation in the appearances themselves, was doubtless greatly recommended to adoption by its conformity to this
prejudice.

§ 7. From the fundamental error of the scientific inquirers
of antiquity, we pass, by a natural association, to a scarcely
less fundamental one of their great rival and successor, Bacon.
It has excited the surprise of philosophers that the detailed
system of inductive logic, which this extraordinary man
laboured to construct, has been turned to so little direct use
by subsequent inquirers, having neither continued, except in a
few of its generalities, to be recognised as a theory, nor having
conducted in practice to any great scientific results.   But this,
though not unfrequently remarked, has scarcely received any
plausible explanation ; and some, indeed, have preferred to
assert that all rules of induction are useless, rather than suppose that Bacon's rules are grounded on an insufficient analysis
of the inductive process.   Such, however, will be seen to be
the fact, as soon as it is considered, that Bacon entirely overlooked Plurality of Causes.   All his rules tacitly imply the
assumption, so contrary to all we now know of nature, that a
phenomenon cannot have more than one cause.

When he is inquiring into what he terms the *forma calidi
aut frigidi, gravis aut levis, sicci aut humidi*, and the like, he
never for an instant doubts that there is some one thing, some
invariable condition or set of conditions, which is present in
all cases of heat, or cold, or whatever other phenomenon he is
considering ; the only difficulty being to find what it is ;
which accordingly he tries to do by a process of elimination,

---

* *Novum Organum*, Aph. 75.

rejecting or excluding, by negative instances, whatever is not
the *forma* or cause, in order to arrive at what is. But, that
this *forma* or cause is *one* thing, and that it is the same in all
hot objects, he has no more doubt of, than another person has
that there is always some cause *or other*. In the present state
of knowledge it could not be necessary, even if we had not
already treated so fully of the question, to point out how
widely this supposition is at variance with the truth. It is
particularly unfortunate for Bacon that, falling into this error,
he should have fixed almost exclusively upon a class of inquiries
in which it was especially fatal; namely, inquiries into the
causes of the sensible qualities of objects. For his assumption,
groundless in every case, is false in a peculiar degree with
respect to those sensible qualities. In regard to scarcely any
of them has it been found possible to trace any unity of
cause, any set of conditions invariably accompanying the
quality. The conjunctions of such qualities with one another
constitute the variety of Kinds, in which, as already remarked,
it has not been found possible to trace any law. Bacon was
seeking for what did not exist. The phenomenon of which he
sought for the one cause has oftenest no cause at all, and when
it has, depends (as far as hitherto ascertained) on an unassignable variety of distinct causes.

And on this rock every one must split, who represents to
himself as the first and fundamental problem of science to
ascertain what is the cause of a given effect, rather than what
are the effects of a given cause. It was shown, in an early
stage of our inquiry into the nature of Induction,* how much
more ample are the resources which science commands for the
latter than for the former inquiry, since it is upon the latter
only that we can throw any direct light by means of experiment; the power of artificially producing an effect, implying
a previous knowledge of at least one of its causes. If we
discover the causes of effects, it is generally by having previously discovered the effects of causes: the greatest skill in
devising crucial instances for the former purpose may only

---

* Supra, book iii. ch. vii, § 4.

end, as Bacon's physical inquiries did, in no result at all. Was it that his eagerness to acquire the power of producing for man's benefit effects of practical importance to human life, rendering him impatient of pursuing that end by a circuitous route, made even him, the champion of experiment, prefer the direct mode, though one of mere observation, to the indirect, in which alone experiment was possible? Or had even Bacon not entirely cleared his mind from the notion of the ancients, that "rerum cognoscere *causas*" was the sole object of philosophy, and that to inquire into the *effects* of things belonged to servile and mechanical arts?

It is worth remarking that, while the only efficient mode of cultivating speculative science was missed from an undue contempt of manual operations, the false speculative views thus engendered gave in their turn a false direction to such practical and mechanical aims as were suffered to exist. The assumption universal among the ancients and in the middle ages, that there were *principles* of heat and cold, dryness and moisture, &c., led directly to a belief in alchemy; in a transmutation of substances, a change from one Kind into another. Why should it not be possible to make gold? Each of the characteristic properties of gold has its *forma*, its essence, its set of conditions, which if we could discover, and learn how to realize, we could superinduce that particular property upon any other substance, upon wood, or iron, or lime, or clay. If, then, we could effect this with respect to every one of the essential properties of the precious metal, we should have converted the other substance into gold. Nor did this, if once the premises were granted, appear to transcend the real powers of mankind. For daily experience showed that almost every one of the distinctive sensible properties of any object, its consistence, its colour, its taste, its smell, its shape, admitted of being totally changed by fire, or water, or some other chemical agent. The *formæ* of all those qualities seeming, therefore, to be within human power either to produce or to annihilate, not only did the transmutation of substances appear abstractedly possible, but the employment of the power, at our choice, for practical ends, seemed by no means hopeless.*

A prejudice, universal in the ancient world, and from which Bacon was so far from being free, that it pervaded and vitiated the whole practical part of his system of logic, may with good reason be ranked high in the order of Fallacies of which we are now treating.

§ 8.   There remains one *à priori* fallacy or natural prejudice, the most deeply-rooted, perhaps, of all which we have enumerated : one which not only reigned supreme in the ancient world, but still possesses almost undisputed dominion over many of the most cultivated minds ; and some of the most remarkable of the numerous instances by which I shall think it necessary to exemplify it, will be taken from recent thinkers.   This is, that the conditions of a phenomenon must, or at least probably will, resemble the phenomenon itself.

Conformably to what we have before remarked to be of frequent occurrence, this fallacy might without much impropriety have been placed in a different class, among Fallacies of Generalization : for experience does afford a certain degree of countenance to the assumption.   The cause does, in very many cases, resemble its effect ; like produces like.   Many phenomena have a direct tendency to perpetuate their own existence, or to give rise to other phenomena similar to themselves.   Not to mention forms actually moulded on one another, as impressions on wax and the like, in which the closest resemblance between the effect and its cause is the very law of the phenomenon ; all motion tends to continue itself, with its own velocity, and in its own original direction ; and the motion of one body tends to set others in motion, which is indeed the most common of the modes in which the

---

* It is hardly needful to remark that nothing is here intended to be said against the possibility at some future period of making gold ; by first discovering it to be a compound, and putting together its different elements or ingredients.   But this is a totally different idea from that of the seekers of the grand arcanum.

motions of bodies originate. We need scarcely refer to con-
tagion, fermentation, and the like; or to the production of
effects by the growth or expansion of a germ or rudiment
resembling on a smaller scale the completed phenomenon, as
in the growth of a plant or animal from an embryo, that
embryo itself deriving its origin from another plant or animal
of the same kind. Again, the thoughts, or reminiscences,
which are effects of our past sensations, resemble those
sensations; feelings produce similar feelings by way of sympathy; acts produce similar acts by involuntary or voluntary
imitation. With so many appearances in its favour, no wonder
if a presumption naturally grew up, that causes must *necessarily* resemble their effects, and that like could *only* be
produced by like.

This principle of fallacy has usually presided over the
fantastical attempts to influence the course of nature by
conjectural means, the choice of which was not directed by
previous observation and experiment. The guess almost
always fixed upon some means which possessed features of
real or apparent resemblance to the end in view. If a charm
was wanted, as by Ovid's Medea, to prolong life, all long-lived
animals, or what were esteemed such, were collected and brewed
into a broth :—

> . . . . nec defuit illic
> Squamea Cinyphii tenuis membrana chelydri
> Vivacisque jecur cervi : quibus insuper addit
> Ora caputque novem cornicis secula passæ.

A similar notion was embodied in the celebrated medical
theory called the " Doctrine of Signatures," " which is no
less," says Dr. Paris,* " than a belief that every natural
substance which possesses any medicinal virtue indicates by
an obvious and well-marked external character the disease for
which it is a remedy, or the object for which it should be
employed." This outward character was generally some
feature of resemblance, real or fantastical, either to the effect

---

* *Pharmacologia*, pp. 43-5.

it was supposed to produce, or to the phenomenon over which its power was thought to be exercised. "Thus the lungs of a fox must be a specific for asthma, because that animal is remarkable for its strong powers of respiration. Turmeric has a brilliant yellow colour, which indicates that it has the power of curing the jaundice; for the same reason, poppies must relieve diseases of the head ; Agaricus those of the bladder; *Cassia fistula* the affections of the intestines, and Aristolochia the disorders of the uterus: the polished surface and stony hardness which so eminently characterize the seeds of the Lithospermum officinale (common gromwell) were deemed a certain indication of their efficacy in calculous and gravelly disorders; for a similar reason, the roots of the Saxifraga granulata (white saxifrage) gained reputation in the cure of the same disease; and the Euphrasia (eye-bright) acquired fame, as an application in complaints of the eye, because it exhibits a black spot in its corolla resembling the pupil. The blood-stone, the Heliotropium of the ancients, from the occasional small specks or points of a blood-red colour exhibited on its green surface, is even at this very day employed in many parts of England and Scotland, to stop a bleeding from the nose; and nettle tea continues a popular remedy for the cure of *Urticaria*. It is also asserted that some substances bear the *signatures* of the humours, as the petals of the red rose that of the blood, and the roots of rhubarb and the flowers of saffron that of the bile."

The early speculations respecting the chemical composition of bodies were rendered abortive by no circumstance more, than by their invariably taking for granted that the properties of the elements must resemble those of the compounds which were formed from them.

To descend to more modern instances; it was long thought, and was stoutly maintained by the Cartesians and even by Leibnitz against the Newtonian system, (nor did Newton himself, as we have seen, contest the assumption, but eluded it by an arbitrary hypothesis), that nothing (of a physical nature at least) could account for motion, except previous motion; the impulse or impact of some other body. It was very long before the scientific world could prevail upon itself to admit attraction and repulsion (*i. e.* spontaneous tendencies of particles to approach or recede from one another) as ultimate laws, no more requiring to be accounted for than impulse itself, if indeed the latter were not, in truth, resolvable into the former. From the same source arose the innumerable hypotheses devised to explain those classes of motions which appeared more mysterious than others because there was no obvious mode of attributing them to impulse, as for example the voluntary motions of the human body. Such were the interminable systems of vibrations propagated along the nerves, or animal spirits rushing up and down between the muscles and the brain; which, if the facts could have been proved, would have been an important addition to our knowledge of physiological laws; but the mere invention, or arbitrary supposition of them, could not unless by the strongest delusion be supposed to render the phenomena of animal life more comprehensible, or less mysterious. Nothing, however, seemed satisfactory, but to make out that motion was caused by motion; by something like itself. If it was not one kind of motion, it must be another. In like manner it was supposed that the physical qualities of objects must arise from some similar quality, or perhaps only some quality bearing the same name, in the particles or atoms of which the objects were composed; that a sharp taste, for example, must arise from sharp particles. And reversing the inference, the effects produced by a phenomenon must, it was supposed, resemble in their physical attributes the phenomenon itself. The influences of the planets were supposed to be analogous to their visible peculiarities: Mars, being of a red colour, portended fire and slaughter; and the like.

Passing from physics to metaphysics, we may notice among the most remarkable fruits of this *à priori* fallacy, two closely analogous theories, employed in ancient and modern times to bridge over the chasm between the world of mind and that of matter: the *species sensibiles* of the Epicureans, and the modern doctrine of perception by means of ideas. These theories are indeed, probably, indebted for their exis-

tence not solely to the fallacy in question, but to that fallacy
combined with another natural prejudice already adverted to,
that a thing cannot act where it is not. In both doctrines it
is assumed that the phenomenon which takes place *in us* when
we see or touch an object, and which we regard as an effect of
that object, or rather as its presence to our organs, must of
necessity resemble very closely the outward object itself. To
fulfil this condition, the Epicureans supposed that objects were
constantly projecting in all directions impalpable images of
themselves, which entered at the eyes and penetrated to the
mind; while modern metaphysicians, though they rejected
this hypothesis, agreed in deeming it necessary to suppose
that not the thing itself, but a mental image or representation
of it, was the direct object of perception. Dr. Reid had to
employ a world of argument and illustration to familiarize
people with the truth, that the sensations or impressions on
our minds need not necessarily be copies of, or bear any
resemblance to, the causes which produce them; in opposition
to the natural prejudice which led people to assimilate the
action of bodies upon our senses, and through them upon our
minds, to the transfer of a given form from one object to
another by actual moulding. The works of Dr. Reid are even
now the most effectual course of study for detaching the mind
from the prejudice of which this was an example. And the
value of the service which he thus rendered to popular philosophy, is not much diminished although we may hold, with
Brown, that he went too far in imputing the "ideal theory"
as an actual tenet, to the generality of the philosophers who
preceded him, and especially to Locke and Hume: for if they
did not themselves consciously fall into the error, unquestionably they often led their readers into it.

The prejudice, that the conditions of a phenomenon must
resemble the phenomenon, is occasionally exaggerated, at least
verbally, into a still more palpable absurdity; the conditions
of the thing are spoken of as if they *were* the very thing
itself. In Bacon's model-inquiry, which occupies so great a
space in the *Novum Organum*, the *inquisitio in formam calidi*,
the conclusion which he favours is that heat is a kind of

motion; meaning of course not the feeling of heat, but the conditions of the feeling; meaning, therefore, only that wherever there is heat, there must first be a particular kind of motion; but he makes no distinction in his language between these two ideas, expressing himself as if heat, and the conditions of heat, were one and the same thing. So Darwin, in the beginning of his *Zoonomia*, says, "The word *idea* has various meanings in the writers of metaphysic: it is here used simply for those notions of external things which our organs of sense bring us acquainted with originally," (thus far the proposition, though vague, is unexceptionable in meaning,) "and is defined a contraction, a motion, or configuration, of the fibres which constitute the immediate organ of sense." Our *notions*, a configuration of the fibres! What kind of logician must he be who thinks that a phenomenon is *defined* to *be* the condition on which he supposes it to depend? Accordingly he says soon after, not that our ideas are caused by, or consequent on, certain organic phenomena, but "our ideas *are* animal motions of the organs of sense." And this confusion runs through the four volumes of the *Zoonomia;* the reader never knows whether the writer is speaking of the effect, or of its supposed cause; of the idea, a state of mental consciousness, or of the state of the nerves and brain which he considers it to presuppose.

I have given a variety of instances in which the natural prejudice, that causes and their effects must resemble one another, has operated in practice so as to give rise to serious errors. I shall now go further, and produce from writings even of the present or very recent times, instances in which this prejudice is laid down as an established principle. M. Victor Cousin, in the last of his celebrated lectures on Locke, enunciates the maxim in the following unqualified terms. "Tout ce qui est vrai de l'effet est vrai de la cause." A doctrine to which, unless in some peculiar and technical meaning of the words cause and effect, it is not to be imagined that any person would literally adhere: but he who could so write must be far enough from seeing, that the very reverse might be the fact; that there is nothing impossible in the

supposition that no one property which is true of the effect
might be true of the cause. Without going quite so far in
point of expression, Coleridge, in his *Biographia Literaria,*[*]
affirms as an "evident truth," that "the law of causality
holds only between homogeneous things, *i. e.* things having
some common property," and therefore "cannot extend from
one world into another, its opposite:" hence, as mind and
matter have no common property, mind cannot act upon
matter, nor matter upon mind. What is this but the *à priori*
fallacy of which we are speaking? The doctrine, like many
others of Coleridge, is taken from Spinoza, in the first book of
whose *Ethica* (*De Deo*) it stands as the Third Proposition,
" Quæ res nihil commune inter se habent, earum una alterius
causa esse non potest," and is there proved from two so-called
axioms, equally gratuitous with itself: but Spinoza, ever
systematically consistent, pursued the doctrine to its inevitable
consequence, the materiality of God.

The same conception of impossibility led the ingenious
and subtle mind of Leibnitz to his celebrated doctrine of a
pre-established harmony. He, too, thought that mind could
not act upon matter, nor matter upon mind, and that the two,
therefore, must have been arranged by their Maker like two
clocks, which, though unconnected with one another, strike
simultaneously, and always point to the same hour. Malebranche's equally famous theory of Occasional Causes was
another form of the same conception: instead of supposing
the clocks originally arranged to strike together, he held that
when the one strikes, God interposes, and makes the other
strike in correspondence with it.

Descartes, in like manner, whose works are a rich mine of
almost every description of *à priori* fallacy, says that the
Efficient Cause must at least have all the perfections of the
effect, and for this singular reason: " Si enim ponamus aliquid
in ideâ reperiri quod non fuerit in ejus causâ, hoc igitur habet
a nihilo ;" of which it is scarcely a parody to say, that if there

---

* Vol. i. chap. 8.

be pepper in the soup there must be pepper in the cook who made it, since otherwise the pepper would be without a cause. A similar fallacy is committed by Cicero in his second book *De Finibus*, where, speaking in his own person against the Epicureans, he charges them with inconsistency in saying that the pleasures of the mind had their origin from those of the body, and yet that the former were more valuable, as if the effect could surpass the cause. "Animi voluptas oritur propter voluptatem corporis, et major est animi voluptas quam corporis? ita fit ut gratulator lætior sit quam is cui gratulatur." Even that, surely, is not an impossibility: a person's good fortune has often given more pleasure to others than it gave to the person himself.

Descartes, with no less readiness, applies the same principle the converse way, and infers the nature of the effects from the assumption that they must, in this or that property or in all their properties, resemble their cause. To this class belong his speculations, and those of so many others after him, tending to infer the order of the universe, not from observation, but by *à priori* reasoning from supposed qualities of the Godhead. This sort of inference was probably never carried to a greater length than it was in one particular instance by Descartes, when, as a proof of one of his physical principles, that the quantity of motion in the universe is invariable, he had recourse to the immutability of the Divine Nature. Reasoning of a very similar character is however nearly as common now as it was in his time, and does duty largely as a means of fencing off disagreeable conclusions. Writers have not yet ceased to oppose the theory of divine benevolence to the evidence of physical facts, to the principle of population for example. And people seem in general to think that they have used a very powerful argument, when they have said, that to suppose some proposition true, would be a reflection on the goodness or wisdom of the Deity. Put into the simplest possible terms, their argument is, "If it had depended on me, I would not have made the proposition true, therefore it is not true." Put into other words it stands thus: "God is perfect, therefore (what I think) perfection must obtain in nature." But since in reality every one feels that nature is very far from perfect, the doctrine is never applied consistently. It furnishes an argument which (like many others of a similar character) people like to appeal to when it makes for their own side. Nobody is convinced by it, but each appears to think that it puts religion on his side of the question, and that it is a useful weapon of offence for wounding an adversary.

Although several other varieties of *à priori* fallacy might probably be added to those here specified, these are all against which it seems necessary to give any special caution. Our object is to open, without attempting or affecting to exhaust, the subject. Having illustrated, therefore, this first class of Fallacies at sufficient length, I shall proceed to the second.

CHAPTER IV.

§ 1. From the fallacies which are properly Prejudices, or presumptions antecedent to, and superseding, proof, we pass to those which lie in the incorrect performance of the proving process. And as Proof, in its widest extent, embraces one or more, or all, of three processes, Observation, Generalization, and Deduction; we shall consider in their order the errors capable of being committed in these three operations. And first, of the first mentioned.

A fallacy of misobservation may be either negative or positive; either Non-observation or Mal-observation. It is non-observation, when all the error consists in overlooking, or neglecting, facts or particulars which ought to have been observed. It is mal-observation, when something is not simply unseen, but seen wrong; when the fact or phenomenon, instead of being recognised for what it is in reality, is mistaken for something else.

§ 2. Non-observation may either take place by overlooking instances, or by overlooking some of the circumstances of a given instance. If we were to conclude that a fortune-teller was a true prophet, from not adverting to the cases in which his predictions had been falsified by the event, this would be non-observation of instances; but if we overlooked or remained ignorant of the fact that in cases where the predictions had been fulfilled, he had been in collusion with some one who had given him the information on which they were grounded, this would be non-observation of circumstances.

one great constitutive law, in the light of which dwell
dominion and the power of prophecy; if these discoveries,
instead of having been, as they really were, preconcerted by
meditation, and evolved out of his own intellect, had occurred
by a set of lucky *accidents* to the illustrious father and founder
of philosophic alchemy; if they had presented themselves to
Professor Davy exclusively in consequence of his *luck* in
possessing a particular galvanic battery; if this battery, as far
as Davy was concerned, had itself been an *accident*, and not
(as in point of fact it was) desired and obtained by him for
the purpose of ensuring the testimony of experience to his
principles, and in order to bind down material nature under
the inquisition of reason, and force from her, as by torture,
unequivocal answers to *prepared* and *preconceived* questions,—
yet still they would not have been talked of or described as
instances of *luck*, but as the natural results of his admitted
genius and known skill. · But should an accident have disclosed similar discoveries to a mechanic at Birmingham or
Sheffield, and if the man should grow rich in consequence,
and partly by the envy of his neighbours and partly with
good reason, be considered by them as a man *below par* in the
general powers of his understanding; then, 'O what a lucky
fellow! Well, Fortune *does* favour fools—that's for certain!
—It is always so!' And forthwith the exclaimer relates half
a dozen similar instances. Thus accumulating the one sort of
facts and never collecting the other, we do, as poets in their
diction, and quacks of all denominations do in their reasoning,
put a part for the whole."

This passage very happily sets forth the manner in which,
under the loose mode of induction which proceeds *per
enumerationem simplicem*, not seeking for instances of such a
kind as to be decisive of the question, but generalizing from
any which occur, or rather which are remembered, opinions
grow up with the apparent sanction of experience, which have
no foundation in the laws of nature at all. "Itaque recte
respondit ille," (we may say with Bacon,[*]) " qui cum suspensa

---

[*] *Nov. Org.*, Aph. 46.

tabula in templo ei monstraretur eorum, qui vota solverant, quod naufragii periculo elapsi sint, atque interrogando premeretur, anne tum quidem Deorum numen agnosceret, quæsivit denuo, *At ubi sunt illi depicti qui post vota nuncupata perierunt?* Eadem ratio est fere omnis superstitionis, ut in Astrologicis, in Somniis, Ominibus, Nemesibus, et hujusmodi; in quibus, homines delectati hujusmodi vanitatibus, advertunt eventus, ubi implentur; ast ubi fallunt, licet multo frequentius, tamen negligunt, et prætereunt." And he proceeds to say, that independently of the love of the marvellous, or any other bias in the inclinations, there is a natural tendency in the intellect itself to this kind of fallacy; since the mind is more moved by affirmative instances, though negative ones are of most use in philosophy; "Is tamen humano intellectui error est proprius et perpetuus, ut magis moveatur et excitetur Affirmativis quam Negativis; cum rite et ordine æquum se utrique præbere debeat; quin contra, in omni Axiomate vero constituendo, major vis est instantiæ negativæ."

But the greatest of all causes of non-observation is a preconceived opinion. This it is which, in all ages, has made the whole race of mankind, and every separate section of it, for the most part unobservant of all facts, however abundant, even when passing under their own eyes, which are contradictory to any first appearance, or any received tenet. It is worth while to recal occasionally to the oblivious memory of mankind some of the striking instances in which opinions that the simplest experiment would have shown to be erroneous, continued to be entertained because nobody ever thought of trying that experiment. One of the most remarkable of these was exhibited in the Copernican controversy. The opponents of Copernicus argued that the earth did not move, because if it did, a stone let fall from the top of a high tower would not reach the ground at the foot of the tower, but at a little distance from it, in a contrary direction to the earth's course; in the same manner (said they) as, if a ball is let drop from the mast-head while the ship is in full sail, it does not fall exactly at the foot of the mast, but nearer to the stern of the vessel. The Copernicans would have silenced these objectors at once if they had *tried* dropping a ball from the mast-head, since they would have found that it does fall exactly at the foot, as the theory requires: but no; they admitted the spurious fact, and struggled vainly to make out a difference between the two cases. "The ball was no *part* of the ship—and the motion forward was not *natural*, either to the ship or to the ball. The stone, on the other hand, let fall from the top of the tower, was a *part* of the earth; and therefore, the diurnal and annular revolutions which were *natural* to the earth, were also *natural* to the stone: the stone would, therefore, retain the same motion with the tower, and strike the ground precisely at the bottom of it."[*]

Other examples, scarcely less striking, are recorded by Dr. Whewell,[†] where imaginary laws of nature have continued to be received as real, merely because no person had steadily looked at facts which almost every one had the opportunity of observing. "A vague and loose mode of looking at facts very easily observable, left men for a long time under the belief that a body ten times as heavy as another falls ten times as fast; that objects immersed in water are always magnified, without regard to the form of the surface; that the magnet exerts an irresistible force; that crystal is always found associated with ice; and the like. These and many others are examples how blind and careless man can be even in observation of the plainest and commonest appearances; and they show us that the mere faculties of perception, although constantly exercised upon innumerable objects, may long fail in leading to any exact knowledge."

If even on physical facts, and these of the most obvious character, the observing faculties of mankind can be to this degree the passive slaves of their preconceived impressions, we need not be surprised that this should be so lamentably true as all experience attests it to be, on things more nearly connected with their stronger feelings—on moral, social, and religious subjects. The information which an ordinary traveller brings back from a foreign country, as the result of the evidence of

---

[*] Playfair's *Dissertation*, sect. 4.　　　[†] *Nov. Org. Renov.*, p. 61.

his senses, is almost always such as exactly confirms the opinions with which he set out. He has had eyes and ears for such things only as he expected to see. Men read the sacred books of their religion, and pass unobserved therein, multitudes of things utterly irreconcileable with even their own notions of moral excellence. With the same authorities before them, different historians, alike innocent of intentional misrepresentation, see only what is favourable to Protestants or Catholics, royalists or republicans, Charles I. or Cromwell; while others, having set out with the preconception that extremes must be in the wrong, are incapable of seeing truth and justice when these are wholly on one side.

The influence of a preconceived theory is well exemplified in the superstitions of barbarians respecting the virtues of medicaments and charms. The negroes, among whom coral, as of old among ourselves, is worn as an amulet, affirm, according to Dr. Paris,* that its colour " is always affected by the state of health of the wearer, it becoming paler in disease." On a matter open to universal observation, a general proposition which has not the smallest vestige of truth is received as a result of experience; the preconceived opinion preventing, it would seem, any observation whatever on the subject.

§ 4. For illustration of the first species of non-observation, that of Instances, what has now been stated may suffice. But there may also be non-observation of some material circumstances, in instances which have not been altogether overlooked—nay, which may be the very instances on which the whole superstructure of a theory has been founded. As, in the cases hitherto examined, a general proposition was too rashly adopted, on the evidence of particulars, true indeed, but insufficient to support it; so in the cases to which we now turn, the particulars themselves have been imperfectly observed, and the singular propositions on which the generalization is grounded, or some at least of those singular propositions, are false.

---

* *Pharmacologia*, p. 21.

Such, for instance, was one of the mistakes committed in the celebrated phlogistic theory; a doctrine which accounted for combustion by the extrication of a substance called phlogiston, supposed to be contained in all combustible matter. The hypothesis accorded tolerably well with superficial appearances: the ascent of flame naturally suggests the escape of a substance; and the visible residuum of ashes, in bulk and weight, generally falls extremely short of the combustible material. The error was, non-observation of an important portion of the actual residue, namely, the gaseous products of combustion. When these were at last noticed and brought into account, it appeared to be an universal law, that all substances gain instead of losing weight by undergoing combustion; and, after the usual attempt to accommodate the old theory to the new fact by means of an arbitrary hypothesis (that phlogiston had the quality of positive levity instead of gravity), chemists were conducted to the true explanation, namely, that instead of a substance separated, there was on the contrary a substance absorbed.

Many of the absurd practices which have been deemed to possess medicinal efficacy, have been indebted for their reputation to non-observance of some accompanying circumstance which was the real agent in the cures ascribed to them. Thus, of the sympathetic powder of Sir Kenelm Digby: "Whenever any wound had been inflicted, this powder was applied to the weapon that had inflicted it, which was, moreover, covered with ointment, and dressed two or three times a day. The wound itself, in the meantime, was directed to be brought together, and carefully bound up with clean linen rags, but *above all, to be let alone* for seven days, at the end of which period the bandages were removed, when the wound was generally found perfectly united. The triumph of the cure was decreed to the mysterious agency of the sympathetic powder which had been so assiduously applied to the weapon, whereas it is hardly necessary to observe that the promptness of the cure depended on the total exclusion of air from the wound, and upon the sanative operations of nature not having received any disturbance from the officious interference of art. The result, beyond all doubt, furnished the first hint which led surgeons to the improved practice of healing wounds by what is technically called the *first intention*."[*] "In all records," adds Dr. Paris, " of extraordinary cures performed by mysterious agents, there is a great desire to conceal the remedies and other curative means which were simultaneously administered with them; thus Oribasius commends in high terms a necklace of Pæony root for the cure of epilepsy; but we learn that he always took care to accompany its use with copious evacuations, although he assigns to them no share of credit in the cure. In later times we have a good specimen of this species of deception, presented to us in a work on Scrofula by Mr. Morley, written, as we are informed, for the sole purpose of restoring the much injured character and use of the Vervain; in which the author directs the root of this plant to be tied with a yard of white satin riband around the neck, where it is to remain until the patient is cured; but mark—during this interval he calls to his aid the most active medicines in the materia medica."[†]

In other cases the cures really produced by rest, regimen, and amusement, have been ascribed to the medicinal, or occasionally to the supernatural, means which were put in requisition. " The celebrated John Wesley, while he commemorates the triumph of sulphur and supplication over his bodily infirmity, forgets to appreciate the resuscitating influence of four months' repose from his apostolic labours; and such is the disposition of the human mind to place confidence in the operation of mysterious agents, that we find him more disposed to attribute his cure to a brown paper plaister of egg and brimstone, than to Dr. Fothergill's salutary prescription of country air, rest, asses' milk, and horse exercise."[‡]

In the following example, the circumstance overlooked was of a somewhat different character. " When the yellow fever raged in America, the practitioners trusted exclusively to the copious use of mercury; at first this plan was deemed so universally efficacious, that, in the enthusiasm of the

* *Pharmacologia*, pp. 23-4.   † Ibid. p. 28.   ‡ Ibid. p. 62.

moment, it was triumphantly proclaimed that death never took place after the mercury had evinced its effect upon the system: all this was very true, but it furnished no proof of the efficacy of that metal, since the disease in its aggravated form was so rapid in its career, that it swept away its victims long before the system could be brought under mercurial influence, while in its milder shape it passed off equally well without any assistance from art."*

In these examples the circumstance overlooked was cognizable by the senses. In other cases, it is one the knowledge of which could only be arrived at by reasoning; but the fallacy may still be classed under the head to which, for want of a more appropriate name, we have given the appellation Fallacies of Non-observation. It is not the nature of the faculties which ought to have been employed, but the non-employment of them, which constitutes this Natural Order of Fallacies. Wherever the error is negative, not positive; wherever it consists especially in *overlooking*, in being ignorant or unmindful of some fact which, if known and attended to, would have made a difference in the conclusion arrived at; the error is properly placed in the Class which we are considering. In this Class, there is not, as in all other fallacies there is, a positive mis-estimate of evidence actually had. The conclusion would be just, if the portion which is seen of the case were the whole of it; but there is another portion overlooked, which vitiates the result.

For instance, there is a remarkable doctrine which has occasionally found a vent in the public speeches of unwise legislators, but which only in one instance that I am aware of has received the sanction of a philosophical writer, namely M. Cousin, who, in his preface to the *Gorgias* of Plato, contending that punishment must have some other and higher justification than the prevention of crime, makes use of this argument—that if punishment were only for the sake of example, it would be indifferent whether we punished the innocent or the guilty, since the punishment, considered as an

---

* *Pharmacologia*, pp. 61-2.

example, is equally efficacious in either case. Now we must,
in order to go along with this reasoning, suppose, that the
person who feels himself under temptation, observing some-
body punished, concludes himself to be in danger of being
punished likewise, and is terrified accordingly. But it is forgotten that if the person punished is supposed to be innocent,
or even if there be any doubt of his guilt, the spectator will
reflect that his own danger, whatever it may be, is not contingent on his guiltiness, but threatens him equally if he
remains innocent, and how therefore is he deterred from guilt
by the apprehension of such punishment? M. Cousin supposes that people will be dissuaded from guilt by whatever
renders the condition of the guilty more perilous, forgetting
that the condition of the innocent (also one of the elements in
the calculation) is, in the case supposed, made perilous in precisely an equal degree.   This is a fallacy of overlooking; or of
non-observation, within the intent of our classification.

Fallacies of this description are the great stumbling-block
to correct thinking in political economy. The economical
workings of society afford numerous cases in which the effects
of a cause consist of two sets of phenomena : the one immediate, concentrated, obvious to all eyes, and passing, in common
apprehension, for the whole effect; the other widely diffused,
or lying deeper under the surface, and which is exactly contrary to the former.   Take, for instance, the common notion
so plausible at the first glance, of the encouragement given to
industry by lavish expenditure. A, who spends his whole
income; and even his capital, in expensive living, is supposed
to give great employment to labour. B, who lives on a small
portion, and invests the remainder in the funds, is thought to
give little or no employment. For everybody sees the gains
which are made by A's tradesmen, servants, and others, while
his money is spending. B's savings, on the contrary, pass
into the hands of the person whose stock he purchased, who
with it pays a debt he owed to some banker, who lends it again
to some merchant or manufacturer ; and the capital being laid
out in hiring spinners and weavers, or carriers and the crews
of merchant vessels, not only gives immediate employment to

at least as much industry as A employs during the whole of his career, but coming back with increase by the sale of the goods which have been manufactured or imported, forms a fund for the employment of the same and perhaps a greater quantity of labour in perpetuity. But the observer does not see, and therefore does not consider, what becomes of B's money; he does see what is done with A's: he observes the amount of industry which A's profusion feeds; he observes not the far greater quantity which it prevents from being fed; and thence the prejudice, universal to the time of Adam Smith, that prodigality encourages industry, and parsimony is a discouragement to it.

The common argument against free trade was a fallacy of the same nature. The purchaser of British silk encourages British industry; the purchaser of Lyons silk encourages only French; the former conduct is patriotic, the latter ought to be interdicted by law. The circumstance is overlooked, that the purchaser of any foreign commodity necessarily causes, directly or indirectly, the export of an equivalent value of some article of home production (beyond what would otherwise be exported), either to the same foreign country or to some other; which fact, though from the complication of the circumstances it cannot always be verified by specific observation, no observation can possibly be brought to contradict, while the evidence of reasoning on which it rests is irrefragable. The fallacy is, therefore, the same as in the preceding case, that of seeing a part only of the phenomena, and imagining that part to be the whole: and may be ranked among Fallacies of Non-observation.

§ 5. To complete the examination of the second of our five classes, we have now to speak of Mal-observation; in which the error does not lie in the fact that something is unseen, but that something seen is seen wrong.

Perception being infallible evidence of whatever is really perceived, the error now under consideration can be committed no otherwise than by mistaking for conception what is in fact inference. We have formerly shown how intimately the two are blended in almost everything which is called observation, and still more in every Description.* What is actually on any occasion perceived by our senses being so minute in amount, and generally so unimportant a portion of the state of facts which we wish to ascertain or to communicate; it would be absurd to say that either in our observations, or in conveying their result to others, we ought not to mingle inference with fact; all that can be said is, that when we do so we ought to be aware of what we are doing, and to know what part of the assertion rests on consciousness, and is therefore indisputable, what part on inference, and is therefore questionable.

One of the most celebrated examples of an universal error produced by mistaking an inference for the direct evidence of the senses, was the resistance made, on the ground of common sense, to the Copernican system. People fancied they *saw* the sun rise and set, the stars revolve in circles round the pole. We now know that they saw no such thing; what they really saw was a set of appearances, equally reconcileable with the theory they held and with a totally different one. It seems strange that such an instance as this, of the testimony of the senses pleaded with the most entire conviction in favour of something which was a mere inference of the judgment, and, as it turned out, a false inference, should not have opened the eyes of the bigots of common sense, and inspired them with a more modest distrust of the competency of mere ignorance to judge the conclusions of cultivated thought.

In proportion to any person's deficiency of knowledge and mental cultivation, is generally his inability to discriminate between his inferences and the perceptions on which they were grounded. Many a marvellous tale, many a scandalous anecdote, owes its origin to this incapacity. The narrator relates, not what he saw or heard, but the impression which he derived from what he saw or heard, and of which perhaps the greater part consisted of inference, though the whole is related not as inference but as matter-of-fact. The difficulty of inducing

---

* Supra, p. 182.

witnesses to restrain within any moderate limits the inter-mixture of their inferences with the narrative of their percep-tions, is well known to experienced cross-examiners; and still more is this the case when ignorant persons attempt to describe any natural phenomenon. "The simplest narrative," says Dugald Stewart,* "of the most illiterate observer involves more or less of hypothesis; nay, in general, it will be found that, in proportion to his ignorance, the greater is the number of conjectural principles involved in his statements. A village apothecary (and, if possible, in a still greater degree, an expe-rienced nurse) is seldom able to describe the plainest case, without employing a phraseology of which every word is a theory: whereas a simple and genuine specification of the phenomena which mark a particular disease; a specification unsophisticated by fancy, or by preconceived opinions, may be regarded as unequivocal evidence of a mind trained by long and successful study to the most difficult of all arts, that of the faithful *interpretation* of nature."

The universality of the confusion between perceptions and the inferences drawn from them, and the rarity of the power to discriminate the one from the other, ceases to surprise us when we consider that in the far greater number of instances the actual perceptions of our senses are of no importance or interest to us except as marks from which we infer something beyond them. It is not the colour and superficial extension perceived by the eye that are important to us, but the object, of which those visible appearances testify the presence; and where the sensation itself is indifferent, as it generally is, we have no motive to attend particularly to it, but acquire a habit of passing it over without distinct consciousness, and going on at once to the inference. So that to know what the sensa-tion actually was, is a study in itself, to which painters, for example, have to train themselves by special and long-con-tinued discipline and application. In things further removed from the dominion of the outward senses, no one who has not great experience in psychological analysis is competent to

* *Elements of the Philosophy of the Mind*, vol. ii. ch. 4, sect. 5.

break this intense association; and when such analytic habits do not exist in the requisite degree, it is hardly possible to mention any of the habitual judgments of mankind on subjects of a high degree of abstraction, from the being of a God and the immortality of the soul down to the multiplication table, which are not, or have not been, considered as matter of direct intuition. So strong is the tendency to ascribe an intuitive character to judgments which are mere inferences, and often false ones. No one can doubt that many a deluded visionary has actually believed that he was directly inspired from Heaven, and that the Almighty had conversed with him face to face; which yet was only, on his part, a conclusion drawn from appearances to his senses, or feelings in his internal consciousness, which afforded no warrant for any such belief. A caution, therefore, against this class of errors, is not only needful but indispensable; though to determine whether, on any of the great questions of metaphysics, such errors are actually committed, belongs not to this place, but, as I have so often said, to a different science.

CHAPTER V.

§ 1. THE class of Fallacies of which we are now to speak, is the most extensive of all; embracing a greater number and variety of unfounded inferences than any of the other classes, and which it is even more difficult to reduce to sub-classes or species. If the attempt made in the preceding books to define the principles of well-grounded generalization has been successful, all generalizations not conformable to those principles might, in a certain sense, be brought under the present class: when however the rules are known and kept in view, but a casual lapse committed in the application of them, this is a blunder, not a fallacy. To entitle an error of generalization to the latter epithet, it must be committed on principle; there must lie in it some erroneous general conception of the inductive process; the legitimate mode of drawing conclusions from observation and experiment must be fundamentally misconceived.

Without attempting anything so chimerical as an exhaustive classification of all the misconceptions which can exist on the subject, let us content ourselves with noting, among the cautions which might be suggested, a few of the most useful and needful.

§ 2. In the first place, there are certain kinds of generalization which, if the principles already laid down be correct, *must* be groundless: experience cannot afford the necessary conditions for establishing them by a correct induction. Such, for instance, are all inferences from the order of nature existing on the earth, or in the solar system, to that which may exist in remote parts of the universe; where the phenomena, for aught we know, may be entirely different, or may succeed one another according to different laws, or even according to no fixed law at all. Such, again, in matters dependent on causation, are all universal negatives, all propositions that assert impossibility. The non-existence of any given phenomenon, however uniformly experience may as yet have testified to the fact, proves at most that no cause, adequate to its production, has yet manifested itself; but that no such causes exist in nature can only be inferred if we are so foolish as to suppose that we know all the forces in nature. The supposition would at least be premature while our acquaintance with some even of those which we do know is so extremely recent. And however much our knowledge of nature may hereafter be extended, it is not easy to see how that knowledge could ever be complete, or how, if it were, we could ever be assured of its being so.

The only laws of nature which afford sufficient warrant for attributing impossibility (even with reference to the existing order of nature, and to our own region of the universe), are first, those of number and extension, which are paramount to the laws of the succession of phenomena, and not exposed to the agency of counteracting causes; and secondly, the universal law of causality itself. That no variation in any effect or consequent will take place while the whole of the antecedents remain the same, may be affirmed with full assurance. But, that the addition of some new antecedent might not entirely alter and subvert the accustomed consequent, or that antecedents competent to do this do not exist in nature, we are in no case empowered positively to conclude.

§ 3. It is next to be remarked that all generalizations which profess, like the theories of Thales, Democritus, and others of the early Greek speculators, to resolve all things into some one element, or like many modern theories, to resolve phenomena radically different into the same, are necessarily false. By radically different phenomena I mean impressions on our senses which differ in quality, and not merely in degree.   On this subject what appeared necessary was said in the chapter on the Limits to the Explanation of Laws of Nature; but as the fallacy is even in our own times a common one, I shall touch on it somewhat further in this place.   •

When we say that the force which retains the planets in their orbits is resolved into gravity, or that the force which makes substances combine chemically is resolved into electricity, we assert in the one case what is, and in the other case what might, and probably will ultimately, be a legitimate result of induction.   In both these cases, motion is resolved into motion.   The assertion is, that a case of motion, which was supposed to be special, and to follow a distinct law of its own, conforms to and is included in the general law which regulates another class of motions.   But, from these and similar generalizations, countenance and currency have been given to attempts to resolve, not motion into motion, but heat into motion, light into motion, sensation itself into motion ; states of consciousness into states of the nervous system, as in the ruder forms of the materialist philosophy; vital phenomena into mechanical or chemical processes, as in some schools of physiology.

Now I am far from pretending that it may not be capable of proof, or that it will not be an important addition to our knowledge if proved, that certain motions in the particles of bodies are among the *conditions* of the production of heat or light; that certain assignable physical modifications of the nerves may be the *conditions* not only of our sensations or emotions, but even of our thoughts ; that certain mechanical and chemical conditions may, in the order of nature, be sufficient to determine to action the physiological laws of life. All I insist upon, in common with every thinker who entertains any clear idea of the logic of science, is, that it shall not be supposed that by proving these things one step would be made towards a real explanation of heat, light, or sensation ; or that the generic peculiarity of those phenomena can be in the least degree evaded by any such discoveries, however

well established. Let it be shown, for instance, that the
most complex series of physical causes and effects succeed
one another in the eye and in the brain to produce a sensation of colour; rays falling on the eye, refracted, converging,
crossing one another, making an inverted image on the
retina, and after this a motion—let it be a vibration, or a
rush of nervous fluid, or whatever else you are pleased to
suppose, along the optic nerve—a propagation of this motion
to the brain itself, and as many more different motions as
you choose; still, at the end of these motions, there is
something which is not motion, there is a feeling or sensation
of colour. Whatever number of motions we may be able to
interpolate, and whether they be real or imaginary, we shall
still find, at the end of the series, a motion antecedent and a
colour consequent. The mode in which any one of the
motions produces the next, might possibly be susceptible of
explanation by some general law of motion; but the mode in
which the last motion produces the sensation of colour, cannot
be explained by any law of motion; it is the law of colour:
which is, and must always remain, a peculiar thing. Where
our consciousness recognises between two phenomena an
inherent distinction; where we are sensible of a difference
which is not merely of degree, and feel that no adding
one of the phenomena to itself would produce the other;
any theory which attempts to bring either under the laws
of the other must be false; though a theory which merely
treats the one as a cause or condition of the other, may possibly be true.

§ 4. Among the remaining forms of erroneous generalization, several of those most worthy of and most requiring
notice have fallen under our examination in former places,
where, in investigating the rules of correct induction, we have
had occasion to advert to the distinction between it and some
common mode of the incorrect. In this number is what I
have formerly called the natural Induction of uninquiring
minds, the Induction of the ancients, which proceeds *per*

enumerationem simplicem : " This, that, and the other A are
B, I cannot think of any A which is not B, therefore every A
is B." As a final condemnation of this rude and slovenly
mode of generalization, I will quote Bacon's emphatic denun-
ciation of it ; the most important part, as I have more than
once ventured to assert, of the permanent service rendered by
him to philosophy. "Inductio quæ procedit per enumerationem simplicem, res puerilis est, et precario concludit" (concludes only *by your leave*, or provisionally,) " et periculo ex-
ponitur ab instantiâ contradictoriâ, et plerumque secundum
pauciora quam par est, et *ex his tantummodo quæ præsto sunt
pronunciat.* At Inductio quæ ad inventionem et demonstra-
tionem Scientiarum et Artium erit utilis, Naturam separare
debet, per rejectiones et exclusiones debitas; ac deinde post
negativas tot quot sufficiunt, super affirmativas concludere."

I have already said that the mode of Simple Enumeration
is still the common and received method of Induction in whatever relates to man and society. Of this a very few instances,
more by way of memento than of instruction, may suffice.
What, for example, is to be thought of all the " common-
sense" maxims for which the following may serve as the universal formula, " Whatsoever has never been, will never be."
As for example : negroes have never been as civilized as
whites sometimes are, therefore it is impossible they should be
so. Women, as a class, are supposed not to have hitherto
been equal in intellect to men, therefore they are necessarily
inferior. Society cannot prosper without this or the other
institution; *e.g.* in Aristotle's time, without slavery; in later
times, without an established priesthood, without artificial
distinctions of rank, &c. One poor person in a thousand,
educated, while the nine hundred and ninety-nine remain
uneducated, has usually aimed at raising himself out of his
class, therefore education makes people dissatisfied with the
condition of a labourer. Bookish men, taken from speculative
pursuits and set to work on something they know nothing
about, have generally been found or thought to do it ill ; therefore philosophers are unfit for business, &c. &c. All these
are inductions by simple enumeration. Reasons having some

reference to the canons of scientific investigation have been attempted to be given, however unsuccessfully, for some of these propositions; but to the multitude of those who parrot them, the *enumeratio simplex, ex his tantummodo quæ præsto sunt pronuncians*, is the sole evidence. Their fallacy consists in this, that they are inductions without elimination: there has been no real comparison of instances, nor even ascertainment of the material facts in any given instance. There is also the further error, of forgetting that such generalizations, even if well established, could not be ultimate truths, but must be results of laws much more elementary; and therefore, until deduced from such, could at most be admitted as empirical laws, holding good within the limits of space and time by which the particular observations that suggested the generalization were bounded.

This error, of placing mere empirical laws, and laws in which there is no direct evidence of causation, on the same footing of certainty as laws of cause and effect, an error which is at the root of perhaps the greater number of bad inductions, is exemplified only in its grossest form in the kind of generalizations to which we have now referred. These, indeed, do not possess even the degree of evidence which pertains to a well-ascertained empirical law; but admit of refutation on the empirical ground itself, without ascending to causal laws. A little reflection, indeed, will show that mere negations can only form the ground of the lowest and least valuable kind of empirical law. A phenomenon has never been noticed; this only proves that the conditions of that phenomenon have not yet occurred in experience, but does not prove that they may not occur hereafter. There is a better kind of empirical law than this, namely, when a phenomenon which is observed presents within the limits of observation a series of gradations, in which a regularity, or something like a mathematical law, is perceptible: from which, therefore, something may be rationally presumed as to those terms of the series which are beyond the limits of observation. But in negation there are no gradations, and no series: the generalizations, therefore, which deny the possibility of any given condition of man and society merely because it has never yet been witnessed, cannot possess this higher degree of validity even as empirical laws. What is more, the minuter examination which that higher order of empirical laws presupposes, being applied to the subject-matter of these, not only does not confirm but actually refutes them. For in reality the past history of Man and Society, instead of exhibiting them as immovable, unchangeable, incapable of ever presenting new phenomena, shows them on the contrary to be, in many most important particulars, not only changeable, but actually undergoing a progressive change. The empirical law, therefore, best expressive, in most cases, of the genuine result of observation, would be, not that such and such a phenomenon will continue unchanged, but that it will continue to change in some particular manner.

Accordingly, while almost all generalizations relating to Man and Society, antecedent to the last fifty years, have erred in the gross way which we have attempted to characterize, namely, by implicitly assuming that human nature and society will for ever revolve in the same orbit, and exhibit essentially the same phenomena; which is also the vulgar error of the ostentatiously practical, the votaries of so-called common sense, in our day, especially in Great Britain; the more thinking minds of the present age, having applied a more minute analysis to the past records of our race, have for the most part adopted a contrary opinion, that the human species is in a state of necessary progression, and that from the terms of the series which are past we may infer positively those which are yet to come. Of this doctrine, considered as a philosophical tenet, we shall have occasion to speak more fully in the concluding Book. If not, in all its forms, free from error, it is at least free from the gross and stupid error which we previously exemplified. But, in all except the most eminently philosophical minds, it is infected with precisely the same *kind* of fallacy as that is. For we must remember that even this other and better generalization, the progressive change in the condition of the human species, is, after all, but an empirical law: to which, too, it is not difficult to point out exceedingly large exceptions; and even if these could be got rid of, either by disputing the facts or by explaining and limiting the theory, the general objection remains valid against the supposed law, as applicable to any other than what, in our third book, were termed Adjacent Cases. For not only is it no ultimate, but not even a causal law. Changes do indeed take place in human affairs, but every one of those changes depends on determinate causes; the "progressiveness of the species" is not a cause, but a summary expression for the general result of all the causes. So soon as, by a quite different sort of induction, it shall be ascertained what causes have produced these successive changes, from the beginning of history, in so far as they have really taken place, and by what causes of a contrary tendency they have been occasionally checked or entirely counteracted, we may then be prepared to predict the future with reasonable foresight; we may be in possession of the real *law* of the future; and may be able to declare on what circumstances the continuance of the same onward movement will eventually depend. But this it is the error of many of the more advanced thinkers, in the present age, to overlook; and to imagine that the empirical law collected from a mere comparison of the condition of our species at different past times, is a real law, is *the* law of its changes, not only past but also to come. The truth is, that the causes on which the phenomena of the moral world depend, are in every age, and almost in every country, combined in some different proportion; so that it is scarcely to be expected that the general result of them all should conform very closely, in its details at least, to any uniformly progressive series. And all generalizations which affirm that mankind have a tendency to grow better or worse, richer or poorer, more cultivated or more barbarous, that population increases faster than subsistence, or subsistence than population, that inequality of fortune has a tendency to increase or to break down, and the like, propositions of considerable value as empirical laws within certain (but generally rather narrow) limits, are in reality true or false according to times and circumstances.

What we have said of empirical generalizations from times

past to times still to come, holds equally true of similar
generalizations from present times to times past; when persons whose acquaintance with moral and social facts is confined to their own age, take the men and the things of that age
for the type of men and things in general, and apply without
scruple to the interpretation of the events of history, the
empirical laws which represent sufficiently for daily guidance
the common phenomena of human nature at that time and in
that particular state of society. If examples are wanted,
almost every historical work, until a very recent period,
abounded in them. The same may be said of those who
generalize empirically from the people of their own country to
the people of other countries, as if human beings felt, judged,
and acted, everywhere in the same manner.

§ 5. In the foregoing instances, the distinction is confounded between empirical laws, which express merely the
customary order of the succession of effects, and the laws of
causation on which the effects depend. There may, however,
be incorrect generalization when this mistake is not committed; when the investigation takes its proper direction, that
of causes, and the result erroneously obtained purports to be
a really causal law.

The most vulgar form of this fallacy is that which is commonly called *post hoc, ergo propter hoc*, or, *cum hoc, ergo propter
hoc*. As when it was inferred that England owed her industrial
pre-eminence to her restrictions on commerce: as when the
old school of financiers, and some speculative writers, maintained that the national debt was one of the causes of national
prosperity: as when the excellence of the Church, of the
Houses of Lords and Commons, of the procedure of the law
courts, &c., were inferred from the mere fact that the country
had prospered under them. In such cases as these, if it
can be rendered probable by other evidence that the supposed
causes have some tendency to produce the effect ascribed to
them, the fact of its having been produced, though only in
one instance, is of some value as a verification by specific
experience: but in itself it goes scarcely any way at all

towards establishing such a tendency, since, admitting the effect, a hundred other antecedents could show an equally strong title of *that* kind to be considered as the cause.

In these examples we see bad generalization *à posteriori*, or empiricism properly so called : causation inferred from casual conjunction, without either due elimination, or any presumption arising from known properties of the supposed agent. But bad generalization *à priori* is fully as common : which is properly called false theory; conclusions drawn, by way of deduction, from properties of some one agent which is known or supposed to be present, all other coexisting agents being overlooked. As the former is the error of sheer ignorance, so the latter is especially that of semi-instructed minds; and is mainly committed in attempting to explain complicated phenomena by a simpler theory than their nature admits of. As when one school of physicians sought for the universal principle of all disease in " lentor and morbid viscidity of the blood," and imputing most bodily derangements to mechanical obstructions, thought to cure them by mechanical remedies;* while another, the chemical school, " acknowledged no source of disease but the presence of some hostile acid or alkali, or some deranged condition in the chemical composition of the fluid or solid parts," and conceived, therefore, that " all remedies must act by producing chemical changes in the body. We find Tournefort busily engaged in testing every vegetable juice, in order to discover in it some traces of an acid or alkaline ingredient, which might confer upon it medicinal activity. The fatal errors into

---

* "Thus Fourcroy," says Dr. Paris, "explained the operation of mercury by its specific gravity, and the advocates of this doctrine favoured the general introduction of the preparations of iron, especially in scirrhus of the spleen or liver, upon the same hypothetical principle; for, say they, whatever is most forcible in removing the obstruction must be the most proper instrument of cure ; such is steel, which, besides the attenuating power with which it is furnished, has still a greater force in this case from the gravity of its particles, which, being seven times specifically heavier than any vegetable, acts in proportion with a stronger impulse, and therefore is a more powerful deobstruent. This may be taken as a specimen of the style in which these mechanical physicians reasoned and practised."—*Pharmacologia*, pp. 38-9.

which such an hypothesis was liable to betray the practitioner, received an awful illustration in the history of the memorable fever that raged at Leyden in the year 1699, and which consigned two-thirds of the population of that city to an untimely grave; an event which in a great measure depended upon the Professor Sylvius de la Boe, who having just embraced the chemical doctrines of Van Helmont, assigned the origin of the distemper to a prevailing acid, and declared that its cure could alone [only] be effected by the copious administration of absorbent and testaceous medicines."[*]

These aberrations in medical theory have their exact parallels in politics. All the doctrines which ascribe absolute goodness to particular forms of government, particular social arrangements, and even to particular modes of education, without reference to the state of civilization and the various distinguishing characters of the society for which they are intended, are open to the same objection—that of assuming one class of influencing circumstances to be the paramount rulers of phenomena which depend in an equal or greater degree on many others. But on these considerations it is the less necessary that we should now dwell, as they will occupy our attention more largely in the concluding Book.

§ 6. The last of the modes of erroneous generalization to which I shall advert, is that to which we may give the name of False Analogies. This Fallacy stands distinguished from those already treated of by the peculiarity, that it does not even simulate a complete and conclusive induction, but consists in the misapplication of an argument which is at best only admissible as an inconclusive presumption, where real proof is unattainable.

An argument from analogy, is an inference that what is true in a certain case, is true in a case known to be somewhat similar, but not known to be exactly parallel, that is, to be similar in all the material circumstances. An object has the property B: another object is not known to have that pro-

---

* *Pharmacologia*, pp. 39, 40.

perty, but resembles the first in a property A, not known to be connected with B; and the conclusion to which the analogy points, is that this object has the property B also. As, for example, that the planets are inhabited, because the earth is so. The planets resemble the earth in describing elliptical orbits round the sun, in being attracted by it and by one another, in being nearly spherical, revolving on their axes, &c.; but it is not known that any of these properties, or all of them together, are the conditions on which the possession of inhabitants is dependent, or are marks of those conditions. Nevertheless, so long as we do not know what the conditions are, they *may* be connected by some law of nature with those common properties; and to the extent of that possibility the planets are more likely to be inhabited, than if they did not resemble the earth at all. This non-assignable and generally small increase of probability, beyond what would otherwise exist, is all the evidence which a conclusion can derive from analogy. For if we have the slightest reason to suppose any real connexion between the two properties A and B, the argument is no longer one of analogy. If it had been ascertained (I purposely put an absurd supposition) that there was a connexion by causation between the fact of revolving on an axis and the existence of animated beings, or if there were any reasonable ground for even suspecting such a connexion, a probability would arise of the existence of inhabitants in the planets, which might be of any degree of strength, up to a complete induction; but we should then infer the fact from the ascertained or presumed law of causation, and not from the analogy of the earth.

The name analogy, however, is sometimes employed by extension to denote those arguments of an inductive character but not amounting to a real induction, which are employed to strengthen the argument drawn from a simple resemblance. Though A, the property common to the two cases, cannot be shown to be the cause or effect of B, the analogical reasoner will endeavour to show that there is some less close degree of connexion between them; that A is one of a set of conditions from which, when all united, B would result; or is an occasional effect of some cause which has been known also to produce B; and the like. Any of which things, if shown, would render the existence of B by so much more probable, than if there had not been even that amount of known connexion between B and A.

Now an error or fallacy of analogy may occur in two ways. Sometimes it consists in employing an argument of either of the above kinds with correctness indeed, but overrating its probative force. This very common aberration is sometimes supposed to be particularly incident to persons distinguished for their imagination; but in reality it is the characteristic intellectual vice of those whose imaginations are barren, either from want of exercise, natural defect, or the narrowness of their range of ideas. To such minds objects present themselves clothed in but few properties; and as, therefore, few analogies between one object and another occur to them, they almost invariably overrate the degree of importance of those few: while one whose fancy takes a wider range, perceives and remembers so many analogies tending to conflicting conclusions, that he is much less likely to lay undue stress on any of them. We always find that those are the greatest slaves to metaphorical language, who have but one set of metaphors.

But this is only one of the modes of error in the employment of arguments of analogy. There is another, more properly deserving the name of fallacy; namely, when resemblance in one point is inferred from resemblance in another point, though there is not only no evidence to connect the two circumstances by way of causation, but the evidence tends positively to disconnect them. This is properly the Fallacy of False Analogies.

As a first instance, we may cite that favourite argument in defence of absolute power, drawn from the analogy of paternal government in a family, which government, however much in need of control, is not and cannot be controlled by the children themselves, while they remain children. Paternal government, says the argument, works well; therefore, despotic government in a state will work well. I wave, as not pertinent in this place, all that could be said in qualification of the alleged excellence of paternal government. However this might be, the argument from the family to the state would not the less proceed on a false analogy; implying that the beneficial working of parental government depends, in the family, on the only point which it has in common with political despotism, namely, irresponsibility. Whereas it depends, when real, not on that but on two other circumstances of the case, the affection of the parent for the children, and the superiority of the parent in wisdom and experience; neither of which properties can be reckoned on, or are at all likely to exist, between a political despot and his subjects; and when either of these circumstances fails even in the family, and the influence of the irresponsibility is allowed to work uncorrected, the result is anything but good government. This, therefore, is a false analogy.

Another example is the not uncommon *dictum*, that bodies politic have youth, maturity, old age, and death, like bodies natural: that after a certain duration of prosperity, they tend spontaneously to decay. This also is a false analogy, because the decay of the vital powers in an animated body can be distinctly traced to the natural progress of those very changes of structure which, in their earlier stages, constitute its growth to maturity: while in the body politic the progress of those changes cannot, generally speaking, have any effect but the still further continuance of growth: it is the stoppage of that progress, and the commencement of retrogression, that alone would constitute decay. Bodies politic die, but it is of disease, or violent death: they have no old age.

The following sentence from Hooker's *Ecclesiastical Polity* is an instance of a false analogy from physical bodies to what are called bodies politic. "As there could be in natural bodies no motion of anything unless there were some which moveth all things, and continueth immovable: even so in politic societies there must be some unpunishable, or else no man shall suffer punishment." There is a double fallacy here, for not only the analogy, but the premise from which it is drawn, is untenable. The notion that there must be something immovable which moves all other things, is the old scholastic error of a *primum mobile.*

The following instance I quote from Archbishop Whately's *Rhetoric:* " It would be admitted that a great and permanent diminution in the quantity of some useful commodity, such as corn, or coal, or iron, throughout the world, would be a serious and lasting loss; and again, that if the fields and coal mines yielded regularly double quantities, with the same labour, we should be so much the richer; hence it might be inferred, that if the quantity of gold and silver in the world were diminished one-half, or were doubled, like results would follow; the utility of these ·metals, for the purposes of coin, being very great. Now there are many points of resemblance and many of difference, between the precious metals on the one hand, and corn, coal, &c., on the other; but the important circumstance to the supposed argument is, that the *utility* of gold and silver (as coin, which is far the chief) *depends on their value,* which is regulated by their scarcity; or rather, to speak strictly, by the difficulty of obtaining them; whereas, if corn and coal were ten times as abundant (*i.e.* more easily obtained), a bushel of either would still be as useful as now. But if it were twice as easy to procure gold as it is, a sovereign would be twice as large; if only half as easy it would be of the size of a half-sovereign, and this (besides the trifling circumstance of the cheapness or dearness of gold ornaments) would be all the difference. The analogy, therefore, fails in the point essential to the argument."

The same author notices, after Bishop Copleston, the case of False Analogy which consists in inferring from the similarity in many respects between the metropolis of a country and the heart of the animal body, that the increased size of the metropolis is a disease.

Some of the false analogies on which systems of physics were confidently grounded in the time of the Greek philosophers, are such as we now call fanciful, not that the resemblances are not often real, but that it is long since any one has been inclined to draw from them the inferences which were then drawn. Such, for instance, are the curious speculations of the Pythagoreans on the subject of numbers. Finding that the distances of the planets bore or seemed to bear to one another a proportion not varying' much from that of the divisions of the monochord, they inferred from it the existence of an inaudible music, that of the spheres: as if the music of a harp had depended solely on the numerical proportions, and not on the material, nor even on the existence of any material, any strings at all. It has been similarly imagined that certain combinations of numbers, which were found to prevail in some natural phenomena, must run through the whole of nature: as that there must be four elements, because there are four possible combinations of hot and cold, wet and dry; that there must be seven planets, because there were seven metals, and even because there were seven days of the week. Kepler himself thought that there could be only six planets because there were only five regular solids. With these we may class the reasonings, so common in the speculations of the ancients, founded on a supposed *perfection* in nature: meaning by nature the customary order of events as they take place of themselves without human interference. This also is a rude guess at an analogy supposed to pervade all phenomena, however dissimilar. Since what was thought to be perfection appeared to obtain in some phenomena, it was inferred (in opposition to the plainest evidence) to obtain in all. "We always suppose that which is better to take place in nature, if it be possible," says Aristotle: and the vaguest and most heterogeneous qualities being confounded together under the notion of being *better*, there was no limit to the wildness of the inferences. Thus, because the heavenly bodies were "perfect," they must move in circles and uniformly. For "they" (the Pythagoreans) "would not allow," says Geminus,[*] "of any such disorder among divine and eternal things, as that they should sometimes move quicker and sometimes slower, and sometimes stand still; for no one would tolerate such anomaly in the movements even of a man, who was decent and orderly. The occasions of life, however, are often reasons for men going

---

[*] I quote from Dr. Whewell's *Hist. Ind. Sc.* 3rd ed. i. 129.

quicker or slower; but in the incorruptible nature of the stars, it is not possible that any cause can be alleged of quickness or slowness." It is seeking an argument of analogy very far, to suppose that the stars must observe the rules of decorum in gait and carriage, prescribed for themselves by the long-bearded philosophers satirized by Lucian.

As late as the Copernican controversy it was urged as an argument in favour of the true theory of the solar system, that it placed the fire, the noblest element, in the centre of the universe. This was a remnant of the notion that the order of nature must be perfect, and that perfection consisted in conformity to rules of precedency in dignity, either real or conventional. Again, reverting to numbers: certain numbers were *perfect*, therefore those numbers must obtain in the great phenomena of nature. Six was a perfect number, that is, equal to the sum of all its factors; an additional reason why there must be exactly six planets. The Pythagoreans, on the other hand, attributed perfection to the number ten; but agreed in thinking that the perfect number must be somehow realized in the heavens; and knowing only of nine heavenly bodies, to make up the enumeration, they asserted "that there was an *antichthon* or counter-earth, on the other side of the sun, invisible to us."[*] Even Huygens was persuaded that when the number of the heavenly bodies had reached twelve, it could not admit of any further increase. Creative power could not go beyond that sacred number.

Some curious instances of false analogy are to be found in the arguments of the Stoics to prove the equality of all crimes, and the equal wretchedness of all who had not realized their idea of perfect virtue. Cicero, towards the end of his Fourth Book *De Finibus*, states some of these as follows. "Ut, inquit, in fidibus plurimis, si nulla earum ita contenta numeris sit, ut concentum servare possit, omnes æque incontentæ sunt; sic peccata, quia discrepant, æque discrepant; paria sunt igitur." To which Cicero himself aptly answers, "æque contingit omnibus fidibus, ut incontentæ sint;

---

illud non continuo, ut æque incontentæ." The Stoic resumes:
"Ut enim, inquit, gubernator æque peccat, si palearum
navem evertit, et si auri ; item æque peccat qui parentem, et
qui servum, injuriâ verberat ;" assuming, that because the
magnitude of the interest at stake makes no difference in the
mere defect of skill, it can make none in the moral defect: a
false analogy. Again, "Quis ignorat, si plures ex alto emer-
gere velint, propius fore eos quidem ad respirandum, qui ad
summam jam aquam appropinquant, sed nihilo magis respirare
posse, quam eos, qui sunt in profundo ? Nihil ergo adjuvat
procedere, et progredi in virtute, quominus miserrimus sit,
antequam ad eam pervenerit, quoniam in aquâ nihil adjuvat :
et quoniam catuli, qui jam despecturi sunt, cæci æque, et ii qui
modo nati ; Platonem quoque necesse est, quoniam nondum
videbat sapientiam, æque cæcum animo, ac Phalarim fuisse."
Cicero, in his own person, combats these false analogies by
other analogies tending to an opposite conclusion. "Ista
similia non sunt, Cato. . . . . Illa sunt similia ; hebes acies
est cuipiam oculorum: corpore alius languescit : hi curatione
adhibitâ levantur in dies : alter valet plus quotidie : alter
videt. Hi similes sunt omnibus, qui virtuti student ; levantur
vitiis, levantur erroribus."

§ 7. In these and all other arguments drawn from remote
analogies, and from metaphors, which are cases of analogy, it
is apparent (especially when we consider the extreme facility
of raising up contrary analogies and conflicting metaphors)
that so far from the metaphor or analogy proving anything,
the applicability of the metaphor is the very thing to be made
out. It has to be shown that in the two cases asserted to be
analogous, the same law is really operating ; that between the
known resemblance and the inferred one there is some connexion by means of causation. Cicero and Cato might have
bandied opposite analogies for ever ; it rested with each of
them to prove by just induction, or at least to render probable,
that the case resembled the one set of analogous cases and
not the other, in the circumstances on which the disputed
question really hinged. Metaphors, for the most part, there-

fore, assume the proposition which they are brought to prove: their use is, to aid the apprehension of it ; to make clearly and vividly comprehended what it is that the person who employs the metaphor is proposing to make out ; and sometimes also, by what media he proposes to do so. For an apt metaphor, though it cannot prove, often suggests the proof.

For instance, when D'Alembert (I believe) remarked that in certain governments, only two creatures find their way to the highest places, the eagle and the serpent ; the metaphor not only conveys with great vividness the assertion intended, but contributes towards substantiating it, by suggesting, in a lively manner, the means by which the two opposite characters thus typified effect their rise. When it is said that a certain person misunderstands another because the lesser of two objects cannot comprehend the greater, the application of what is true in the literal sense of the word *comprehend,* to its metaphorical sense, points to the fact which is the ground and justification of the assertion, viz. that one mind cannot thoroughly understand another unless it can contain it in itself, that is, unless it possesses all that is contained in the other. When it is urged as an argument for education, that if the soil is left uncultivated, weeds will spring up, the metaphor, though no proof, but a statement of the thing to be proved, states it in terms which, by suggesting a parallel case, put the mind upon the track of the real proof. For, the reason why weeds grow in an uncultivated soil, is that the seeds of worthless products exist everywhere, and can germinate and grow in almost all circumstances, while the reverse is the case with those which are valuable ; and this being equally true of mental products, this mode of conveying an argument, independently of its rhetorical advantages, has a logical value ; since it not only suggests the grounds of the conclusion, but points to another case in which those grounds have been found, or at least deemed to be, sufficient.

On the other hand, when Bacon, who is equally conspicuous in the use and abuse of figurative illustration, says that the stream of time has brought down to us only the least valuable part of the writings of the ancients, as a river carries froth and straws floating on its surface, while more weighty objects sink to the bottom; this, even if the assertion illustrated by it were true, would be no good illustration, there being no parity of cause. The levity by which substances float on a stream, and the levity which is synonymous with worthlessness, have nothing in common except the name; and (to show how little value there is in the metaphor) we need only change the word into *buoyancy*, to turn the semblance of argument involved in Bacon's illustration against himself.

A metaphor, then, is not to be considered as an argument, but as an assertion that an argument exists; that a parity subsists between the case from which the metaphor is drawn and that to which it is applied. This parity may exist though the two cases be apparently very remote from one another; the only resemblance existing between them may be a resemblance of relations, an analogy in Ferguson's and Archbishop Whately's sense: as in the preceding instance, in which an illustration from agriculture was applied to mental cultivation.

§ 8. To terminate the subject of Fallacies of Generalization, it remains to be said, that the most fertile source of them is bad classification: bringing together in one group, and under one name, things which have no common properties, or none but such as are too unimportant to allow general propositions of any considerable value to be made respecting the class. The misleading effect is greatest, when a word which in common use expresses some definite fact, is extended by slight links of connexion to cases in which that fact does not exist, but some other or others, only slightly resembling it. Thus Bacon,* in speaking of the *Idola* or Fallacies arising from notions *temere et inæqualiter à rebus abstractæ*, exemplifies them by the notion of Humidum or Wet, so familiar in the physics of antiquity and of the middle ages. "Invenietur verbum istud, Humidum, nihil aliud quam nota confusa diversarum

---

* *Nov. Org.* Aph. 60.

actionum, quæ nullam constantiam aut reductionem patiuntur.
Significat enim, et quod circa aliud corpus facile se circum-
fundit; et quod in se est indeterminabile, nec consistere
potest; et quod facile cedit undique; et quod facile se dividit
et dispergit; et quod facile se unit et colligit; et quod facile
fluit, et in motu ponitur; et quod alteri corpori facile adhæret,
idque madefacit; et quod facile reducitur in liquidum, sive
colliquatur, cum antea consisteret. Itaque quum ad hujus
nominis prædicationem et impositionem ventum sit; si alia
accipias, flamma humida est; si alia accipias, aer humidus non
est; si alia, pulvis minutus humidus est; si alia, vitrum
humidum est: ut facile appareat, istam notionem ex aquâ
tantum, et communibus et vulgaribus liquoribus, absque ullâ
debitâ verificatione, temere abstractam esse."

Bacon himself is not exempt from a similar accusation
when inquiring into the nature of heat: where he occasionally
proceeds like one who seeking for the cause of hardness, after
examining that quality in iron, flint,¹ and diamond, should
expect to find that it is something which can be traced also in
hard water, a hard knot, and a hard heart.

The word κίνησις in the Greek philosophy, and the words
Generation and Corruption both then and long afterwards,
denoted such a multitude of heterogeneous phenomena, that
any attempt at philosophizing in which those words were
used was almost as necessarily abortive as if the word *hard*
had been taken to denote a class including all the things
mentioned above. Κίνησις, for instance, which properly
signified motion, was taken to denote not only all motion
but even all change: ἀλλοίωσις being recognised as one of
the modes of κίνησις. The effect was, to connect with every
form of ἀλλοίωσις or change, ideas drawn from motion in the
proper and literal sense, and which had no real connexion
with any other kind of κίνησις than that. Aristotle and Plato
laboured under a continual embarrassment from this misuse
of terms. But if we proceed further in this direction we shall
encroach upon the Fallacy of Ambiguity, which belongs to a
different class, the last in order of our classification, Fallacies
of Confusion.

CHAPTER VI.

FALLACIES OF RATIOCINATION.

§ 1. WE have now, in our progress through the classes of Fallacies, arrived at those to which, in the common books of logic, the appellation is in general exclusively appropriated; those which have their seat in the ratiocinative or deductive part of the investigation of truth. On these fallacies it is the less necessary for us to insist at any length, as they have been most satisfactorily treated in a work familiar to almost all, in this country at least, who feel any interest in these speculations, Archbishop Whately's *Logic*. Against the more obvious forms of this class of fallacies, the rules of the syllogism are a complete protection. Not (as we have so often said) that the ratiocination cannot be good unless it be in the form of a syllogism; but that, by showing it in that form, we are sure to discover if it be bad, or at least if it contain any fallacy of this class.

§ 2. Among Fallacies of Ratiocination, we ought perhaps to include the errors committed in processes which have the appearance only, not the reality, of an inference from premises; the fallacies connected with the conversion and æquipollency of propositions. I believe errors of this description to be far more frequently committed than is generally supposed, or than their extreme obviousness might seem to admit of. For example, the simple conversion of an universal affirmative proposition, All A are B, therefore all B are A, I take to be a very common form of error: though committed, like many other fallacies, oftener in the silence of thought than in express words, for it can scarcely be clearly enunciated without being detected. And so with another form of fallacy, not

substantially different from the preceding: the erroneous conversion of an hypothetical proposition. The proper converse
of an hypothetical proposition is this: If the consequent be
false, the antecedent is false; but this, If the consequent be
true, the antecedent is true, by no means holds good, but
is an error corresponding to the simple conversion of an
universal affirmative. Yet hardly anything is more common
than for people, in their private thoughts, to draw this inference. As when the conclusion is accepted, which it so often
is, for proof of the premises. That the premises cannot be
true if the conclusion is false, is the unexceptionable foundation of the legitimate mode of reasoning called a *reductio ad
absurdum*. But people continually think and express themselves, as if they also believed that the premises cannot be
false if the conclusion is true. The truth, or supposed truth,
of the inferences which follow from a doctrine, often enables
it to find acceptance in spite of gross absurdities in it. How
many philosophical systems which had scarcely any intrinsic
recommendation, have been received by thoughtful men because they were supposed to lend additional support to religion, morality, some favourite view of politics, or some other
cherished persuasion: not merely because their wishes were
thereby enlisted on its side, but because its leading to what
they deemed sound conclusions appeared to them a strong
presumption in favour of its truth: though the presumption,
when viewed in its true light, amounted only to the absence
of that particular evidence of falsehood, which would have
resulted from its leading by correct inference to something
already known to be false.

Again, the very frequent error in conduct, of mistaking
reverse of wrong for right, is the practical form of a logical
error with respect to the Opposition of Propositions. It is
committed for want of the habit of distinguishing the *contrary*
of a proposition from the *contradictory* of it, and of attending
to the logical canon, that contrary propositions, though they
cannot both be true, may both be false. If the error were to
express itself in words, it would run distinctly counter to this
canon. It generally, however, does not so express itself, and

to compel it to do so is the most effectual method of detecting and exposing it.

§ 3. Among Fallacies of Ratiocination are to be ranked in the first place, all the cases of vicious syllogism laid down in the books. These generally resolve themselves into having more than three terms to the syllogism, either avowedly, or in the covert mode of an undistributed middleterm, or an *illicit process* of one of the two extremes. It is not, indeed, very easy fully to convict an argument of falling under any one of these vicious cases in particular; for the reason already more than once referred to, that the premises are seldom formally set out: if they were, the fallacy would impose upon nobody; and while they are not, it is almost always to a certain degree optional in what manner the suppressed link shall be filled up. The rules of the syllogism are rules for compelling a person to be aware of the whole of what he must undertake to defend if he persists in maintaining his conclusion. He has it almost always in his power to make his syllogism good by introducing a false premise; and hence it is scarcely ever possible decidedly to affirm that any argument involves a bad syllogism: but this detracts nothing from the value of the syllogistic rules, since it is by them that a reasoner is compelled distinctly to make his election what premises he is prepared to maintain. The election made, there is generally so little difficulty in seeing whether the conclusion follows from the premises set out, that we might without much logical impropriety have merged this fourth class of fallacies in the fifth, or Fallacies of Confusion.

§ 4. Perhaps, however, the commonest, and certainly the most dangerous fallacies of this class, are those which do not lie in a single syllogism, but slip in between one syllogism and another in a chain of argument, and are committed by *changing the premises*. A proposition is proved, or an acknowledged truth laid down, in the first part of an argumentation, and in the second a further argument is founded not on the same proposition, but on some other, resembling it sufficiently to be mistaken for it. Instances of this fallacy will be found in almost all the argumentative discourses of unprecise thinkers; and we need only here advert to one of the obscurer forms of it, recognised by the schoolmen as the fallacy *à dicto secundum quid ad dictum simpliciter*. This is committed when, in the premises, a proposition is asserted with a qualification, and the qualification lost sight of in the conclusion; or oftener, when a limitation or condition, though not asserted, is necessary to the truth of the proposition, but is forgotten when that proposition comes to be employed as a premise. Many of the bad arguments in vogue belong to this class of error. The premise is some admitted truth, some common maxim, the reasons or evidence for which have been forgotten, or are not thought of at the time, but if they had been thought of would have shown the necessity of so limiting the premise that it would no longer have supported the conclusion drawn from it.

Of this nature is the fallacy in what is called, by Adam Smith and others, the Mercantile Theory in Political Economy. That theory sets out from the common maxim, that whatever brings in money enriches; or that every one is rich in proportion to the quantity of money he obtains. From this it is concluded that the value of any branch of trade, or of the trade of the country altogether, consists in the balance of money it brings in; that any trade which carries more money out of the country than it draws into it is a losing trade; that therefore money should be attracted into the country and kept there, by prohibitions and bounties: and a train of similar corollaries. All for want of reflecting that if the riches of an individual are in proportion to the quantity of money he can command, it is because that is the measure of his power of purchasing money's worth; and is therefore subject to the proviso that he is not debarred from employing his money in such purchases. The premise, therefore, is only true *secundum quid;* but the theory assumes it to be true absolutely, and infers that increase of money is increase of riches, even when produced by means subversive of the condition under which alone money can be riches.

A second instance is, the argument by which it used to be contended, before the commutation of tithe, that tithes fell on the landlord, and were a deduction from rent; because the rent of tithe-free land was always higher than that of land of the same quality, and the same advantages of situation, subject to tithe. Whether it be true or not that a tithe falls on rent, a treatise on Logic is not the place to examine; but it is certain that this is no proof of it. Whether the proposition be true or false, tithe-free land must, by the necessity of the case, pay a higher rent. For if tithes do not fall on rent, it must be because they fall on the consumer; because they raise the price of agricultural produce. But if the produce be raised in price, the farmer of tithe-free as well as the farmer of tithed land gets the benefit. To the latter the rise is but a compensation for the tithe he pays; to the first, who pays none, it is clear gain, and therefore enables him, and if there be freedom of competition forces him, to pay so much more rent to his landlord. The question remains, to what class of fallacies this belongs. The premise is, that the owner of tithed land receives less rent than the owner of tithe-free land; the conclusion is, that therefore he receives less than he himself would receive if tithe were abolished. But the premise is only true conditionally; the owner of tithed land receives less than what the owner of tithe-free land is enabled to receive *when other lands are tithed;* while the conclusion is applied to a state of circumstances in which that condition fails, and in which, by consequence, the premise would not be true. The fallacy, therefore, is *à dicto secundum quid ad dictum simpliciter.*

A third example is the opposition sometimes made to legitimate interferences of government in the economical affairs of society, grounded on a misapplication of the maxim, that an individual is a better judge than the government, of what is for his own pecuniary interest. This objection was urged to Mr. Wakefield's principle of colonization; the concentration of the settlers, by fixing such a price on unoccupied land as may preserve the most desirable proportion between the quantity of land in culture, and the labouring population. Against this it was argued, that if individuals found it for their advantage to occupy extensive tracts of land, they, being better judges of their own interest than the legislature (which can only proceed on general rules) ought not to be restrained from doing so. But in this argument it was forgotten that the fact of a person's taking a large tract of land is evidence only that it is his interest to take as much as other people, but not that it might not be for his interest to content himself with less, if he could be assured that other people would do so too; an assurance which nothing but a government regulation can give. If all other people took much, and he only a little, he would reap none of the advantages derived from the concentration of the population and the consequent possibility of procuring labour for hire, but would have placed himself, without equivalent, in a situation of voluntary inferiority. The proposition, therefore, that the quantity of land which people will take when left to themselves is that which is most for their interest to take, is true only *secundum quid:* it is only their interest while they have no guarantee for the conduct of one another. But the arrangement disregards the limitation, and takes the proposition for true *simpliciter.*

One of the conditions oftenest dropped, when what would otherwise be a true proposition is employed as a premise for proving others, is the condition of *time.* It is a principle of political economy that prices, profits, wages, &c. " always find their level;" but this is often interpreted as if it meant that they are always, or generally, *at* their level; while the truth is, as Coleridge epigrammatically expresses it, that they are always *finding* their level, "which might be taken as a paraphrase or ironical definition of a storm."

Under the same head of fallacy (*à dicto secundum quid ad dictum simpliciter*) might be placed all the errors which are vulgarly called misapplications of abstract truths: that is, where a principle, true (as the common expression is) *in the abstract,* that is, all modifying causes being supposed absent, is reasoned on as if it were true absolutely, and no modifying circumstance could ever by possibility exist. This very com-

mon form of error it is not requisite that we should exemplify
here, as it will be particularly treated of hereafter in its application to the subjects on which it is most frequent and most
fatal, those of politics and society.*

* "An advocate," says Mr. De Morgan (*Formal Logic*, p. 270), "is sometimes guilty of the argument *à dicto secundum quid ad dictum simpliciter :* it is
his business to do for his client all that his client might *honestly* do for himself.
Is not the word in italics frequently omitted ? *Might* any man honestly try to
do for himself all that counsel frequently try to do for him ? We are often
reminded of the two men who stole the leg of mutton ; one could swear he had
not got it, the other that he had not taken it. The counsel is doing his duty
by his client, the client has left the matter to his counsel. Between the
unexecuted intention of the client, and the unintended execution of the counsel, there may be a wrong done, and, if we are to believe the usual maxims, no
wrong-doer."

The same writer justly remarks (p. 251) that there is a converse fallacy, *à
dicto simpliciter ad dictum secundum quid,* called by the scholastic logicians,
fallacia accidentis ; and another which may be called *à dicto secundum quid ad
dictum secundum alterum quid* (p. 265). For apt instances of both, I must
refer the reader to Mr. De Morgan's able chapter on Fallacies.

CHAPTER VII.

FALLACIES OF CONFUSION.

§ 1. UNDER this fifth and last class it is convenient to arrange all those fallacies, in which the source of error is not so much a false estimate of the probative force of known evidence, as an indistinct, indefinite, and fluctuating conception of what the evidence is.

At the head of these stands that multitudinous body of fallacious reasonings, in which the source of error is the ambiguity of terms: when something which is true if a word be used in a particular sense, is reasoned on as if it were true in another sense. In such a case there is not a mal-estimation of evidence, because there is not properly any evidence to the point at all; there is evidence, but to a different point, which from a confused apprehension of the meaning of the terms used, is supposed to be the same. This error will naturally be oftener committed in our ratiocinations than in our direct inductions, because in the former we are deciphering our own or other people's notes, while in the latter we have the things themselves present, either to the senses or to the memory. Except, indeed, when the induction is not from individual cases to a generality, but from generalities to a still higher generalization; in that case the fallacy of ambiguity may affect the inductive process as well as the ratiocinative. It occurs in ratiocination in two ways: when the middleterm is ambiguous, or when one of the terms of the syllogism is taken in one sense in the premises, and in another sense in the conclusion.

Some good exemplifications of this fallacy are given by Archbishop Whately. "One case," says he, "which may be regarded as coming under the head of Ambiguous Middle, is (what I believe logical writers mean by ' *Fallacia Figuræ*

Dictionis,') the fallacy built on the grammatical structure of language, from men's usually taking for granted that *paronymous* (or *conjugate*) words, *i.e.* those belonging to each other, as the substantive, adjective, verb, &c. of the same root, have a precisely corresponding meaning; which is by no means universally the case. Such a fallacy could not indeed be even exhibited in strict logical form, which would preclude even the attempt at it, since it has two middleterms in sound as well as sense. But nothing is more common in practice than to vary continually the terms employed, with a view to grammatical convenience; nor is there anything unfair in such a practice, as long as the *meaning* is preserved unaltered; *e. g.* 'murder should be punished with death; this man is a murderer, therefore he deserves to die,' &c. Here we proceed on the assumption (in this case just) that to commit murder, and to be a murderer,—to deserve death, and to be one who ought to die, are, respectively, equivalent expressions; and it would frequently prove a heavy inconvenience to be debarred this kind of liberty; but the abuse of it gives rise to the Fallacy in question: e. g. *projectors* are unfit to be trusted; this man has formed a *project*, therefore he is unfit to be trusted: here the sophist proceeds on the hypothesis that he who forms a *project* must be a *projector*: whereas the bad sense that commonly attaches to the latter word, is not at all implied in the former. This fallacy may often be considered as lying not in the Middle, but in one of the terms of the Conclusion; so that the conclusion drawn shall not be, in reality, at all warranted by the premises, though it will appear to be so, by means of the grammatical affinity of the words: *e. g.* to be acquainted with the guilty is a *presumption* of guilt; this man is so acquainted, therefore we may *presume* that he is guilty: this argument proceeds on the supposition of an exact correspondence between *presume* and *presumption*, which, however, does not really exist; for 'presumption' is commonly used to express a kind of *slight suspicion;* whereas, 'to presume' amounts to actual belief. There are innumerable instances of a non-correspondence in paronymous words, similar to that above instanced; as between *art* and *artful*, *design* and *design-*

ing, faith and *faithful*, &c. ; and the more slight the variation
of the meaning, the more likely is the fallacy to be successful ;
for when the words have become so widely removed in sense
as 'pity' and 'pitiful,' every one would perceive such a fallacy, nor could it be employed but in jest.*

"The present Fallacy is nearly allied to, or rather, perhaps,
may be regarded as a branch of, that founded on *etymology ;*
viz. when a term is used, at one time in its customary, and at
another in its etymological sense. Perhaps no example of this
can be found that is more extensively and mischievously employed than in the case of the word *representative* : assuming
that its right meaning must correspond exactly with the strict
and original sense of the verb 'represent,' the sophist persuades the multitude, that a member of the House of Commons is bound to be guided in all points by the opinion of his
constituents ; and, in short, to be merely their *spokesman ;*
whereas law and custom, which in this case may be considered
as fixing the meaning of the term, require no such thing, but
enjoin the representative to act according to the best of his
own judgment, and on his own responsibility."

The following are instances of great practical importance, in which arguments are habitually founded on a verbal
ambiguity.

The mercantile public are frequently led into this fallacy
by the phrase, "scarcity of money." In the language of commerce "money" has two meanings: *currency*, or the circulating medium ; and *capital seeking investment*, especially
investment on loan. In this last sense the word is used when
the "money market" is spoken of, and when the "value of
money" is said to be high or low, the rate of interest being

---

* An example of this fallacy is the popular error that *strong* drink must
be a cause of *strength*. There is here fallacy within fallacy; for granting that
the words "strong" and "strength" were not (as they are) applied in a totally
different sense to fermented liquors and to the human body, there would still
be involved the error of supposing that an effect must be like its cause ; that
the conditions of a phenomenon are likely to resemble the phenomenon itself;
which we have already treated of as an *à priori* fallacy of the first rank. As
well might it be supposed that a strong poison will make the person who takes
it, strong.

meant. The consequence of this ambiguity is, that as soon as scarcity of money in the latter of these senses begins to be felt,—as soon as there is difficulty of obtaining loans, and the rate of interest is high,—it is concluded that this must arise from causes acting upon the quantity of money in the other and more popular sense; that the circulating medium must have diminished in quantity, or ought to be increased. I am aware that, independently of the double meaning of the term, there are in facts themselves some peculiarities, giving an apparent support to this error; but the ambiguity of the language stands on the very threshold of the subject, and intercepts all attempts to throw light upon it.

Another ambiguous expression which continually meets us in the political controversies of the present time, especially in those which relate to organic changes, is the phrase " influence of property :" which is sometimes used for the influence of respect for superior intelligence, or gratitude for the kind offices which persons of large property have it so much in their power to bestow; at other times for the influence of fear; fear of the worst sort of power, which large property also gives to its possessor, the power of doing mischief to dependents. To confound these two, is the standing fallacy of ambiguity brought against those who seek to purify the electoral system from corruption and intimidation. Persuasive influence, acting through the conscience of the voter, and carrying his heart and mind with it, is beneficial—therefore (it is pretended) coercive influence, which compels him to forget that he is a moral agent, or to act in opposition to his moral convictions, ought not to be placed under restraint.

Another word which is often turned into an instrument of the fallacy of ambiguity, is Theory. In its most proper acceptation, theory means the completed result of philosophical induction from experience. In that sense, there are erroneous as well as true theories, for induction may be incorrectly performed, but theory of some sort is the necessary result of knowing anything of a subject, and having put one's knowledge into the form of general propositions for the guidance of practice. In this, the proper sense of the word, Theory is the explanation of practice. In another and a more vulgar sense, theory means any mere fiction of the imagination, endeavouring to conceive how a thing may possibly have been produced, instead of examining how it was produced. In this sense only are theory, and theorists, unsafe guides; but because of this, ridicule or discredit is attempted to be attached to theory in its proper sense, that is, to legitimate generalization, the end and aim of all philosophy; and a conclusion is represented as worthless, just because that has been done, which if done correctly, constitutes the highest worth that a principle for the guidance of practice can possess, namely, to comprehend in a few words the real law on which a phenomenon depends, or some property or relation which is universally true of it.

"The Church" is sometimes understood to mean the clergy alone, sometimes the whole body of believers, or at least of communicants. The declamations respecting the inviolability of church property are indebted for the greater part of their apparent force to this ambiguity. The clergy, being called the church, are supposed to be the real owners of what is called church property; whereas they are in truth only the managing members of a much larger body of proprietors, and enjoy on their own part a mere usufruct, not extending beyond a life interest.

The following is a Stoical argument taken from Cicero *De Finibus*, book the third: "Quod est bonum, omne laudabile est. Quod autem laudabile est, omne honestum est. Bonum igitur quod est, honestum est." Here the ambiguous word is *laudabile*, which in the minor premise means anything which mankind are accustomed, on good grounds, to admire or value; as beauty, for instance, or good fortune: but in the major, it denotes exclusively moral qualities. In much the same manner the Stoics endeavoured logically to justify as philosophical truths, their figurative and rhetorical expressions of ethical sentiment: as that the virtuous man is alone free, alone beautiful, alone a king, &c. Whoever has virtue has Good (because it has been previously determined not to call anything else good); but, again, Good necessarily includes freedom, beauty, and even kingship, all these being good things; therefore whoever has virtue has all these.

The following is an argument of Descartes to prove, in his *à priori* manner, the being of a God. The conception, says he, of an infinite Being proves the real existence of such a being. For if there is not really any such being, *I* must have made the conception; but if I could make it, I can also unmake it; which evidently is not true; therefore there must be, externally to myself, an archetype, from which the conception was derived. In this argument (which, it may be observed, would equally prove the real existence of ghosts and of witches) the ambiguity is in the pronoun *I*, by which, in one place, is to be understood my *will*, in another the *laws of my nature*. If the conception, existing as it does in my mind, had no original without, the conclusion would unquestionably follow that *I* made it; that is, the laws of my nature must have somehow evolved it: but that my *will* made it, would not follow. Now when Descartes afterwards adds that I cannot unmake the conception, he means that I cannot get rid of it by an act of my will: which is true, but is not the proposition required. I can as much unmake this conception as I can any other: no conception which I have once had, can I ever dismiss by mere volition: but what some of the laws of my nature have produced, other laws, or those same laws in other circumstances, may, and often do, subsequently efface.

Analogous to this are some of the ambiguities in the free-will controversy; which, as they will come under special consideration in the concluding Book, I only mention *memoriæ causâ*. In that discussion, too, the word *I* is often shifted from one meaning to another, at one time standing for my volitions, at another time for the actions which are the consequences of them, or the mental dispositions from which they proceed. The latter ambiguity is exemplified in an argument of Coleridge (in his *Aids to Reflection*), in support of the freedom of the will. It is not true, he says, that a man is governed by motives; "the man makes the motive, not the motive the man ;" the proof being that " what is a strong motive to one man is no motive at all to another." The premise is true, but only amounts to this, that different persons have different degrees of susceptibility to the same motive ; as they have also to the same intoxicating liquid, which however does not prove that they are free to be drunk or not drunk, whatever quantity of the fluid they may drink. What is proved is, that certain mental conditions in the person himself, must co-operate, in the production of the act, with the external inducement : but those mental conditions also are the effect of causes ; and there is nothing in the argument to prove that they can arise without a cause—that a spontaneous determination of the will, without any cause at all, ever takes place, as the free-will doctrine supposes.

The double use, in the free-will controversy, of the word Necessity, which sometimes stands only for Certainty, at other times for Compulsion ; sometimes for what *cannot* be prevented, at other times only for what we have reason to be assured *will* not ; we shall have occasion hereafter to pursue to some of its ulterior consequences.

A most important ambiguity, both in common and in metaphysical language, is thus pointed out by Archbishop Whately in the Appendix to his Logic : "*Same* (as well as *One, Identical*, and other words derived from them,) is used frequently in a sense very different from its primary one, as applicable to a *single* object ; being employed to denote great *similarity*. When several objects are undistinguishably alike, *one single description* will apply equally to any of them ; and thence they are said to be all of *one and the same* nature, appearance, &c. As, *e. g.* when we say ' this house is built of the *same* stone with such another,' we only mean that the stones are undistinguishable in their qualities ; not that the one building was pulled down, and the other constructed with the materials. Whereas *sameness*, in the primary sense, does not even necessarily imply similarity ; for if we say of any man that he is greatly altered since such a time, we understand, and indeed imply by the very expression, that he is *one person*, though different in several qualities. It is worth observing also, that Same, in the secondary sense, admits, according to popular usage, of degrees: we speak of two things being *nearly* the same, but not entirely: personal identity does not admit of degrees. Nothing, perhaps, has contributed more to the error of Realism than inattention to this ambiguity. When several persons are said to have *one and the same* opinion, thought, or idea, many men, overlooking the true simple statement of the case, which is, that they are *all thinking alike*, look for something more abstruse and mystical, and imagine there must be some *One Thing*, in the primary sense, though not an individual, which is present at once in the mind of each of these persons; and thence readily sprung Plato's theory of Ideas, each of which was, according to him, one real, eternal object, existing entire and complete in each of the individual objects that are known by one name."

It is, indeed, not a matter of inference but of authentic history, that Plato's doctrine of Ideas, and the Aristotelian doctrine (in this respect similar to the Platonic) of substantial forms and second substances, grew up in the precise way here pointed out; from the supposed necessity of finding, in things which were said to have the *same* nature, or the *same* qualities, something which was the *same* in the very sense in which a man is the same as himself. All the idle speculations respecting τὸ ὄν, τὸ ἕν, τὸ ὅμοιον, and similar abstractions, so common in the ancient and in some modern schools of thought, sprang from the same source. The Aristotelian logicians saw, however, one case of the ambiguity, and provided against it with their peculiar felicity in the invention of technical language, when they distinguished things which differed both *specie* and *numero*, from those which differed *numero tantum*, that is, which were exactly alike (in some particular respect at least) but were distinct individuals. An extension of this distinction to the two meanings of the word Same, namely, things which are the same *specie tantum*, and a thing which is the same *numero* as well as *specie*, would have prevented the confusion which has been a source of so much darkness and such an abundance of positive error in metaphysical philosophy.

One of the most singular examples of the length to which a thinker of eminence may be led away by an ambiguity of language, is afforded by this very case. I refer to the famous argument by which Bishop Berkeley flattered himself that he had for ever put an end to "scepticism, atheism, and irreligion." It is briefly as follows. I thought of a thing yesterday; I ceased to think of it; I think of it again to-day. I had, therefore, in my mind yesterday an *idea* of the object; I have also an idea of it to-day; this idea is evidently not another, but the very same idea. Yet an intervening time elapsed in which I had it not. Where was the idea during this interval? It must have been somewhere; it did not cease to exist; otherwise the idea I had yesterday could not be the *same* idea; no more than the man I see alive to-day can be the same whom I saw yesterday, if the man has died in the meanwhile. Now an idea cannot be conceived to exist anywhere except in a mind; and hence there must exist an Universal Mind, in which all ideas have their permanent residence, during the intervals of their conscious presence in our own minds.

It is evident that Berkeley here confounded sameness *numero* with sameness *specie*, that is, with exact resemblance, and assumed the former where there was only the latter; not perceiving that when we say we have the same thought to-day which we had yesterday, we do not mean the same individual thought, but a thought exactly similar: as we say that we have the same illness which we had last year, meaning only the same sort of illness.

In one remarkable instance the scientific world was divided into two furiously hostile parties by an ambiguity of language affecting a branch of science which, more completely than most others, enjoys the advantage of a precise and well-defined terminology. I refer to the famous dispute respecting the *vis viva*, the history of which is given at large in Professor Playfair's Dissertation. The question was, whether the *force* of a moving body was proportional (its mass being given) to its velocity simply, or to the square of its velocity: and the ambiguity was in the word Force. " One of the effects," says

Playfair, "produced by a moving body is proportional to the square of the velocity, while another is proportional to the velocity simply:" from whence clearer thinkers were subsequently led to establish a double measure of the efficiency of a moving power, one being called *vis viva*, and the other *momentum*. About the facts, both parties were from the first agreed: the only question was, with which of the two effects the term *force* should be, or could most conveniently be, associated. But the disputants were by no means aware that this was all; they thought that force was one thing, the production of effects another; and the question, by which set of effects the force which produced both the one and the other should be measured, was supposed to be a question not of terminology but of fact.

The ambiguity of the word Infinite is the real fallacy in the amusing logical puzzle of Achilles and the Tortoise, a puzzle which has been too hard for the ingenuity or patience of many philosophers, and which no less a thinker than Sir William Hamilton considered as insoluble; as a sound argument, though leading to a palpable falsehood. The fallacy, as Hobbes hinted, lies in the tacit assumption that whatever is infinitely divisible is infinite; but the following solution, (to the invention of which I have no claim,) is more precise and satisfactory.

The argument is, let Achilles run ten times as fast as the tortoise, yet if the tortoise has the start, Achilles will never overtake him. For suppose them to be at first separated by an interval of a thousand feet: when Achilles has run these thousand feet, the tortoise will have got on a hundred; when Achilles has run those hundred, the tortoise will have run ten, and so on for ever: therefore Achilles may run for ever without overtaking the tortoise.

Now, the "for ever," in the conclusion, means, for any length of time that can be supposed; but in the premises "ever" does not mean any *length* of time: it means any *number of subdivisions* of time. It means that we may divide a thousand feet by ten, and that quotient again by ten, and so on as often as we please; that there never needs be an end to the subdivisions of the distance, nor consequently to those of the time in which it is performed. But an unlimited number of subdivisions may be made of that which is itself limited. The argument proves no other infinity of duration than may be embraced within five minutes. As long as the five minutes are not expired, what remains of them may be divided by ten, and again by ten, as often as we like, which is perfectly compatible with their being only five minutes altogether. It proves, in short, that to pass through this finite space requires a time which is infinitely divisible, but not an infinite time: the confounding of which distinction Hobbes had already seen to be the gist of the fallacy.

The following ambiguities of the word *right* (in addition to the obvious and familiar one of *a* right and the *adjective* right) are extracted from a forgotten paper of my own, in a periodical:—

"Speaking morally, you are said to have a right to do a thing, if all persons are morally bound not to hinder you from doing it. But, in another sense, to have a right to do a thing is the opposite of having *no* right to do it, *i.e.* of being under a moral obligation to forbear doing it. In this sense, to say that you have a right to do a thing, means that you may do it without any breach of duty on your part; that other persons not only ought not to hinder you, but have no cause to think worse of you for doing it. This is a perfectly distinct proposition from the preceding. The right which you have by virtue of a duty incumbent upon other persons, is obviously quite a different thing from a right consisting in the absence of any duty incumbent upon yourself. Yet the two things are perpetually confounded. Thus a man will say he has a right to publish his opinions; which may be true in this sense, that it would be a breach of duty in any other person to interfere and prevent the publication: but he assumes thereupon, that in publishing his opinions, he himself violates no duty; which may either be true or false, depending, as it does, on his having taken due pains to satisfy himself, first, that the opinions are true, and next, that their publication in this manner, and at this particular juncture, will probably be beneficial to the interests of truth on the whole.

"The second ambiguity is that of confounding a right of any kind, with a right to enforce that right by resisting or punishing a violation of it. People will say, for example, that they have a right to good government, which is undeniably true, it being the moral duty of their governors to govern them well. But in granting this, you are supposed to have admitted their right or liberty to turn out their governors, and perhaps to punish them, for having failed in the performance of this duty; which, far from being the same thing, is by no means universally true, but depends on an immense number of varying circumstances," requiring to be conscientiously weighed before adopting or acting on such a resolution. This last example is (like others which have been cited) a case of fallacy within fallacy; it involves not only the second of the two ambiguities pointed out, but the first likewise.

One not unusual form of the Fallacy of Ambiguous Terms, is known technically as the Fallacy of Composition and Division: when the same term is collective in the premises, distributive in the conclusion, or *vice versâ*: or when the middle term is collective in one premise, distributive in the other. As if one were to say (I quote from Archbishop Whately) "All the angles of a triangle are equal to two right angles: ABC is an angle of a triangle; therefore ABC is equal to two right angles. . . . . There is no fallacy more common, or more likely to deceive, than the one now before us. The form in which it is most usually employed is to establish some truth, separately, concerning *each single* member of a certain class, and thence to infer the same of the *whole collectively*." As in the argument one sometimes hears, to prove that the world could do without great men. If Columbus (it is said) had never lived, America would still have been discovered, at most only a few years later; if Newton had never lived, some other person would have discovered the law of gravitation; and so forth. Most true: these things would have been

done, but in all probability not until some one had again
been found with the qualities of Columbus or Newton.
Because any one great man might have had his place supplied
by other great men, the argument concludes that all great
men could have been dispensed with. The term "great
men" is distributive in the premises and collective in the
conclusion.

"Such also is the fallacy which probably operates on most
adventurers in lotteries; *e. g.* 'the gaining of a high prize is
no uncommon occurrence; and what is no uncommon occurrence may reasonably be expected; therefore the gaining of a
high prize may reasonably be expected:' the conclusion when
applied to the individual (as in practice it is) must be understood in the sense of 'reasonably expected *by a certain individual*;' therefore for the major premise to be true, the middle
term must be understood to mean, 'no uncommon occurrence
to some one *particular* person;' whereas for the minor (which
has been placed first) to be true, you must understand it of
'no uncommon occurrence to *some one or other*;' and thus you
will have the Fallacy of Composition.

"This is a Fallacy with which men are extremely apt to
deceive *themselves*; for when a multitude of particulars are
presented to the mind, many are too weak or too indolent to
take a comprehensive view of them, but confine their attention
to each single point, by turns; and then decide, infer, and act,
accordingly: *e. g.* the imprudent spendthrift, finding that he is
able to afford this, *or* that, *or* the other expense, forgets that
all of them together will ruin him." The debauchee destroys
his health by successive acts of intemperance, because no *one*
of those acts would be of itself sufficient to do him any serious
harm. A sick person reasons with himself, "one, and another,
and another, of my symptoms, do not prove that I have a fatal
disease;" and practically concludes that all taken together do
not prove it.

§ 2. We have now sufficiently exemplified one of the
principal Genera in this Order of Fallacies; where, the source
of error being the ambiguity of terms, the premises are ver-

bally what is required to support the conclusion, but not
really so. In the second great Fallacy of Confusion they are
neither verbally nor really sufficient, though, from their
multiplicity and confused arrangement, and still oftener
from defect of memory, they are not seen to be what they
are. The fallacy I mean is that of Petitio Principii, or
begging the question; including the more complex and not
uncommon variety of it, which is termed Reasoning in a
Circle.

Petitio Principii, as defined by Archbishop Whately, is the
fallacy "in which the premise either appears manifestly to be
the same as the conclusion, or is actually proved from the
conclusion, or is such as would naturally and properly so be
proved." By the last clause I presume is meant, that it is
not susceptible of any other proof; for otherwise, there would
be no fallacy. To deduce from a proposition, propositions
from which it would itself more naturally be deduced, is often
an allowable deviation from the usual didactic order; or at most,
what, by an adaptation of a phrase familiar to mathematicians,
may be called a logical *inelegance*.*

The employment of a proposition to prove that on which
it is itself dependent for proof, by no means implies the degree
of mental imbecility which might at first be supposed. The
difficulty of comprehending how this fallacy could possibly be
committed, disappears when we reflect that all persons, even
the instructed, hold a great number of opinions without
exactly recollecting how they came by them. Believing that
they have at some former time verified them by sufficient
evidence, but having forgotten what the evidence was, they
may easily be betrayed into deducing from them the very propositions which are alone capable of serving as premises for

* In his later editions, Archbishop Whately confines the name of Petitio
Principii "to those cases in which one of the premises either is manifestly the
same in sense with the conclusion, or is actually proved from it, or is such as
the persons you are addressing are not likely to know, or to admit,' except as an
inference from the conclusion: as, *e.g.* if any one should infer the authenticity
of a certain history, from its recording such and such facts, the reality of which
rests on the evidence of that history."

their establishment. "As if," says Archbishop Whately, "one should attempt to prove the being of a God from the authority of Holy Writ;" which might easily happen to one with whom both doctrines, as fundamental tenets of his religious creed, stand on the same ground of familiar and traditional belief.

Arguing in a circle, however, is a stronger case of the fallacy, and implies more than the mere passive reception of a premise by one who does not remember how it is to be proved. It implies an actual attempt to prove two propositions reciprocally from one another; and is seldom resorted to, at least in express terms, by any person in his own speculations, but is committed by those who, being hard pressed by an adversary, are forced into giving reasons for an opinion of which, when they began to argue, they had not sufficiently considered the grounds. As in the following example from Archbishop Whately: "Some mechanicians attempt to prove (what they ought to lay down as a probable but doubtful hypothesis*) that every particle of matter gravitates equally: 'why?' 'because those bodies which contain more particles ever gravitate more strongly, *i.e.* are heavier:' 'but, (it may be urged,) those which are heaviest are not always more bulky;' 'no, but they contain more particles, though more closely condensed:' 'how do you know that?' 'because they are heavier:' 'how does that prove it?' 'because all particles of matter gravitating equally, that mass which is specifically the heavier must needs have the more of them in the same space.'" It appears to me that the fallacious reasoner, in his private thoughts, would not be likely to proceed beyond the first step. He would acquiesce in the sufficiency of the reason first given, "bodies which contain more particles are heavier." It is when he finds this questioned, and is called upon to

---

* No longer even a probable hypothesis, since the establishment of the atomic theory; it being now certain that the integral particles of different substances gravitate unequally. It is true that these particles, though real *minima* for the purposes of chemical combination, may not be the ultimate particles of the substance; and this doubt alone renders the hypothesis admissible, even as an hypothesis.

prove it, without knowing how, that he tries to establish his premise by supposing proved what he is attempting to prove by it. The most effectual way, in fact, of exposing a Petitio Principii, when circumstances allow of it, is by challenging the reasoner to prove his premises; which if he attempts to do, he is necessarily driven into arguing in a circle.

It is not uncommon, however, for thinkers, and those not of the lowest description, to be led, even in their own thoughts, not indeed into formally proving each of two propositions from the other, but into admitting propositions which can only be so proved. In the preceding example the two together form a complete and consistent, though hypothetical, explanation of the facts concerned. And the tendency to mistake mutual coherency for truth; to trust one's safety to a strong chain though it has no point of support; is at the bottom of much which, when reduced to the strict forms of argumentation, can exhibit itself no otherwise than as reasoning in a circle. All experience bears testimony to the enthralling effect of neat concatenation in a system of doctrines, and the difficulty with which people admit the persuasion that anything which holds so well together can possibly fall.

Since every case where a conclusion which can only be proved from certain premises is used for the proof of those premises, is a case of *petitio principii*, that fallacy includes a very great proportion of all incorrect reasoning. It is necessary, for completing our view of the fallacy, to exemplify some of the disguises under which it is accustomed to mask itself, and to escape exposure.

A proposition would not be admitted by any person in his senses as a corollary from itself, unless it were expressed in language which made it seem different. One of the commonest modes of so expressing it, is to present the proposition itself in abstract terms, as a proof of the same proposition expressed in concrete language. This is a very frequent mode, not only of pretended proof, but of pretended explanation; and is parodied when Molière makes

one of his absurd physicians say, "l'opium endormit
parcequ'il a une vertu soporifique," or, in the equivalent
doggrel,

Mihi à docto doctore,
Domandatur causam et rationem quare
Opium facit dormire.
A quoi respondeo,
Quia est in eo
Virtus dormitiva,
Cujus est natura
Sensus assoupire.

The words Nature and Essence are grand instruments of
this mode of begging the question. As in the well-known
argument of the scholastic theologians, that the mind thinks
always, because the *essence* of the mind is to think. Locke
had to point out, that if by essence is here meant some property which must manifest itself by actual exercise at all
times, the premise is a direct assumption of the conclusion;
while if it only means that to think is the distinctive property
of a mind, there is no connexion between the premise and the
conclusion, since it is not necessary that a distinctive property
should be perpetually in action.

The following is one of the modes in which these abstract
terms, Nature and Essence, are used as instruments of this
fallacy. Some particular properties of a thing are selected,
more or less arbitrarily, to be termed its nature or essence;
and when this has been done, these properties are supposed to
be invested with a kind of indefeasibleness; to have become
paramount to all the other properties of the thing, and incapable of being prevailed over or counteracted by them. As
when Aristotle, in a passage already cited, "decides that
there is no void on such arguments as this: in a void there
could be no difference of up and down; for as in nothing
there are no differences, so there are none in a privation or
negation; but a void is merely a privation or negation of
matter; therefore, in a void, bodies could not move up and
down, which it is in their *nature* to do."* In other words;
it is in the *nature* of bodies to move up and down, *ergo* any

* *Hist. Ind. Sc.* i. 34.

physical fact which supposes them not so to move, cannot be authentic. This mode of reasoning, by which a bad generalization is made to overrule all facts which contradict it, is *petitio principii* in one of its most palpable forms.

None of the modes of assuming what should be proved are in more frequent use than what are termed by Bentham "question-begging appellatives;" names which beg the question under the disguise of stating it. The most potent of these are such as have a laudatory or vituperative character. For instance, in politics, the word Innovation. The dictionary meaning of this term being merely "a change to something new," it is difficult for the defenders even of the most salutary improvement to deny that it is an innovation; yet the word having acquired in common usage a vituperative connotation in addition to its dictionary meaning, the admission is always construed as a large concession to the disadvantage of the thing proposed.

The following passage from the argument in refutation of the Epicureans, in the second book of Cicero *de Finibus*, affords a fine example of this sort of fallacy. "Et quidem illud ipsum non nimium probo (et tantum patior) philosophum loqui de cupiditatibus finiendis. An potest cupiditas finiri? tollenda est, atque extrahenda radicitus. Quis est enim, in quo sit cupiditas, quin recte cupidus dici possit? Ergo et avarus erit, sed finite: adulter, verum habebit modum: et luxuriosus eodem modo. Qualis ista philosophia est, quæ non interitum afferat pravitatis, sed sit contenta mediocritate vitiorum?" The question was, whether certain desires, when kept within bounds, are vices or not; and the argument decides the point by applying to them a word (*cupiditas*) which *implies* vice. It is shown, however, in the remarks which follow, that Cicero did not intend this as a serious argument, but as a criticism on what he deemed an inappropriate expression. "Rem ipsam prorsus probo: elegantiam desidero. Appellet hæc *desideria naturæ;* cupiditatis nomen servet alio," &c. But many persons, both ancient and modern, have employed this, or something equivalent to it, as a real and conclusive argument. We may remark that the passage respecting *cupiditas* and *cupidus* is also an example of another fallacy already noticed, that of Paronymous Terms.

Many more of the arguments of the ancient moralists, and especially of the Stoics, fall within the definition of Petitio Principii. In the *De Finibus*, for example, which I continue to quote as being probably the best extant exemplification at once of the doctrines and the methods of the schools of philosophy existing at that time; of what value as arguments are such pleas as those of Cato in the third book: That if virtue were not happiness, it could not be a thing to *boast* of: That if death or pain were evils, it would be impossible not to fear them, and it could not, therefore, be laudable to despise them, &c. In one way of viewing these arguments, they may be regarded as appeals to the authority of the general sentiment of mankind, which had stamped its approval upon certain actions and characters by the phrases referred to; but that such could have been the meaning intended is very unlikely, considering the contempt of the ancient philosophers for vulgar opinion. In any other sense they are clear cases of Petitio Principii, since the word laudable, and the idea of boasting, imply principles of conduct; and practical maxims can only be proved from speculative truths, namely from the properties of the subject matter, and cannot, therefore, be employed to prove those properties. As well might it be argued that a government is good because we ought to support it, or that there is a God because it is our duty to pray to him.

It is assumed by all the disputants in the *De Finibus* as the foundation of the inquiry into the *summum bonum*, that " sapiens semper beatus est." Not simply that wisdom gives the best chance of happiness, or that wisdom consists in knowing what happiness is, and by what things it is promoted; these propositions would not have been enough for them:—but that the sage always is, and must of necessity be, happy. The idea that wisdom could be consistent with unhappiness, was always rejected as inadmissible: the reason assigned by one of the interlocutors, near the beginning of the third book, being, that if the wise could be unhappy, there was little use in pursuing wisdom. But by unhappiness they did not mean pain or suffering; to that, it was granted that the wisest person was liable in common with others: he was happy, because in possessing wisdom he had the most valuable of all possessions, the most to be sought and prized of all things, and to possess the most valuable thing was to be the most happy. By laying it down, therefore, at the commencement of the inquiry, that the sage must be happy, the disputed question respecting the *summum bonum* was in fact begged; with the further assumption, that pain and suffering, so far as they can coexist with wisdom, are not unhappiness, and are no evil.

The following are additional instances of Petitio Principii, under more or less of disguise.

Plato, in the *Sophistes*, attempts to prove that things may exist which are incorporeal, by the argument that justice and wisdom are incorporeal, and justice and wisdom must be something. Here, if by *something* be meant, as Plato did in fact mean, a thing capable of existing in and by itself, and not as a quality of some other thing, he begs the question in asserting that justice and wisdom must be something: if he means anything else, his conclusion is not proved. This fallacy might also be classed under ambiguous middleterm: *something*, in the one premise, meaning some substance, in the other merely some object of thought, whether substance or attribute.

It was formerly an argument employed in proof of what is now no longer a popular doctrine, the infinite divisibility of matter, that every portion of matter, however small, must at least have an upper and an under surface. Those who used this argument did not see that it assumed the very point in dispute, the impossibility of arriving at a minimum of thickness; for if there be a minimum, its upper and under surface will of course be one: it will be itself a surface, and no more. The argument owes its very considerable plausibility to this, that the premise does actually seem more obvious than the conclusion, though really identical with it. As expressed in the premise, the proposition appeals directly and in concrete language to the incapacity of the human imagination for conceiving a minimum. Viewed in this light, it becomes a case of the *à priori* fallacy or natural prejudice, that whatever cannot be conceived cannot exist. Every fallacy of Confusion (it is almost unnecessary to repeat) will, if cleared up, become a fallacy of some other sort; and it will be found of deductive or ratiocinative fallacies generally, that when they mislead, there is mostly, as in this case, a fallacy of some other description lurking under them, by virtue of which chiefly it is that the verbal juggle, which is the outside or body of this kind of fallacy, passes undetected.

Euler's Algebra, a book otherwise of great merit, but full, to overflowing, of logical errors in respect to the foundation of the science, contains the following argument to prove that *minus* multiplied by *minus* gives *plus*, a doctrine the opprobrium of all mere mathematicians, and which Euler had not a glimpse of the true method of proving. He says, *minus* multiplied by *minus* cannot give *minus*; for *minus* multiplied by *plus* gives *minus*, and *minus* multiplied by *minus* cannot give the same product as *minus* multiplied by *plus*. Now one is obliged to ask, why minus multiplied by minus must give any product at all? and if it does, why its product cannot be the same as that of minus multiplied by plus? for this would seem, at the first glance, not more absurd than that minus by minus should give the same as plus by plus, the proposition which Euler prefers to it. The premise requires proof, as much as the conclusion: nor can it be proved, except by that more comprehensive view of the nature of multiplication, and of algebraic processes in general, which would also supply a far better proof of the mysterious doctrine which Euler is here endeavouring to demonstrate.

A striking instance of reasoning in a circle is that of some ethical writers, who first take for their standard of moral truth what, being the general, they deem to be the natural or instinctive sentiments and perceptions of mankind, and then explain away the numerous instances of divergence from their assumed standard, by representing them as cases in which the perceptions are unhealthy. Some particular mode of conduct or feeling is affirmed to be *unnatural;* why? because it is abhorrent to the universal and natural sentiments of mankind. Finding no such sentiment in yourself, you question the fact; and the answer is (if your antagonist is polite), that you are an exception, a peculiar case. But neither (say you) do I find in the people of some other country, or of some former age, any such feeling of abhorrence; " ay, but their feelings were sophisticated and unhealthy."

One of the most notable specimens of reasoning in a circle is the doctrine of Hobbes, Rousseau, and others, which rests the obligations by which human beings are bound as members of society, on a supposed social compact. I wave the consideration of the fictitious nature of the compact itself; but when Hobbes, through the whole Leviathan, elaborately deduces the obligation of obeying the sovereign, not from the necessity or utility of doing so, but from a promise supposed to have been made by our ancestors, on renouncing savage life and agreeing to establish political society, it is impossible not to retort by the question, why are we bound to keep a promise made for us by others? or why bound to keep a promise at all? No satisfactory ground can be assigned for the obligation, except the mischievous consequences of the absence of faith and mutual confidence among mankind. We are, therefore, brought round to the interests of society, as the ultimate ground of the obligation of a promise; and yet those interests are not admitted to be a sufficient justification for the existence of government and law. Without a promise it is thought that we should not be bound to that which is implied in all modes of living in society, namely, to yield a general obedience to the laws therein established; and so necessary is the promise deemed, that if none has actually been made, some additional safety is supposed to be given to the foundations of society by feigning one.

§ 3. Two principal subdivisions of the class of Fallacies of Confusion having been disposed of; there remains a third, in which the confusion is not, as in the Fallacy of Ambiguity, in misconceiving the import of the premises, nor, as in *Petitio Principii*, in forgetting what the premises are, but in mistaking the conclusion which is to be proved. This is the fallacy of *Ignoratio Elenchi*, in the widest sense of the phrase; also called by Archbishop Whately the Fallacy of Irrelevant Conclusion. His examples and remarks are highly worthy of citation.

"Various kinds of propositions are, according to the occasion, substituted for the one of which proof is required : sometimes the particular for the universal; sometimes a proposition with different terms; and various are the contrivances employed to effect and to conceal this substitution, and to make the conclusion which the sophist has drawn, answer practically the same purpose as the one he ought to have established. We say, 'practically the same purpose,' because it will very often happen that some *emotion* will be excited, some sentiment impressed on the mind, (by a dexterous employment of this fallacy), such as shall bring men into the *disposition* requisite for your purpose; though they may not have assented to, or even stated distinctly in their own minds, the *proposition* which it was your business to establish. Thus if a sophist has to defend one who has been guilty of some *serious* offence, which he wishes to extenuate, though he is unable distinctly to prove that it is not such, yet if he can succeed in *making the audience laugh* at some casual matter, he has gained practically the same point. So also if any one has pointed out the extenuating circumstances in some particular case of offence, so as to show that it differs widely from the generality of the same class, the sophist if he find himself unable to disprove these circumstances, may do away the force of them, by simply *referring the action to that very class,* which no one can deny that it belongs to, and the very name of which will excite a feeling of disgust sufficient to counteract the extenuation ; *e. g.* let it be a case of peculation, and that many *mitigating* circumstances have been brought forward which cannot be denied; the sophistical opponent will reply, 'Well, but after all, the man is a *rogue,* and there is an end of it ;' now in reality this was (by hypothesis) never the question ; and the mere assertion of what was never denied, *ought* not, in fairness, to be regarded as decisive: but, practically, the odiousness of the word, arising in great measure from the association of those very circumstances which belong to most of the class, but which we have supposed to be *absent* in *this particular* instance, excites precisely that feeling of disgust, which in effect destroys the force of the defence. In like manner we may refer to this head all cases of improper appeal to the passions, and everything else which is mentioned by Aristotle as extraneous to the matter in hand (ἔξω τοῦ πράγματος)."

Again, "instead of proving that 'this prisoner has committed an atrocious fraud,' you prove that the fraud he is accused of is atrocious: instead of proving (as in the well-known tale of Cyrus and the two coats) that the taller boy had a right to force the other boy to exchange coats with him, you prove that the exchange would have been advantageous to both: instead of proving that the poor ought to be relieved in this way rather than in that, you prove that the poor ought to be relieved: instead of proving that the irrational agent—whether a brute or a madman—can never be deterred from any act by apprehension of punishment (as for instance a dog from sheep-biting, by fear of being beaten), you prove that the beating of one dog does not operate as an *example* to *other* dogs, &c.

"It is evident that *ignoratio elenchi* may be employed as well for the apparent refutation of your opponent's proposition, as for the apparent establishment of your own; for it is substantially the same thing, to prove what was not denied or to disprove what was not asserted. The latter practice is not less common, and it is more offensive, because it frequently amounts to a personal affront, in attributing to a person, opinions, &c., which he perhaps holds in abhorrence. Thus, when in a discussion one party vindicates, on the ground of general expediency, a particular instance of resistance to government in a case of intolerable oppression, the opponent may gravely maintain, 'that we ought not to do evil that good may come;' a proposition which of course had never been denied, the point in dispute being, 'whether

resistance in this particular case *were* doing evil or not.'   Or
again, by way of disproving the assertion of the right of
private judgment in religion, one may hear a grave argument
to prove that ' it is impossible every one can be *right in his
judgment.*' "

The works of controversial writers are seldom free from
this fallacy. The attempts, for instance, to disprove the
population doctrines of Malthus, have been mostly cases of
ignoratio elenchi. Malthus has been supposed to be refuted
if it could be shown that in some countries or ages population has been nearly stationary; as if he had asserted that
population always increases in a given ratio, or had not
expressly declared that it increases only in so far as it is not
restrained by prudence, or kept down by poverty and disease.
Or, perhaps, a collection of facts is produced to prove that
in some one country the people are better off with a dense
population than they are in another country with a thin one;
or that the people have become more numerous and better
off at the same time. As if the assertion were that a dense
population could not possibly be well off: as if it were not part
of the very doctrine, and essential to it, that where there is a
more abundant capital there may be a greater population
without any increase of poverty, or even with a diminution
of it.

The favourite argument against Berkeley's theory of the
non-existence of matter, and the most popularly effective, next
to a " grin"*—an argument, moreover, which is not confined to " coxcombs," nor to men like Samuel Johnson,
whose greatly overrated ability certainly did not lie in the
direction of metaphysical speculation, but is the stock argument of the Scotch school of metaphysicians—is a palpable
ignoratio elenchi. The argument is perhaps as frequently
expressed by gesture as by words, and one of its commonest
forms consists in knocking a stick against the ground. This
short and easy confutation overlooks the fact, that in denying
matter, Berkeley did not deny anything to which our senses

---

* " And coxcombs vanquish Berkeley with a grin."

bear witness, and therefore cannot be answered by any appeal to them. His scepticism related to the supposed substratum, or hidden cause of the appearances perceived by our senses: the evidence of which, whatever may be thought of its conclusiveness, is certainly not the evidence of sense. And it will always remain a signal proof of the want of metaphysical profundity of Reid, Stewart, and, I am sorry to add, of Brown, that they should have persisted in asserting that Berkeley, if he believed his own doctrine, was bound to walk into the kennel, or run his head against a post. As if persons who do not recognise an occult cause of their sensations, could not possibly believe that a fixed order subsists among the sensations themselves. Such a want of comprehension of the distinction between a thing and its sensible manifestation, or, in metaphysical language, between the noumenon and the phenomenon, would be impossible to even the dullest disciple of Kant or Coleridge.

It would be easy to add a greater number of examples of this fallacy, as well as of the others which I have attempted to characterize. But a more copious exemplification does not seem to be necessary; and the intelligent reader will have little difficulty in adding to the catalogue from his own reading and experience. We shall therefore here close our exposition of the general principles of logic, and proceed to the supplementary inquiry which is necessary to complete our design.

BOOK VI.

—

ON THE LOGIC OF THE MORAL SCIENCES.

"Si l'homme peut prédire, avec une assurance presque entière, les phénomènes dont il connaît les lois ; si lors même qu'elles lui sont inconnues, il peut,
d'après l'expérience, prévoir avec une grande probabilité les événemens de
l'avenir ; pourquoi regarderait-on comme une entreprise chimérique, celle de
tracer avec quelque vraisemblance le tableau des destinées futures de l'espèce
humaine, d'après les résultats de son histoire ? Le seul fondement de croyance
dans les sciences naturelles, est cette idée, que les lois générales, connues ou
ignorées, qui règlent les phénomènes de l'univers, sont nécessaires et constantes ;
et par quelle raison ce principe serait-il moins vrai pour le développement des
facultés intellectuelles et morales de l'homme, que pour les autres opérations de
la nature ? Enfin, puisque des opinions formées d'après l'expérience . . . sont
la seule règle de la conduite des hommes les plus sages, pourquoi interdirait-on
au philosophe d'appuyer ses conjectures sur cette même base, pourvu qu'il ne
leur attribue pas une certitude supérieure à celle qui peut naître du nombre,
de la constance, de l'exactitude des observations ?"—CONDORCET, *Esquisse d'un
Tableau Historique des Progrès de l'Esprit Humain.*

CHAPTER I.

INTRODUCTORY REMARKS.

§ 1. PRINCIPLES of Evidence and Theories of Method are not to be constructed à *priori*. The laws of our rational faculty, like those of every other natural agency, are only learnt by seeing the agent at work. The earlier achievements of science were made without the conscious observance of any Scientific Method; and we should never have known by what process truth is to be ascertained, if we had not previously ascertained many truths. But it was only the easier problems which could be thus resolved: natural sagacity, when it tried its strength against the more difficult ones, either failed altogether, or if it succeeded here and there in obtaining a solution, had no sure means of convincing others that its solution was correct. In scientific investigation, as in all other works of human skill, the way of obtaining the end is seen as it were instinctively by superior minds in some comparatively simple case, and is then, by judicious generalization, adapted to the variety of complex cases. We learn to do a thing in difficult circumstances, by attending to the manner in which we have spontaneously done the same thing in easier ones.

This truth is exemplified by the history of the various branches of knowledge which have successively, in the ascending order of their complication, assumed the character of sciences; and will doubtless receive fresh confirmation from those, of which the final scientific constitution is yet to come, and which are still abandoned to the uncertainties of vague and popular discussion. Although several other sciences have emerged from this state at a comparatively recent date, none now remain in it except those which relate to man himself, the most complex and most difficult subject of study on which the human mind can be engaged.

Concerning the physical nature of man, as an organized being,—though there is still much uncertainty and much controversy, which can only be terminated by the general acknowledgment and employment of stricter rules of induction than are commonly recognised,—there is, however, a considerable body of truths which all who have attended to the subject consider to be fully established; nor is there now any radical imperfection in the method observed in this department of science by its most distinguished modern teachers. But the laws of Mind, and, in even a greater degree, those of Society, are so far from having attained a similar state of even partial recognition, that it is still a controversy whether they are capable of becoming subjects of science in the strict sense of the term: and among those who are agreed on this point, there reigns the most irreconcileable diversity on almost every other. Here, therefore, if anywhere, the principles laid down in the preceding Books may be expected to be useful.

If, on matters so much the most important with which human intellect can occupy itself, a more general agreement is ever to exist among thinkers; if what has been pronounced "the proper study of mankind" is not destined to remain the only subject which Philosophy cannot succeed in rescuing from Empiricism; the same process through which the laws of many simpler phenomena have by general acknowledgment been placed beyond dispute, must be consciously and deliberately applied to those more difficult inquiries. If there are some subjects on which the results obtained have finally received the unanimous assent of all who have attended to the proof, and others on which mankind have not yet been equally successful; on which the most sagacious minds have occupied themselves from the earliest date, and have never succeeded in establishing any considerable body of truths, so as to be beyond denial or doubt; it is by generalizing the methods successfully followed in the former inquiries, and adapting them to the latter, that we may hope to remove this blot on the face of science. The remaining chapters are an endeavour to facilitate this most desirable object.

§ 2. In attempting this, I am not unmindful how little can be done towards it in a mere treatise on Logic, or how vague and unsatisfactory all precepts of Method must necessarily appear, when not practically exemplified in the establishment of a body of doctrine. Doubtless, the most effectual mode of showing how the sciences of Ethics and Politics may be constructed, would be to construct them: a task which, it needs scarcely be said, I am not about to undertake. But even if there were no other examples, the memorable one of Bacon would be sufficient to demonstrate, that it is sometimes both possible and useful to point out the way, though without being oneself prepared to adventure far into it. And if more were to be attempted, this at least is not a proper place for the attempt.

In substance, whatever can be done in a work like this for the Logic of the Moral Sciences, has been or ought to have been accomplished in the five preceding Books; to which the present can be only a kind of supplement or appendix, since the methods of investigation applicable to moral and social science must have been already described, if I have succeeded in enumerating and characterizing those of science in general. It remains, however, to examine which of those methods are more especially suited to the various branches of moral inquiry; under what peculiar facilities or difficulties they are there employed; how far the unsatisfactory state of those inquiries is owing to a wrong choice of methods, how far to want of skill in the application of right ones; and what degree of ultimate success may be attained or hoped for, by a better choice or more careful employment of logical processes appropriate to the case. In other words, whether moral sciences exist, or can exist; to what degree of perfection they are susceptible of being carried; and by what selection or adaptation of the methods brought to view in the previous part of this work, that degree of perfection is attainable.

At the threshold of this inquiry we are met by an objection, which, if not removed, would be fatal to the attempt to treat human conduct as a subject of science. Are the actions of human beings, like all other natural events, subject to invariable laws? Does that constancy of causation, which is the foundation of every scientific theory of successive phenomena, really obtain among them? This is often denied; and for the sake of systematic completeness, if not from any very urgent practical necessity, the question should receive a deliberate answer in this place. We shall devote to the subject a chapter apart.

CHAPTER II.

§ 1. THE question, whether the law of causality applies in the same strict sense to human actions as to other phenomena, is the celebrated controversy concerning the freedom of the will: which, from at least as far back as the time of Pelagius, has divided both the philosophical and the religious world. The affirmative opinion is commonly called the doctrine of Necessity, as asserting human volitions and actions to be necessary and inevitable. The negative maintains that the will is not determined, like other phenomena, by antecedents, but determines itself; that our volitions are not, properly speaking, the effects of causes, or at least have no causes which they uniformly and implicitly obey.

I have already made it sufficiently apparent that the former of these opinions is that which I consider the true one; but the misleading terms in which it is often expressed, and the indistinct manner in which it is usually apprehended, have both obstructed its reception, and perverted its influence when received. The metaphysical theory of free will, as held by philosophers, (for the practical feeling of it, common in a greater or less degree to all mankind, is in no way inconsistent with the contrary theory,) was invented because the supposed alternative of admitting human actions to be *necessary*, was deemed inconsistent with every one's instinctive consciousness, as well as humiliating to the pride and even degrading to the moral nature of man. Nor do I deny that the doctrine, as sometimes held, is open to these imputations; for the misapprehension in which I shall be able to show that they originate, unfortunately is not confined to the opponents of the doctrine, but participated in by many, perhaps we might say by most, of its supporters.

§ 2. Correctly conceived, the doctrine called Philosophical Necessity is simply this: that, given the motives which are present to an individual's mind, and given likewise the character and disposition of the individual, the manner in which he will act might be unerringly inferred: that if we knew the person thoroughly, and knew all the inducements which are acting upon him, we could foretell his conduct with as much certainty as we can predict any physical event. This proposition I take to be a mere interpretation of universal experience, a statement in words of what every one is internally convinced of. No one who believed that he knew thoroughly the circumstances of any case, and the characters of the different persons concerned, would hesitate to foretell how all of them would act. Whatever degree of doubt he may in fact feel, arises from the uncertainty whether he really knows the circumstances, or the character of some one or other of the persons, with the degree of accuracy required: but by no means from thinking that if he did know these things, there could be any uncertainty what the conduct would be. Nor does this full assurance conflict in the smallest degree with what is called our feeling of freedom. We do not feel ourselves the less free, because those to whom we are intimately known are well assured how we shall will to act in a particular case. We often, on the contrary, regard the doubt what our conduct will be, as a mark of ignorance of our character, and sometimes even resent it as an imputation. The religious metaphysicians who have asserted the freedom of the will, have always maintained it to be consistent with divine foreknowledge of our actions: and if with divine, then with any other foreknowledge. We may be free, and yet another may have reason to be perfectly certain what use we shall make of our freedom. It is not, therefore, the doctrine that our volitions and actions are invariable consequents of our antecedent states of mind, that is either contradicted by our consciousness, or felt to be degrading.

But the doctrine of causation, when considered as obtaining between our volitions and their antecedents, is almost universally conceived as involving more than this. Many do not believe, and very few practically feel, that there is nothing in causation but invariable, certain, and unconditional sequence. There are few to whom mere constancy of succession appears a sufficiently stringent bond of union for so peculiar a relation as that of cause and effect. Even if the reason repudiates, the imagination retains, the feeling of some more intimate connexion, of some peculiar tie, or mysterious constraint exercised by the antecedent over the consequent. Now this it is which, considered as applying to the human will, conflicts with our consciousness, and revolts our feelings. We are certain that, in the case of our volitions, there is not this mysterious constraint. We know that we are not compelled, as by a magical spell, to obey any particular motive. We feel, that if we wished to prove that we have the power of resisting the motive, we could do so, (that wish being, it needs scarcely be observed, a *new antecedent;*) and it would be humiliating to our pride, and (what is of more importance) paralysing to our desire of excellence if we thought otherwise. But neither is any such mysterious compulsion now supposed, by the best philosophical authorities, to be exercised by any other cause over its effect. Those who think that causes draw their effects after them by a mystical tie, are right in believing that the relation between volitions and their antecedents is of another nature. But they should go farther, and admit that this is also true of all other effects and their antecedents. If such a tie is considered to be involved in the word necessity, the doctrine is not true of human actions; but neither is it then true of inanimate objects. It would be more correct to say that matter is not bound by necessity than that mind is so.

That the free-will metaphysicians, being mostly of the school which rejects Hume's and Brown's analysis of Cause and Effect, should miss their way for want of the light which that analysis affords, cannot surprise us. The wonder is, that the necessarians, who usually admit that philosophical theory, should in practice equally lose sight of it. The very same misconception of the doctrine called Philosophical Necessity, which prevents the opposite party from recognising its truth, I believe to exist more or less obscurely in the minds of most necessarians, however they may in words disavow it. I am much mistaken if they habitually feel that the necessity which they recognise in actions is but uniformity of order, and capability of being predicted. They have a feeling as if there were at bottom a stronger tie between the volitions and their causes: as if, when they asserted that the will is governed by the balance of motives, they meant something more cogent than if they had only said, that whoever knew the motives, and our habitual susceptibilities to them, could predict how we should will to act. They commit, in opposition to their own scientific system, the very same mistake which their adversaries commit in obedience to theirs; and in consequence do really in some instances suffer those depressing consequences, which their opponents erroneously impute to the doctrine itself.

§ 8. I am inclined to think that this error is almost wholly an effect of the associations with a word; and that it would be prevented, by forbearing to employ, for the expression of the simple fact of causation, so extremely inappropriate a term as Necessity. That word, in its other acceptations, involves much more than mere uniformity of sequence: it implies irresistibleness. Applied to the will, it only means that the given cause will be followed by the effect, subject to all possibilities of counteraction by other causes: but in common use it stands for the operation of those causes exclusively, which are supposed too powerful to be counteracted at all. When we say that all human actions take place of necessity, we only mean that they will certainly happen if nothing prevents:—when we say that dying of want, to those who cannot get food, is a necessity, we mean that it will certainly happen whatever may be done to prevent it. The application of the same term to the agencies on which human actions depend, as is used to express those agencies of nature which are really uncontrollable, cannot fail, when habitual, to create a feeling of uncontrollableness in the former also. This however is a mere illusion. There are physical sequences which we call necessary, as death for want of food or air; there are others which, though as much cases of causation as the former, are not said to be necessary, as death from poison, which an antidote, or the use of the stomach-pump, will sometimes avert. It is apt to be forgotten by people's feelings, even if remembered by their understandings, that human actions are in this last predicament: they are never (except in some cases of mania) ruled by any one motive with such absolute sway, that there is no room for the influence of any other. The causes, therefore, on which action depends, are never uncontrollable; and any given effect is only necessary provided that the causes tending to produce it are not controlled. That whatever happens, could not have happened otherwise unless something had taken place which was capable of preventing it, no one surely needs hesitate to admit. But to call this by the name necessity is to use the term in a sense so different from its primitive and familiar meaning, from that which it bears in the common occasions of life, as to amount almost to a play upon words. The associations derived from the ordinary sense of the term will adhere to it in spite of all we can do: and though the doctrine of Necessity, as stated by most who hold it, is very remote from fatalism, it is probable that most necessarians are fatalists, more or less, in their feelings.

A fatalist believes, or half believes (for nobody is a consistent fatalist), not only that whatever is about to happen, will be the infallible result of the causes which produce it, (which is the true necessarian doctrine,) but moreover that there is no use in struggling against it; that it will happen however we may strive to prevent it. Now, a necessarian, believing that our actions follow from our characters, and that our characters follow from our organization, our education, and our circumstances, is apt to be, with more or less of consciousness on his part, a fatalist as to his own actions, and to believe that his nature is such, or that his education and circumstances have so moulded his character, that nothing can now prevent him from feeling and acting in a particular way, or at least that no effort of his own can hinder it. In the words of the sect which in our own day has most perseveringly inculcated and most perversely misunderstood this great doctrine, his cha-

racter is formed *for* him, and not *by* him; therefore his wishing
that it had been formed differently is of no use; he has no
power to alter it. But this is a grand error. He has, to a
certain extent, a power to alter his character. Its being, in
the ultimate resort, formed for him, is not inconsistent with its
being, in part, formed *by* him as one of the intermediate agents.
His character is formed by his circumstances (including among
these his particular organization); but his own desire to mould
it in a particular way, is one of those circumstances, and by
no means one of the least influential. We cannot, indeed,
directly will to be different from what we are. But neither
did those who are supposed to have formed our characters,
directly will that we should be what we are. Their will had
no direct power except over their own actions. They made
us what they did make us, by willing, not the end, but the
requisite means; and we, when our habits are not too inve-
terate, can, by similarly willing the requisite means, make
ourselves different. If they could place us under the influence
of certain circumstances, we, in like manner, can place ourselves under the influence of other circumstances. We are
exactly as capable of making our own character, *if we will*, as
others are of making it for us.

Yes (answers the Owenite), but these words, " if we will,"
surrender the whole point: since the will to alter our own
character is given us, not by any efforts of ours, but by
circumstances which we cannot help; it comes to us either
from external causes, or not at all. Most true: if the Owenite
stops here, he is in a position from which nothing can expel
him. Our character is formed by us as well as for us; but
the wish which induces us to attempt to form it is formed for
us; and how? Not, in general, by our organization, nor
wholly by our education, but by our experience; experience
of the painful consequences of the character we previously
had: or by some strong feeling of admiration or aspiration,
accidentally aroused. But to think that we have no power of
altering our character, and to think that we shall not use our
power unless we desire to use it, are very different things, and
have a very different effect on the mind. A person who does

not wish to alter his character, cannot be the person who is supposed to feel discouraged or paralysed by thinking himself unable to do it. The depressing effect of the fatalist doctrine can only be felt where there *is* a wish to do what that doctrine represents as impossible. It is of no consequence what we think forms our character, when we have no desire of our own about forming it; but it is of great consequence that we should not be prevented from forming such a desire by thinking the attainment impracticable, and that if we have the desire, we should know that the work is not so irrevocably done as to be incapable of being altered.

And indeed, if we examine closely, we shall find that this feeling, of our being able to modify our own character *if we wish*, is itself the feeling of moral freedom which we are conscious of. A person feels morally free who feels that his habits or his temptations are not his masters, but he theirs: who even in yielding to them knows that he could resist; that were he desirous of altogether throwing them off, there would not be required for that purpose a stronger desire than he knows himself to be capable of feeling. It is of course necessary, to render our consciousness of freedom complete, that we should have succeeded in making our character all we have hitherto attempted to make it; for if we have wished and not attained, we have, to that extent, not power over our own character, we are not free. Or at least, we must feel that our wish, if not strong enough to alter our character, is strong enough to conquer our character when the two are brought into conflict in any particular case of conduct. And hence it is said with truth, that none but a person of confirmed virtue is completely free.

The application of so improper a term as Necessity to the doctrine of cause and effect in the matter of human character, seems to me one of the most signal instances in philosophy of the abuse of terms, and its practical consequences one of the most striking examples of the power of language over our associations. The subject will never be generally understood, until that objectionable term is dropped. The free-will doctrine, by keeping in view precisely that portion of the truth

which the word Necessity puts out of sight, namely the power
of the mind to co-operate in the formation of its own character,
has given to its adherents a practical feeling much nearer to the
truth than has generally (I believe) existed in the minds of
necessarians.   The latter may have had a stronger sense of the
importance of what human beings can do to shape the characters
of one another; but the free-will doctrine has, I believe,
fostered in its supporters a much stronger spirit of self-culture.

§ 4.   There is still one fact which requires to be noticed
(in addition to the existence of a power of self-formation)
before the doctrine of the causation of human actions can be
freed from the confusion and misapprehensions which surround it in many minds.   When the will is said to be determined by motives, a motive does not mean always, or solely,
the anticipation of a pleasure or of a pain.   I shall not here
inquire whether it be true that, in the commencement, all our
voluntary actions are mere means consciously employed to
obtain some pleasure, or avoid some pain.   It is at least
certain that we gradually, through the influence of association,
come to desire the means without thinking of the end: the
action itself becomes an object of desire, and is performed
without reference to any motive beyond itself.   Thus far, it
may still be objected, that, the action having through association become pleasurable, we are, as much as before, moved to
act by the anticipation of a pleasure, namely the pleasure of
the action itself.   But granting this, the matter does not end
here.   As we proceed in the formation of habits, and become
accustomed to will a particular act or a particular course of
conduct because it is pleasurable, we at last continue to will it
without any reference to its being pleasurable.   Although,
from some change in us or in our circumstances, we have
ceased to find any pleasure in the action, or perhaps to antici-
pate any pleasure as the consequence of it, we still continue to
desire the action, and consequently to do it.   In this manner
it is that habits of hurtful excess continue to be practised
although they have ceased to be pleasurable; and in this
manner also it is that the habit of willing to persevere in the

course which he has chosen, does not desert the moral hero, even when the reward, however real, which he doubtless receives from the consciousness of well-doing, is anything but an equivalent for the sufferings he undergoes, or the wishes which he may have to renounce.

A habit of willing is commonly called a purpose; and among the causes of our volitions, and of the actions which flow from them, must be reckoned not only likings and aversions, but also purposes. It is only when our purposes have become independent of the feelings of pain or pleasure from which they originally took their rise, that we are said to have a confirmed character. "A character," says Novalis, "is a completely fashioned will:" and the will, once so fashioned, may be steady and constant, when the passive susceptibilities of pleasure and pain are greatly weakened, or materially changed.

With the corrections and explanations now given, the doctrine of the causation of our volitions by motives, and of motives by the desirable objects offered to us, combined with our particular susceptibilities of desire, may be considered, I hope, as sufficiently established for the purposes of this treatise.*

---

* Some arguments and explanations, supplementary to those in the text, will be found in *An Examination of Sir William Hamilton's Philosophy*, chap. xxvi.

CHAPTER III.

§ 1. IT is a common notion, or at least it is implied in
many common modes of speech, that the thoughts, feelings,
and actions of sentient beings are not a subject of science, in
the same strict sense in which this is true of the objects of
outward nature. This notion seems to involve some confusion
of ideas, which it is necessary to begin by clearing up.

Any facts are fitted, in themselves, to be a subject of
science, which follow one another according to constant laws;
although those laws may not have been discovered, nor even
be discoverable by our existing resources. Take, for instance,
the most familiar class of meteorological phenomena, those of
rain and sunshine. Scientific inquiry has not yet succeeded
in ascertaining the order of antecedence and consequence
among these phenomena, so as to be able, at least in our
regions of the earth, to predict them with certainty, or even
with any high degree of probability. Yet no one doubts that
the phenomena depend on laws, and that these must be derivative laws resulting from known ultimate laws, those of heat,
electricity, vaporization, and elastic fluids. Nor can it be
doubted that if we were acquainted with all the antecedent
circumstances, we could, even from those more general laws,
predict (saving difficulties of calculation) the state of the
weather at any future time. Meteorology, therefore, not only
has in itself every natural requisite for being, but actually is, a
science; though, from the difficulty of observing the facts on
which the phenomena depend (a difficulty inherent in the
peculiar nature of those phenomena) the science is extremely
imperfect; and were it perfect, might probably be of little avail

in practice, since the data requisite for applying its principles to particular instances would rarely be procurable.

A case may be conceived, of an intermediate character between the perfection of science, and this its extreme imperfection. It may happen that the greater causes, those on which the principal part of the phenomena depends, are within the reach of observation and measurement; so that if no other causes intervened, a complete explanation could be given not only of the phenomenon in general, but of all the variations and modifications which it admits of. But inasmuch as other, perhaps many other causes, separately insignificant in their effects, co-operate or conflict in many or in all cases with those greater causes; the effect, accordingly, presents more or less of aberration from what would be produced by the greater causes alone. Now if these minor causes are not so constantly accessible, or not accessible at all, to accurate observation; the principal mass of the effect may still, as before, be accounted for, and even predicted; but there will be variations and modifications which we shall not be competent to explain thoroughly, and our predictions will not be fulfilled accurately, but only approximately.

It is thus, for example, with the theory of the tides. No one doubts that Tidology (as Dr. Whewell proposes to call it) is really a science. As much of the phenomena as depends on the attraction of the sun and moon is completely understood, and may in any, even unknown, part of the earth's surface, be foretold with certainty; and the far greater part of the phenomena depends on those causes. But circumstances of a local or casual nature, such as the configuration of the bottom of the ocean, the degree of confinement from shores, the direction of the wind, &c., influence, in many or in all places, the height and time of the tide; and a portion of these circumstances being either not accurately knowable, not precisely measurable, or not capable of being certainly foreseen, the tide in known places commonly varies from the calculated result of general principles by some difference that we cannot explain, and in unknown ones may vary from it by a difference that we are not able to foresee or conjecture. Nevertheless, not only is it certain that these variations depend on causes, and follow their causes by laws of unerring uniformity; not only, therefore, is tidology a science, like meteorology, but it is, what hitherto at least meteorology is not, a science largely available in practice. General laws may be laid down respecting the tides, predictions may be founded on those laws, and the result will in the main, though often not with complete accuracy, correspond to the predictions.

And this is what is or ought to be meant by those who speak of sciences which are not *exact* sciences. Astronomy was once a science, without being an exact science. It could not become exact until not only the general course of the planetary motions, but the perturbations also, were accounted for, and referred to their causes. It has become an exact science, because its phenomena have been brought under laws comprehending the whole of the causes by which the phenomena are influenced, whether in a great or only in a trifling degree, whether in all or only in some cases, and assigning to each of those causes the share of effect which really belongs to it. But in the theory of the tides the only laws as yet accurately ascertained, are those of the causes which affect the phenomenon in all cases, and in a considerable degree; while others which affect it in some cases only, or, if in all, only in a slight degree, have not been sufficiently ascertained and studied to enable us to lay down their laws; still less to deduce the completed law of the phenomenon, by compounding the effects of the greater with those of the minor causes. Tidology, therefore, is not yet an exact science; not from any inherent incapacity of being so, but from the difficulty of ascertaining with complete precision the real derivative uniformities. By combining, however, the exact laws of the greater causes, and of such of the minor ones as are sufficiently known, with such empirical laws or such approximate generalizations respecting the miscellaneous variations as can be obtained by specific observation, we can lay down general propositions which will be true in the main, and on which, with allowance for the degree of their probable inaccuracy, we may safely ground our expectations and our conduct.

§ 2. The science of human nature is of this description. It falls far short of the standard of exactness now realized in Astronomy; but there is no reason that it should not be as much a science as Tidology is, or as Astronomy was when its calculations had only mastered the main phenomena, but not the perturbations.

The phenomena with which this science is conversant being the thoughts, feelings, and actions of human beings, it would have attained the ideal perfection of a science if it enabled us to foretell how an individual would think, feel, or act, throughout life, with the same certainty with which astronomy enables us to predict the places and the occultations of the heavenly bodies. It needs scarcely be stated that nothing approaching to this can be done. The actions of individuals could not be predicted with scientific accuracy, were it only because we cannot foresee the whole of the circumstances in which those individuals will be placed. But further, even in any given combination of (present) circumstances, no assertion, which is both precise and universally true, can be made respecting the manner in which human beings will think, feel, or act. This is not, however, because every person's modes of thinking, feeling, and acting, do not depend on causes; nor can we doubt that if, in the case of any individual, our data could be complete, we even now know enough of the ultimate laws by which mental phenomena are determined, to enable us in many cases to predict, with tolerable certainty, what, in the greater number of supposable combinations of circumstances, his conduct or sentiments would be. But the impressions and actions of human beings are not solely the result of their present circumstances, but the joint result of those circumstances and of the characters of the individuals: and the agencies which determine human character are so numerous and diversified, (nothing which has happened to the person throughout life being without its portion of influence,) that in the aggregate they are never in any two cases exactly similar. Hence, even if our science of human nature were theoretically perfect, that is, if we could calculate any character as we can calculate the orbit of

any planet, *from given data ;* still, as the data are never all
given, nor ever precisely alike in different cases, we could
neither make positive predictions, nor lay down universal
propositions.

Inasmuch, however, as many of those effects which it is
of most importance to render amenable to human foresight
and control are determined, like the tides, in an incomparably
greater degree by general causes, than by all partial causes
taken together; depending in the main on those circumstances
and qualities which are common to all mankind, or at least to
large bodies of them, and only in a small degree on the
idiosyncrasies of organization or the peculiar history of individuals ; it is evidently possible with regard to all such effects,
to make predictions which will *almost* always be verified, and
general propositions which are almost always true.   And
whenever it is sufficient to know how the great majority of
the human race, or of some nation or class of persons, will
think, feel, and act, these propositions are equivalent to universal ones.   For the purposes of political and social science
this *is* sufficient.   As we formerly remarked,* an approximate
generalization is, in social inquiries, for most practical purposes equivalent to an exact one ; that which is only probable
when asserted of individual human beings indiscriminately
selected, being certain when affirmed of the character and
collective conduct of masses.

It is no disparagement, therefore, to the science of Human
Nature, that those of its general propositions which descend
sufficiently into detail to serve as a foundation for predicting
phenomena in the concrete, are for the most part only
approximately true.   But in order to give a genuinely scientific character to the study, it is indispensable that these
approximate generalizations, which in themselves would
amount only to the lowest kind of empirical laws, should be
connected deductively with the laws of nature from which
they result ; should be resolved into the properties of the
causes on which the phenomena depend.   In other words, the

* Supra, p. 137.

science of Human Nature may be said to exist, in proportion as the approximate truths, which compose a practical knowledge of mankind, can be exhibited as corollaries from the universal laws of human nature on which they rest; whereby the proper limits of those approximate truths would be shown, and we should be enabled to deduce others for any new state of circumstances, in anticipation of specific experience.

The proposition now stated is the text on which the two succeeding chapters will furnish the comment.

CHAPTER IV.

OF THE LAWS OF MIND.

§ 1. WHAT the Mind is, as well as what Matter is, or any other question respecting Things in themselves, as distinguished from their sensible manifestations, it would be foreign to the purposes of this treatise to consider. Here, as throughout our inquiry, we shall keep clear of all speculations respecting the mind's own nature, and shall understand by the laws of mind, those of mental Phenomena; of the various feelings or states of consciousness of sentient beings. These, according to the classification we have uniformly followed, consist of Thoughts, Emotions, Volitions, and Sensations; the last being as truly states of Mind as the three former. It is usual indeed to speak of sensations as states of body, not of mind. But this is the common confusion, of giving one and the same name to a phenomenon and to the proximate cause or conditions of the phenomenon. The immediate antecedent of a sensation is a state of body, but the sensation itself is a state of mind. If the word mind means anything, it means that which feels. Whatever opinion we hold respecting the fundamental identity or diversity of matter and mind, in any case the distinction between mental and physical facts, between the internal and the external world, will always remain, as a matter of classification: and in that classification, sensations, like all other feelings, must be ranked as mental phenomena. The mechanism of their production, both in the body itself and in what is called outward nature, is all that can with any propriety be classed as physical.

The phenomena of mind, then, are the various feelings of our nature, both those improperly called physical, and those peculiarly designated as mental: and by the laws of mind, I mean the laws according to which those feelings generate one another.

§ 2. All states of mind are immediately caused either by other states of mind, or by states of body. When a state of mind is produced by a state of mind, I call the law concerned in the case, a law of Mind. When a state of mind is produced directly by a state of body, the law is a law of Body, and belongs to physical science.

With regard to those states of mind which are called sensations, all are agreed that these have for their immediate antecedents, states of body. Every sensation has for its proximate cause some affection of the portion of our frame called the nervous system; whether this affection originate in the action of some external object, or in some pathological condition of the nervous organization itself. The laws of this portion of our nature—the varieties of our sensations, and the physical conditions on which they proximately depend—manifestly belong to the province of Physiology.

Whether the remainder of our mental states are similarly dependent on physical conditions, is one of the *vexatæ questiones* in the science of human nature. It is still disputed whether our thoughts, emotions, and volitions are generated through the intervention of material mechanism; whether we have organs of thought and of emotion, in the same sense in which we have organs of sensation. Many eminent physiologists hold the affirmative. These contend, that a thought (for example) is as much the result of nervous agency, as a sensation: that some particular state of our nervous system, in particular of that central portion of it called the brain, invariably precedes, and is presupposed by, every state of our consciousness. According to this theory, one state of mind is never really produced by another: all are produced by states of body. When one thought seems to call up another by association, it is not really a thought which recals a thought; the association did not exist between the two thoughts, but between the two states of the brain or nerves which preceded the thoughts : one of those states recals the other, each being attended, in its passage, by the particular state of consciousness which is consequent on it. On this theory the uniformities of succession among states of mind would be mere derivative uniformities, resulting from the laws of succession of the bodily states which cause them. There would be no original mental laws, no Laws of Mind in the sense in which I use the term, at all : and mental science would be a mere branch, though the highest and most recondite branch, of the science of physiology. M. Comte, accordingly, claims the scientific cognizance of moral and intellectual phenomena exclusively for physiologists ; and not only denies to Psychology, or Mental Philosophy properly so called, the character of a science, but places it, in the chimerical nature of its objects and pretensions, almost on a par with astrology.

But, after all has been said which can be said, it remains incontestible that there exist uniformities of succession among states of mind, and that these can be ascertained by observation and experiment. Further, that every mental state has a nervous state for its immediate antecedent and proximate cause, though extremely probable, cannot hitherto be said to be proved, in the conclusive manner in which this can be proved of sensations ; and even were it certain, yet every one must admit that we are wholly ignorant of the characteristics of these nervous states ; we know not, and at present have no means of knowing, in what respect one of them differs from another ; and our only mode of studying their successions or coexistences must be by observing the successions and coexistences of the mental states, of which they are supposed to be the generators or causes. The successions, therefore, which obtain among mental phenomena, do not admit of being deduced from the physiological laws of our nervous organization : and all real knowledge of them must continue, for a long time at least, if not always, to be sought in the direct study, by observation and experiment, of the mental successions themselves. Since therefore the order of our mental phenomena must be studied in those phenomena, and not inferred from the laws of any phenomena more general, there is a distinct and separate Science of Mind.

The relations, indeed, of that science to the science of physiology must never be overlooked or undervalued. It must by no means be forgotten that the laws of mind may be derivative laws resulting from laws of animal life, and that their truth therefore may ultimately depend on physical conditions; and the influence of physiological states or physiological changes in altering or counteracting the mental successions, is one of the most important departments of psychological study. But, on the other hand, to reject the resource of psychological analysis, and construct the theory of the mind solely on such data as physiology at present affords, seems to me as great an error in principle, and an even more serious one in practice. Imperfect as is the science of mind, I do not scruple to affirm, that it is in a considerably more advanced state than the portion of physiology which corresponds to it; and to discard the former for the latter appears to me an infringement of the true canons of inductive philosophy, which must produce, and which does produce, erroneous conclusions in some very important departments of the science of human nature.

§ 3. The subject, then, of Psychology, is the uniformities of succession, the laws, whether ultimate or derivative, according to which one mental state succeeds another; is caused by, or at least, is caused to follow, another. Of these laws, some are general, others more special. The following are examples of the most general laws.

First: Whenever any state of consciousness has once been excited in us, no matter by what cause; an inferior degree of the same state of consciousness, a state of consciousness resembling the former, but inferior in intensity, is capable of being reproduced in us, without the presence of any such cause as excited it at first. Thus, if we have once seen or touched an object, we can afterwards think of the object though it be absent from our sight or from our touch. If we have been joyful or grieved at some event, we can think of, or remember our past joy or grief, though no new event of a happy or painful nature has taken place. When a poet has put together a mental picture of an imaginary object, a Castle of Indolence, a Una, or a Hamlet, he can afterwards think of the ideal object he has created, without any fresh act of intellectual combination. This law is expressed by saying, in the language of Hume, that every mental *impression* has its *idea*.

Secondly: These ideas, or secondary mental states, are excited by our impressions, or by other ideas, according to certain laws which are called Laws of Association. Of these laws the first is, that similar ideas tend to excite one another. The second is, that when two impressions have been frequently experienced (or even thought of) either simultaneously or in immediate succession, then whenever one of these impressions, or the idea of it, recurs, it tends to excite the idea of the other. The third law is, that greater intensity in either or both of the impressions, is equivalent, in rendering them excitable by one another, to a greater frequency of conjunction. These are the laws of ideas: on which I shall not enlarge in this place, but refer the reader to works professedly psychological, in particular to Mr. James Mill's *Analysis of the Phenomena of the Human Mind*, where the principal laws of association, along with many of their applications, are copiously exemplified, and with a masterly hand.*

These simple or elementary Laws of Mind have been ascertained by the ordinary methods of experimental inquiry; nor could they have been ascertained in any other manner. But

---

* When this chapter was written, Mr. Bain had not yet published even the first part ("The Senses and the Intellect") of his profound Treatise on the Mind. In this, the laws of association have been more comprehensively stated and more largely exemplified than by any previous writer; and the work, having been completed by the publication of "The Emotions and the Will," may now be referred to as incomparably the most complete analytical exposition of the mental phenomena, on the basis of a legitimate Induction, which has yet been produced.

Many striking applications of the laws of association to the explanation of complex mental phenomena, are also to be found in Mr. Herbert Spencer's "Principles of Psychology."

a certain number of elementary laws having thus been obtained, it is a fair subject of scientific inquiry how far those laws can be made to go in explaining the actual phenomena. It is obvious that complex laws of thought and feeling not only may, but must, be generated from these simple laws. And it is to be remarked, that the case is not always one of Composition of Causes: the effect of concurring causes is not always precisely the sum of the effects of those causes when separate, nor even always an effect of the same kind with them. Reverting to the distinction which occupies so prominent a place in the theory of induction; the laws of the phenomena of mind are sometimes analogous to mechanical, but sometimes also to chemical laws. When many impressions or ideas are operating in the mind together, there sometimes takes place a process, of a similar kind to chemical combination. When impressions have been so often experienced in conjunction, that each of them calls up readily and instantaneously the ideas of the whole group, those ideas sometimes melt and coalesce into one another, and appear not several ideas, but one; in the same manner as, when the seven prismatic colours are presented to the eye in rapid succession, the sensation produced is that of white. But as in this last case it is correct to say that the seven colours when they rapidly follow one another *generate* white, but not that they actually *are* white; so it appears to me that the Complex Idea, formed by the blending together of several simpler ones, should, when it really appears simple, (that is, when the separate elements are not consciously distinguishable in it,) be said to *result from*, or *be generated by,* the simple ideas, not to *consist* of them. Our idea of an orange really *consists* of the simple ideas of a certain colour, a certain form, a certain taste and smell, &c., because we can, by interrogating our consciousness, perceive all these elements in the idea. But we cannot perceive, in so apparently simple a feeling as our perception of the shape of an object by the eye, all that multitude of ideas derived from other senses, without which it is well ascertained that no such visual perception would ever have had existence; nor, in our idea of Extension, can we discover those elementary ideas of resistance, derived from our muscular frame, in which it· has been conclusively shown that the idea originates. These therefore are cases of mental chemistry: in which it is proper to say that the simple ideas generate, rather than that they compose, the complex ones.

With respect to all the other constituents of the mind, its beliefs, its abstruser conceptions, its sentiments, emotions, and volitions; there are some (among whom are Hartley, and the author of the *Analysis*) who think that the whole of these are generated from simple ideas of sensation, by a chemistry similar to that which we have just exemplified. These philosophers have made out a great part of their case, but I am not satisfied that they have established the whole of it. They have shown that there is such a thing as mental chemistry; that the heterogeneous nature of a feeling A, considered in relation to B and C, is no conclusive argument against its being generated from B and C. Having proved this, they proceed to show, that where A is found, B and C were or may have been present, and why therefore, they ask, should not A have been generated from B and C? But even if this evidence were carried to the highest degree of completeness which it admits of; if it were shown (which hitherto it has not, in all cases, been) that certain groups of associated ideas not only might have been, but actually were, present whenever the more recondite mental feeling was experienced; this would amount only to the Method of Agreement, and could not prove causation until confirmed by the more conclusive evidence of the Method of Difference. If the question be whether Belief is a mere case of close association of ideas, it would be necessary to examine experimentally if it be true that any ideas whatever, provided they are associated with the required degree of closeness, give rise to belief. If the inquiry be into the origin of moral feelings, the feeling for example of moral reprobation, it is necessary to compare all the varieties of actions or states of mind which are ever morally disapproved, and see whether in all these cases it can be shown, or reasonably surmised, that the action or state of mind had become connected by association, in the disapproving mind, with some particular class of hateful or dis-

gusting ideas; and the method employed is, thus far, that of
Agreement. But this is not enough. Supposing this proved,
we must try further by the Method of Difference, whether this
particular kind of hateful or disgusting ideas, when it becomes
associated with an action previously indifferent, will render
that action a subject of moral disapproval. If this question
can be answered in the affirmative, it is shown to be a law of
the human mind, that an association of that particular description is the generating cause of moral reprobation. That all
this is the case has been rendered extremely probable, but the
experiments have not been tried with the degree of precision
necessary for a complete and absolutely conclusive induction.[*]

It is further to be remembered, that even if all which this
theory of mental phenomena contends for could be proved,
we should not be the more enabled to resolve the laws of the
more complex feelings into those of the simpler ones. The
generation of one class of mental phenomena from another,
whenever it can be made out, is a highly interesting fact in
psychological chemistry; but it no more supersedes the
necessity of an experimental study of the generated phenomenon, than a knowledge of the properties of oxygen and
sulphur enables us to deduce those of sulphuric acid without
specific observation and experiment. Whatever, therefore, may
be the final issue of the attempt to account for the origin of
our judgments, our desires, or our volitions, from simpler
mental phenomena, it is not the less imperative to ascertain
the sequences of the complex phenomena themselves, by special
study in conformity to the canons of Induction. Thus, in
respect to Belief, psychologists will always have to inquire,
what beliefs we have by direct consciousness, and according
to what laws one belief produces another; what are the laws,

---

[*] In the case of the moral sentiments the place of direct experiment is to a
considerable extent supplied by historical experience, and we are able to trace
with a tolerable approach to certainty the particular associations by which
those sentiments are engendered. This has been attempted, so far as respects
the sentiment of justice, in a little work by the present author, entitled
Utilitarianism.

in virtue of which one thing is recognised by the mind, either
rightly or erroneously, as evidence of another thing. In
regard to Desire, they will have to examine what objects we
desire naturally, and by what causes we are made to desire
things originally indifferent, or even disagreeable to us; and
so forth. It may be remarked, that the general laws of association prevail among these more intricate states of mind, in
the same manner as among the simpler ones. A desire, an
emotion, an idea of the higher order of abstraction, even our
judgments and volitions when they have become habitual, are
called up by association, according to precisely the same laws
as our simple ideas.

§ 4. In the course of these inquiries it will be natural
and necessary to examine, how far the production of one
state of mind by another is influenced by any assignable state
of body. The commonest observation shows that different
minds are susceptible in very different degrees, to the action
of the same psychological causes. The idea, for example,
of a given desirable object, will excite in different minds
very different degrees of intensity of desire. The same subject of meditation, presented to different minds, will excite
in them very unequal degrees of intellectual action. These
differences of mental susceptibility in different individuals
may be, first, original and ultimate facts, or, secondly, they
may be consequences of the previous mental history of those
individuals, or thirdly and lastly, they may depend on varieties
of physical organization. That the previous mental history
of the individuals must have some share in producing or in
modifying the whole of their mental character, is an inevitable
consequence of the laws of mind; but that differences of
bodily structure also co-operate, is the opinion of all physiologists, confirmed by common experience. It is to be re-
gretted that hitherto this experience, being accepted in the
gross, without due analysis, has been made the groundwork
of empirical generalizations most detrimental to the progress
of real knowledge.

It is certain that the natural differences which really exist in the mental predispositions or susceptibilities of different persons, are often not unconnected with diversities in their organic constitution. But it does not therefore follow that these organic differences must in all cases influence the mental phenomena directly and immediately. They often affect them through the medium of their psychological causes. For example, the idea of some particular pleasure may excite in different persons, even independently of habit or education, very different strengths of desire, and this may be the effect of their different degrees or kinds of nervous susceptibility; but these organic differences, we must remember, will render the pleasurable sensation itself more intense in one of these persons than in the other; so that the idea of the pleasure will also be an intenser feeling, and will, by the operation of mere mental laws, excite an intenser desire, without its being necessary to suppose that the desire itself is directly influenced by the physical peculiarity. As in this, so in many cases, such differences in the kind or in the intensity of the physical sensations as must necessarily result from differences of bodily organization, will of themselves account for many differences not only in the degree, but even in the kind, of the other mental phenomena. So true is this, that even different *qualities* of mind, different types of mental character, will naturally be produced by mere differences of intensity in the sensations generally: as is well pointed out in an able essay on Dr. Priestley, mentioned in a former chapter:—

"The sensations which form the elements of all knowledge are received either simultaneously or successively; when several are received simultaneously, as the smell, the taste, the colour, the form, &c. of a fruit, their association together constitutes our idea of an *object;* when received successively, their association makes up the idea of an *event.* Anything, then, which favours the associations of synchronous ideas, will tend to produce a knowledge of objects, a perception of qualities; while anything which favours association in the successive order, will tend to produce a knowledge of events, of the order of occurrences, and of the connexion of cause and effect: in other words, in the one case a perceptive mind, with a discriminate feeling of the pleasurable and painful properties of things, a sense of the grand and the beautiful, will be the result: in the other, a mind attentive to the movements and phenomena, a ratiocinative and philosophic intellect. Now it is an acknowledged principle, that all sensations experienced during the presence of any vivid impression, become strongly associated with it, and with each other; and does it not follow, that the synchronous feelings of a sensitive constitution, (*i.e.* the one which has vivid impressions,) will be more intimately blended than in a differently formed mind? If this suggestion has any foundation in truth, it leads to an inference not unimportant; that where nature has endowed an individual with great original susceptibility, he will probably be distinguished by fondness for natural history, a relish for the beautiful and great, and moral enthusiasm; where there is but a mediocrity of sensibility, a love of science, of abstract truth, with a deficiency of taste and of fervour, is likely to be the result."

We see from this example, that when the general laws of mind are more accurately known, and above all, more skilfully applied to the detailed explanation of mental peculiarities, they will account for many more of those peculiarities than is ordinarily supposed. Unfortunately the reaction of the last and present generation against the philosophy of the eighteenth century has produced a very general neglect of this great department of analytical inquiry; of which, consequently, the recent progress has been by no means proportional to its early promise. The majority of those who speculate on human nature, prefer dogmatically to assume that the mental differences which they perceive, or think they perceive, among human beings, are ultimate facts, incapable of being either explained or altered, rather than take the trouble of fitting themselves, by the requisite processes of thought, for referring those mental differences to the outward causes by which they are for the most part produced, and on the removal of which they would cease to exist. The German school of metaphysical speculation, which has not yet lost its temporary predominance in European thought, has had this among many other injurious influences: and at the opposite extreme of the psychological scale, no writer, either of early or of recent date, is chargeable in a higher degree with this aberration from the true scientific spirit, than M. Comte.

It is certain that, in human beings at least, differences in education and in outward circumstances are capable of affording an adequate explanation of by far the greatest portion of character; and that the remainder may be in great part accounted for by physical differences in the sensations produced in different individuals by the same external or internal cause. There are, however, some mental facts which do not seem to admit of these modes of explanation. Such, to take the strongest case, are the various instincts of animals, and the portion of human nature which corresponds to those instincts. No mode has been suggested, even by way of hypothesis, in which these can receive any satisfactory, or even plausible, explanation from psychological causes alone; and there is great reason to think that they have as positive, and even as direct and immediate, a connexion with physical conditions of the brain and nerves, as any of our mere sensations have. A supposition which (it is perhaps not superfluous to add) in no way conflicts with the indisputable fact, that these instincts may be modified to any extent, or entirely conquered, in human beings at least, by other mental influences, and by education.

Whether organic causes exercise a direct influence over any other classes of mental phenomena, is hitherto as far from being ascertained, as is the precise nature of the organic conditions even in the case of instincts. The physiology, however, of the brain and nervous system is in a state of such rapid advance, and is continually bringing forth such new and interesting results, that if there be really a connexion between mental peculiarities and any varieties cognizable by our senses in the structure of the cerebral and nervous apparatus, the nature of that connexion is now in a fair way of being found out. The latest discoveries in cerebral physiology appear to have proved, that any such connexion which may exist is of a radically different character from that contended for by Gall and his followers, and that whatever may hereafter be found to be the true theory of the subject, phrenology at least is untenable.

CHAPTER V.

OF ETHOLOGY, OR THE SCIENCE OF THE FORMATION OF CHARACTER.

§ 1. THE laws of mind as characterized in the preceding chapter, compose the universal or abstract portion of the philosophy of human nature; and all the truths of common experience, constituting a practical knowledge of mankind, must, to the extent to which they are truths, be results or consequences of these. Such familiar maxims, when collected *à posteriori* from observation of life, occupy among the truths of the science the place of what, in our analysis of Induction, have so often been spoken of under the title of Empirical Laws.

An Empirical Law (it will be remembered) is an uniformity, whether of succession or of coexistence, which holds true in all instances within our limits of observation, but is not of a nature to afford any assurance that it would hold beyond those limits; either because the consequent is not really the effect of the antecedent, but forms part along with it of a chain of effects, flowing from prior causes not yet ascertained; or because there is ground to believe that the sequence (though a case of causation) is resolvable into simpler sequences, and, depending therefore on a concurrence of several natural agencies, is exposed to an unknown multitude of possibilities of counteraction. In other words, an empirical law is a generalization, of which, not content with finding it true, we are obliged to ask, why is it true? knowing that its truth is not absolute, but dependent on some more general conditions, and that it can only be relied on in so far as there is ground of assurance that those conditions are realized.

Now, the observations concerning human affairs collected from common experience, are precisely of this nature. Even if they were universally and exactly true within the bounds of experience, which they never are, still they are not the ultimate laws of human action; they are not the principles of human nature, but results of those principles under the circumstances in which mankind have happened to be placed. When the Psalmist said in his haste that " all men are liars," he enunciated what in some ages and countries is borne out by ample experience; but it is not a law of man's nature to lie; though it is one of the consequences of the laws of human nature, that lying is nearly universal when certain external circumstances exist universally, especially circumstances productive of habitual distrust and fear. When the character of the old is asserted to be cautious, and of the young impetuous, this, again, is but an empirical law; for it is not because of their youth that the young are impetuous, nor because of their age that the old are cautious. It is chiefly, if not wholly, because the old, during their many years of life, have generally had much experience of its various evils, and having suffered or seen others suffer much from incautious exposure to them, have acquired associations favourable to circumspection: while the young, as well from the absence of similar experience as from the greater strength of the inclinations which urge them to enterprise, engage themselves in it more readily. Here, then, is the *explanation* of the empirical law; here are the conditions which ultimately determine whether the law holds good or not. If an old man has not been oftener than most young men in contact with danger and difficulty, he will be equally incautious : if a youth has not stronger inclinations than an old man, he probably will be as little enterprising. The empirical law derives whatever truth it has, from the causal laws of which it is a consequence. If we know those laws, we know what are the limits to the derivative law : while, if we have not yet accounted for the empirical law—if it rest only on observation—there is no safety in applying it far beyond the limits of time, place, and circumstance, in which the observations were made.

The really scientific truths, then, are not these empirical laws, but the causal laws which explain them. The empirical laws of those phenomena which depend on known causes, and of which a general theory can therefore be constructed, have, whatever may be their value in practice, no other function in science than that of verifying the conclusions of theory. Still more must this be the case when most of the empirical laws amount, even within the limits of observation, only to approximate generalizations.

§ 2. This however is not, so much as is sometimes supposed, a peculiarity of the sciences called moral. It is only in the simplest branches of science that empirical laws are ever exactly true; and not always in those. Astronomy, for example, is the simplest of all the sciences which explain, in the concrete, the actual course of natural events. The causes or forces, on which astronomical phenomena depend, are fewer in number than those which determine any other of the great phenomena of nature. Accordingly, as each effect results from the conflict of but few causes, a great degree of regularity and uniformity might be expected to exist among the effects; and such is really the case: they have a fixed order, and return in cycles. But propositions which should express, with absolute correctness, all the successive positions of a planet until the cycle is completed, would be of almost unmanageable complexity, and could be obtained from theory alone. The generalizations which can be collected on the subject from direct observation, even such as Kepler's law, are mere approximations: the planets, owing to their perturbations by one another, do not move in exact ellipses. Thus even in astronomy, perfect exactness in the mere empirical laws is not to be looked for; much less, then, in more complex subjects of inquiry.

The same example shows how little can be inferred against the universality or even the simplicity of the ultimate laws, from the impossibility of establishing any but approximate empirical laws of the effects. The laws of causation according to which a class of phenomena are produced may be very few and simple, and yet the effects themselves may be so various and complicated that it shall be impossible to trace any regularity whatever completely through them. For the phenomena in question may be of an eminently modifiable character; insomuch that innumerable circumstances are capable of influencing the effect, although they may all do it according to a very small number of laws. Suppose that all which passes in the mind of man is determined by a few simple laws: still, if those laws be such that there is not one of the facts surrounding a human being, or of the events which happen to him, that does not influence in some mode or degree his subsequent mental history, and if the circumstances of different human beings are extremely different, it will be no wonder if very few propositions can be made respecting the details of their conduct or feelings, which will be true of all mankind.

Now, without deciding whether the ultimate laws of our mental nature are few or many, it is at least certain that they are of the above description. It is certain that our mental states, and our mental capacities and susceptibilities, are modified, either for a time or permanently, by everything which happens to us in life. Considering therefore how much these modifying causes differ in the case of any two individuals, it would be unreasonable to expect that the empirical laws of the human mind, the generalizations which can be made respecting the feelings or actions of mankind without reference to the causes that determine them, should be anything but approximate generalizations. They are the common wisdom of common life, and as such are invaluable; especially as they are mostly to be applied to cases not very dissimilar to those from which they were collected. But when maxims of this sort, collected from Englishmen, come to be applied to Frenchmen, or when those collected from the present day are applied to past or future generations, they are apt to be very much at fault. Unless we have resolved the empirical law into the laws of the causes on which it depends, and ascertained that those causes extend to the case which we have in view, there can be no reliance placed in our inferences.

For every individual is surrounded by circumstances different from those of every other individual; every nation or generation of mankind from every other nation or generation: and none of these differences are without their influence in forming a different type of character. There is, indeed, also a certain general resemblance; but peculiarities of circumstances are continually constituting exceptions even to the propositions which are true in the great majority of cases.

Although, however, there is scarcely any mode of feeling or conduct which is, in the absolute sense, common to all mankind; and though the generalizations which assert that any given variety of conduct or feeling will be found universally, (however nearly they may approximate to truth within given limits of observation,) will be considered as scientific propositions by no one who is at all familiar with scientific investigation; yet all modes of feeling and conduct met with among mankind have causes which produce them; and in the propositions which assign those causes, will be found the explanation of the empirical laws, and the limiting principle of our reliance on them. Human beings do not all feel and act alike in the same circumstances; but it is possible to determine what makes one person, in a given position, feel or act in one way, another in another; how any given mode of feeling and conduct, compatible with the general laws (physical and mental) of human nature, has been, or may be, formed. In other words, mankind have not one universal character, but there exist universal laws of the Formation of Character. And since it is by these laws, combined with the facts of each particular case, that the whole of the phenomena of human action and feeling are produced, it is on these that every rational attempt to construct the science of human nature in the concrete, and for practical purposes, must proceed.

§ 3. The laws, then, of the formation of character being the principal object of scientific inquiry into human nature; it remains to determine the method of investigation best fitted for ascertaining them. And the logical principles according to which this question is to be decided, must be those which preside over every other attempt to investigate the laws of very complex phenomena. For it is evident that both the character of any human being, and the aggregate of the circumstances by which that character has been formed, are facts of a high order of complexity. Now to such cases we have seen that the Deductive Method, setting out from general laws, and verifying their consequences by specific experience, is alone applicable. The grounds of this great logical doctrine have formerly been stated: and its truth will derive additional support from a brief examination of the specialities of the present case.

There are only two modes in which laws of nature can be ascertained: deductively, and experimentally: including under the denomination of experimental inquiry, observation as well as artificial experiment. Are the laws of the formation of character susceptible of a satisfactory investigation by the method of experimentation? Evidently not; because, even if we suppose unlimited power of varying the experiment, (which is abstractedly possible, though no one but an oriental despot has that power, or if he had, would probably be disposed to exercise it,) a still more essential condition is wanting; the power of performing any of the experiments with scientific accuracy.

The instances requisite for the prosecution of a directly experimental inquiry into the formation of character, would be a number of human beings to bring up and educate, from infancy to mature age. And to perform any one of these experiments with scientific propriety, it would be necessary to know and record every sensation or impression received by the young pupil from a period long before it could speak; including its own notions respecting the sources of all those sensations and impressions. It is not only impossible to do this completely, but even to do so much of it as should constitute a tolerable approximation. One apparently trivial circumstance which eluded our vigilance, might let in a train of impressions and associations sufficient to vitiate the experiment as an authentic exhibition of the effects flowing from given causes. No one who has sufficiently reflected on education is ignorant of this truth: and whoever has not, will find it most instructively illustrated in the writings of Rousseau and Helvetius on that great subject.

Under this impossibility of studying the laws of the formation of character by experiments purposely contrived to elucidate them, there remains the resource of simple observation. But if it be impossible to ascertain the influencing circumstances with any approach to completeness even when we have the shaping of them ourselves, much more impossible is it when the cases are further removed from our observation, and altogether out of our control. Consider the difficulty of the very first step—of ascertaining what actually is the character of the individual, in each particular case that we examine. There is hardly any person living, concerning some essential part of whose character there are not differences of opinion even among his intimate acquaintances: and a single action, or conduct continued only for a short time, goes a very little way towards ascertaining it. We can only make our observations in a rough way, and *en masse;* not attempting to ascertain completely in any given instance, what character has been formed, and still less by what causes; but only observing in what state of previous circumstances it is found that certain marked mental qualities or deficiencies *oftenest* exist. These conclusions, besides that they are mere approximate generalizations, deserve no reliance even as such, unless the instances are sufficiently numerous to eliminate not only chance, but every assignable circumstance in which a number of the cases examined may happen to have resembled one another. So numerous and various, too, are the circumstances which form individual character, that the consequence of any particular combination is hardly ever some definite and strongly marked character, always found where that combination exists, and not otherwise. What is obtained, even after the most extensive and accurate observation, is merely a comparative result; as for example, that in a given number of Frenchmen, taken indiscriminately, there will be found more persons of a particular mental tendency, and fewer of the contrary tendency,

than among an equal number of Italians or English, similarly
taken; or thus: of a hundred Frenchmen and an equal number
of Englishmen, fairly selected, and arranged according to the
degree in which they possess a particular mental characteristic, each number, 1, 2, 3, &c., of the one series, will be
found to possess more of that characteristic than the corresponding number of the other. Since, therefore, the comparison is not one of kinds, but of ratios and degrees; and since
in proportion as the differences are slight, it requires a greater
number of instances to eliminate chance; it cannot often
happen to any one to know a sufficient number of cases with
the accuracy requisite for making the sort of comparison last
mentioned; less than which, however, would not constitute
a real induction. Accordingly there is hardly one current
opinion respecting the characters of nations, classes, or descriptions of persons, which is universally acknowledged as
indisputable.*

And finally, if we could even obtain by way of experiment
a much more satisfactory assurance of these generalizations
than is really possible, they would still be only empirical

---

* The most favourable cases for making such approximate generalizations
are what may be termed collective instances; where we are fortunately enabled
to see the whole class respecting which we are inquiring, in action at once;
and, from the qualities displayed by the collective body, are able to judge what
must be the qualities of the majority of the individuals composing it. Thus
the character of a nation is shown in its acts as a nation; not so much in the
acts of its government, for those are much influenced by other causes; but in
the current popular maxims, and other marks of the general direction of public
opinion; in the character of the persons or writings that are held in permanent
esteem or admiration; in laws and institutions, so far as they are the work of
the nation itself, or are acknowledged and supported by it; and so forth. But
even here there is a large margin of doubt and uncertainty. These things are
liable to be influenced by many circumstances: they are partly determined by
the distinctive qualities of that nation or body of persons, but partly also by
external causes which would influence any other body of persons in the same
manner. In order, therefore, to make the experiment really complete, we
ought to be able to try it without variation upon other nations: to try how
Englishmen would act or feel if placed in the same circumstances in which
we have supposed Frenchmen to be placed; to apply, in short, the Method of
Difference as well as that of Agreement. Now these experiments we cannot
try, nor even approximate to.

laws. They would show, indeed, that there was some connexion between the type of character formed, and the circumstances existing in the case; but not what the precise
connexion was, nor to which of the peculiarities of those
circumstances the effect was really owing. They could only,
therefore, be received as results of causation, requiring to be
resolved into the general laws of the causes: until the determination of which, we could not judge within what limits
the derivative laws might serve as presumptions in cases yet
unknown, or even be depended on as permanent in the very
cases from which they were collected. The French people
had, or were supposed to have, a certain national character:
but they drive out their royal family and aristocracy, alter their
institutions, pass through a series of extraordinary events for
half a century, and at the end of that time are found to be,
in many respects, greatly altered. A long list of mental and
moral differences are observed, or supposed, to exist between
men and women: but at some future, and, it may be hoped,
not distant period, equal freedom and an equally independent
social position come to be possessed by both, and their differences of character are either removed or totally altered.

But if the differences which we think we observe between
French and English, or between men and women, can be connected with more general laws; if they be such as might be
expected to be ˈproduced by the differences of government,
former customs, and physical peculiarities in the two nations,
and by the diversities of education, occupations, personal independence, and social privileges, and whatever original differences there may be in bodily strength and nervous sensibility, between the two sexes; then, indeed, the coincidence
of the two kinds of evidence justifies us in believing that we
have both reasoned rightly and observed rightly. Our observation, though not sufficient as proof, is ample as verification.
And having ascertained not only the empirical laws, but
the causes, of the peculiarities, we need be under no difficulty
in judging how far they may be expected to be permanent, or by what circumstances they would be modified or
destroyed.

§ 4. Since, then, it is impossible to obtain really accurate propositions respecting the formation of character from observation and experiment alone, we are driven perforce to that which, even if it had not been the indispensable, would have been the most perfect, mode of investigation, and which it is one of the principal aims of philosophy to extend; namely, that which tries its experiments not on the complex facts, but on the simple ones of which they are compounded; and after ascertaining the laws of the causes, the composition of which gives rise to the complex phenomena, then considers whether these will not explain and account for the approximate generalizations which have been framed empirically respecting the sequences of those complex phenomena. The laws of the formation of character are, in short, derivative laws, resulting from the general laws of mind; and are to be obtained by deducing them from those general laws; by supposing any given set of circumstances, and then considering what, according to the laws of mind, will be the influence of those circumstances on the formation of character.

A science is thus formed, to which I would propose to give the name of Ethology, or the Science of Character; from ἦθος, a word more nearly corresponding to the term "character" as I here use it, than any other word in the same language. The name is perhaps etymologically applicable to the entire science of our mental and moral nature; but if, as is usual and convenient, we employ the name Psychology for the science of the elementary laws of mind, Ethology will serve for the ulterior science which determines the kind of character produced, in conformity to those general laws, by any set of circumstances, physical and moral. According to this definition, Ethology is the science which corresponds to the art of education; in the widest sense of the term, including the formation of national or collective character as well as individual. It would indeed be vain to expect (however completely the laws of the formation of character might be ascertained) that we could know so accurately the circumstances of any given case as to be able positively to predict the character that would be produced in that case. But we must remember that a degree of knowledge far short of the power of actual prediction, is often of much practical value. There may be great power of influencing phenomena, with a very imperfect knowledge of the causes by which they are in any given instance determined. It is enough that we know that certain means have a *tendency* to produce a given effect, and that others have a tendency to frustrate it. When the circumstances of an individual or of a nation are in any considerable degree under our control, we may, by our knowledge of tendencies, be enabled to shape those circumstances in a manner much more favourable to the ends we desire, than the shape which they would of themselves assume. This is the limit of our power; but within this limit the power is a most important one.

This science of Ethology may be called the Exact Science of Human Nature; for its truths are not, like the empirical laws which depend on them, approximate generalizations, but real laws. It is, however, (as in all cases of complex phenomena) necessary to the exactness of the propositions, that they should be hypothetical only, and affirm tendencies, not facts. They must not assert that something will always, or certainly, happen; but only that such and such will be the effect of a given cause, so far as it operates uncounteracted. It is a scientific proposition, that bodily strength tends to make men courageous; not that it always makes them so: that an interest on one side of a question tends to bias the judgment; not that it invariably does so: that experience tends to give wisdom; not that such is always its effect. These propositions, being assertive only of tendencies, are not the less universally true because the tendencies may be frustrated.

§ 5. While on the one hand Psychology is altogether, or principally, a science of observation and experiment, Ethology, as I have conceived it, is, as I have already remarked, altogether deductive. The one ascertains the simple laws of Mind in general, the other traces their operation in complex combinations of circumstances. Ethology stands to

Psychology in a relation very similar to that in which the various branches of natural philosophy stand to mechanics. The principles of Ethology are properly the middle principles, the *axiomata media* (as Bacon would have said) of the science of mind : as distinguished, on the one hand from the empirical laws resulting from simple observation, and on the other from the highest generalizations.

And this seems a suitable place for a logical remark, which, though of general application, is of peculiar importance in reference to the present subject. Bacon has judiciously observed that the *axiomata media* of every science principally constitute its value. The lowest generalizations, until explained by and resolved into the middle principles of which they are the consequences, have only the imperfect accuracy of empirical laws ; while the most general laws are *too* general, and include too few circumstances, to give sufficient indication of what happens in individual cases, where the circumstances are almost always immensely numerous. In the importance, therefore, which Bacon assigns, in every science, to the middle principles, it is impossible not to agree with him. But I conceive him to have been radically wrong in his doctrine respecting the mode in which these *axiomata media* should be arrived at ; though there is no one proposition laid down in his works for which he has been more extravagantly eulogized. He enunciates as an universal rule, that induction should proceed from the lowest to the middle principles, and from those to the highest, never reversing that order, and consequently leaving no room for the discovery of new principles by way of deduction at all. It is not to be conceived that a man of his sagacity could have fallen into this mistake, if there had existed in his time, among the sciences which treat of successive phenomena, one single instance of a deductive science, such as mechanics, astronomy, optics, acoustics, &c. now are. In those sciences it is evident that the higher and middle principles are by no means derived from the lowest, but the reverse. In some of them the very highest generalizations were those earliest ascertained with any scientific exactness ; as, for example (in mechanics), the laws of motion. Those general laws had not indeed at first the acknowledged universality which they acquired after having been successfully employed to explain many classes of phenomena to which they were not originally seen to be applicable; as when the laws of motion were employed, in conjunction with other laws, to explain deductively the celestial phenomena. Still, the fact remains, that the propositions which were afterwards recognised as the most general truths of the science, were, of all its accurate generalizations, those earliest arrived at. Bacon's greatest merit cannot therefore consist, as we are so often told that it did, in exploding the vicious method pursued by the ancients of flying to the highest generalizations first, and deducing the middle principles from them; since this is neither a vicious nor an exploded, but the universally accredited method of modern science, and that to which it owes its greatest triumphs. The error of ancient speculation did not consist in making the largest generalizations first, but in making them without the aid or warrant of rigorous inductive methods, and applying them deductively without the needful use of that important part of the Deductive Method termed Verification.

The order in which truths of the various degrees of generality should be ascertained, cannot, I apprehend, be prescribed by any unbending rule. I know of no maxim which can be laid down on the subject, but to obtain those first, in respect to which the conditions of a real induction can be first and most completely realized. Now, wherever our means of investigation can reach causes, without stopping at the empirical laws of the effects, the simplest cases, being those in which fewest causes are simultaneously concerned, will be most amenable to the inductive process; and these are the cases which elicit laws of the greatest comprehensiveness. In every science, therefore, which has reached the stage at which it becomes a science of causes, it will be usual as well as desirable first to obtain the highest generalizations, and then deduce the more special ones from them. Nor can I discover any foundation for the Baconian maxim, so much extolled by subsequent writers, except this: That before we attempt to explain deductively from more general laws any new class of phenomena, it is desirable to have gone as far as is practicable in ascertaining the empirical laws of those phenomena; so as to compare the results of deduction, not with one individual instance after another, but with general propositions expressive of the points of agreement which have been found among many instances. For if Newton had been obliged to verify the theory of gravitation, not by deducing from it Kepler's laws, but by deducing all the observed planetary positions which had served Kepler to establish those laws, the Newtonian theory would probably never have emerged from the state of an hypothesis.[*]

The applicability of these remarks to the special case under consideration, cannot admit of question. The science of the formation of character is a science of causes. The subject is one to which those among the canons of induction, by which laws of causation are ascertained, can be rigorously applied. It is, therefore, both natural and advisable to ascertain the simplest, which are necessarily the most general, laws of causation first, and to deduce the middle principles from them. In other words, Ethology, the deductive science, is a system of corollaries from Psychology, the experimental science.

---

[*] "To which," says Dr. Whewell, "we may add, that it is certain from the history of the subject, that in that case the hypothesis would never have been framed at all."

Dr. Whewell (*Philosophy of Discovery*, pp. 277—282) defends Bacon's rule against the preceding strictures. But his defence consists only in asserting and exemplifying a proposition which I had myself stated, viz. that though the largest generalizations may be the earliest made, they are not at first seen in their entire generality, but acquire it by degrees, as they are found to explain one class after another of phenomena. The laws of motion, for example, were not known to extend to the celestial regions, until the motions of the celestial bodies had been deduced from them. This however does not in any way affect the fact, that the middle principles of astronomy, the central force for example, and the law of the inverse square, could not have been discovered, if the laws of motion, which are so much more universal, had not been known first. On Bacon's system of step-by-step generalization, it would be impossible in any science to ascend higher than the empirical laws; a remark which Dr. Whewell's own Inductive Tables, referred to by him in support of his argument, amply bear out.

§ 6. Of these, the earlier alone has been, as yet, really conceived or studied as a science; the other, Ethology, is still to be created. But its creation has at length become practicable. The empirical laws, destined to verify its deductions, have been formed in abundance by every successive age of humanity; and the premises for the deductions are now sufficiently complete. Excepting the degree of uncertainty which still exists as to the extent of the natural differences of individual minds, and the physical circumstances on which these may be dependent, (considerations which are of secondary importance when we are considering mankind in the average, or *en masse*,) I believe most competent judges will agree that the general laws of the different constituent elements of human nature are even now sufficiently understood, to render it possible for a competent thinker to deduce from those laws with a considerable approach to certainty, the particular type of character which would be formed, in mankind generally, by any assumed set of circumstances. A science of Ethology, founded on the laws of Psychology, is therefore possible; though little has yet been done, and that little not at all systematically, towards forming it. The progress of this important but most imperfect science will depend on a double process: first, that of deducing theoretically the ethological consequences of particular circumstances of position, and comparing them with the recognised results of common experience; and secondly, the reverse operation; increased study of the various types of human nature that are to be found in the world; conducted by persons not only capable of analysing and recording the circumstances in which these types severally prevail, but also sufficiently acquainted with psychological laws, to be able to explain and account for the characteristics of the type, by the peculiarities of the circumstances: the residuum alone, when there proves to be any, being set down to the account of congenital predispositions.

For the experimental or *à posteriori* part of this process, the materials are continually accumulating by the observation of mankind. So far as thought is concerned, the great problem of Ethology is to deduce the requisite middle principles from the general laws of Psychology. The subject to be studied is, the origin and sources of all those qualities in human beings which are interesting to us, either as facts to be produced, to be avoided, or merely to be understood: and the object is, to determine, from the general laws of mind, combined with the general position of our species in the universe, what actual or possible combinations of circumstances are capable of promoting or of preventing the production of those qualities. A science which possesses middle principles of this kind, arranged in the order, not of causes, but of the effects which it is desirable to produce or to prevent, is duly prepared to be the foundation of the corresponding Art. And when Ethology shall be thus prepared, practical education will be the mere transformation of those principles into a parallel system of precepts, and the adaptation of these to the sum total of the individual circumstances which exist in each particular case.

It is hardly necessary again to repeat, that, as in every other deductive science, verification *à posteriori* must proceed *pari passu* with deduction *à priori*. The inference given by theory as to the type of character which would be formed by any given circumstances, must be tested by specific experience of those circumstances whenever obtainable; and the conclusions of the science as a whole, must undergo a perpetual verification and correction from the general remarks afforded by common experience respecting human nature in our own age, and by history respecting times gone by. The conclusions of theory cannot be trusted, unless confirmed by observation; nor those of observation, unless they can be affiliated to theory, by deducing them from the laws of human nature, and from a close analysis of the circumstances of the particular situation. It is the accordance of these two kinds of evidence separately taken—the consilience of *à priori* reasoning and specific experience—which forms the only sufficient ground for the principles of any science so "immersed in matter," dealing with such complex and concrete phenomena, as Ethology.

CHAPTER VI.

§ 1. NEXT after the science of individual man, comes the science of man in society : of the actions of collective masses of mankind, and the various phenomena which constitute social life.

If the formation of individual character is already a complex subject of study, this subject must be, in appearance at least, still more complex ; because the number of concurrent causes, all exercising more or less influence on the total effect, is greater, in the proportion in which a nation, or the species at large, exposes a larger surface to the operation of agents, psychological and physical, than any single individual. If it was necessary to prove, in opposition to an existing prejudice, that the simpler of the two is capable of being a subject of science ; the prejudice is likely to be yet stronger against the possibility of giving a scientific character to the study of Politics, and of the phenomena of Society. It is, accordingly, but of yesterday that the conception of a political or social science has existed, anywhere but in the mind of here and there an insulated thinker, generally very ill prepared for its realization : though the subject itself has of all others engaged the most general attention, and been a theme of interested and earnest discussions, almost from the beginning of recorded time.

The condition indeed of politics, as a branch of knowledge, was until very lately, and has scarcely even yet ceased to be, that which Bacon animadverted on, as the natural state of the sciences while their cultivation is abandoned to practitioners ; not being carried on as a branch of speculative inquiry, but only with a view to the exigencies of daily practice, and the *fructifera experimenta*, therefore, being aimed at, almost to the exclusion of the *lucifera*. Such was medical investigation, before physiology and natural history began to be cultivated as branches of general knowledge. The only questions examined were, what diet is wholesome, or what medicine will cure some given disease; without any previous systematic inquiry into the laws of nutrition, and of the healthy and morbid action of the different organs, on which laws the effect of any diet or medicine must evidently depend. And in politics, the questions which engaged general attention were similar:—Is such an enactment, or such a form of government, beneficial or the reverse—either universally, or to some particular community? without any previous inquiry into the general conditions by which the operation of legislative measures, or the effects produced by forms of government, are determined. Students in politics thus attempted to study the pathology and therapeutics of the social body, before they had laid the necessary foundation in its physiology; to cure disease, without understanding the laws of health. And the result was such as it must always be when persons, even of ability, attempt to deal with the complex questions of a science before its simpler and more elementary truths have been established.

No wonder that when the phenomena of society have so rarely been contemplated in the point of view characteristic of science, the philosophy of society should have made little progress; should contain few general propositions sufficiently precise and certain, for common inquirers to recognise in them a scientific character. The vulgar notion accordingly is, that all pretension to lay down general truths on politics and society is quackery; that no universality and no certainty are attainable in such matters. What partly excuses this common notion is, that it is really not without foundation in one particular sense. A large proportion of those who have laid claim to the character of philosophic politicians, have attempted, not to ascertain universal sequences, but to frame universal precepts. They have imagined some one form of government, or system of laws, to fit all cases; a pretension well meriting the ridicule with which it is treated by practitioners, and wholly unsupported by the analogy of the art to which, from the nature of its subject, that of politics must be the most nearly allied. No one now supposes it possible that one remedy can cure all diseases, or even the same disease in all constitutions and habits of body.

It is not necessary even to the perfection of a science, that the corresponding art should possess universal, or even general, rules. The phenomena of society might not only be completely dependent on known causes, but the mode of action of all those causes might be reducible to laws of considerable simplicity, and yet no two cases might admit of being treated in precisely the same manner. So great might be the variety of circumstances on which the results in different cases depend, that the art might not have a single general precept to give, except that of watching the circumstances of the particular case, and adapting our measures to the effects which, according to the principles of the science, result from those circumstances. But although, in so complicated a class of subjects, it is impossible to lay down practical maxims of universal application, it does not follow that the phenomena do not conform to universal laws.

§ 2. All phenomena of society are phenomena of human nature, generated by the action of outward circumstances upon masses of human beings: and if, therefore, the phenomena of human thought, feeling, and action, are subject to fixed laws, the phenomena of society cannot but conform to fixed laws, the consequence of the preceding. There is, indeed, no hope that these laws, though our knowledge of them were as certain and as complete as it is in astronomy, would enable us to predict the history of society, like that of the celestial appearances, for thousands of years to come. But the difference of certainty is not in the laws themselves, it is in the data to which these laws are to be applied. In astronomy the causes influencing the result are few, and change little, and that little according to known laws; we can ascertain what they are now, and thence determine what they will be at any epoch of a distant future. The data, therefore, in astronomy, are as certain as the laws themselves. The circumstances, on the contrary, which influence the condition and progress of society, are innumerable, and perpetually changing; and though they all change in obedience to causes, and therefore to laws, the multitude of the causes is so great as to defy our limited powers of calculation. Not to say that the impossibility of applying precise numbers to facts of such a description, would set an impassable limit to the possibility of calculating them beforehand, even if the powers of the human intellect were otherwise adequate to the task.

But, as before remarked, an amount of knowledge quite insufficient for prediction, may be most valuable for guidance. The science of society would have attained a very high point of perfection, if it enabled us, in any given condition of social affairs, in the condition for instance of Europe or any European country at the present time, to understand by what causes it had, in any and every particular, been made what it was; whether it was tending to any, and to what, changes; what effects each feature of its existing state was likely to produce in the future; and by what means any of those effects might be prevented, modified, or accelerated, or a different class of effects superinduced. There is nothing chimerical in the hope that general laws, sufficient to enable us to answer these various questions for any country or time with the individual circumstances of which we are well acquainted, do really admit of being ascertained; and that the other branches of human knowledge, which this undertaking presupposes, are so far advanced that the time is ripe for its commencement. Such is the object of the Social Science.

That the nature of what I consider the true method of the science may be made more palpable, by first showing what that method is not; it will be expedient to characterize briefly two radical misconceptions of the proper mode of philosophizing on society and government, one or other of which is, either explicitly or more often unconsciously, entertained by almost all who have meditated or argued respecting the logic of politics since the notion of treating it by strict rules, and on Baconian principles, has been current among the more advanced thinkers. These erroneous methods, if the word method can be applied to erroneous tendencies arising from the absence of any sufficiently distinct conception of method, may be termed the Experimental, or Chemical, mode of investigation, and the Abstract, or Geometrical, mode. We shall begin with the former.

CHAPTER VII.

OF THE CHEMICAL, OR EXPERIMENTAL, METHOD IN THE SOCIAL SCIENCE.

§ 1. ¡THE laws of the phenomena of society are, and can be, nothing but the laws of the actions and passions of human beings united together in the social state. Men, however, in a state of society, are still men; their actions and passions are obedient to the laws of individual human nature. Men are not, when brought together, converted into another kind of substance, with different properties; as hydrogen and oxygen are different from water, or as hydrogen, oxygen, carbon, and azote, are different from nerves, muscles, and tendons. Human beings in society have no properties but those which are derived from, and may be resolved into, the laws of the nature of individual man. In social phenomena the Composition of Causes is the universal law.

Now, the method of philosophizing which may be termed chemical overlooks this fact, and proceeds as if the nature of man as an individual were not concerned at all, or were concerned in a very inferior degree, in the operations of human beings in society. All reasoning in political or social affairs, grounded on principles of human nature, is objected to by reasoners of this sort, under such names as "abstract theory." For the direction of their opinions and conduct, they profess to demand, in all cases without exception, specific experience.

This mode of thinking is not only general with practitioners in politics, and with that very numerous class who (on a subject which no one, however ignorant, thinks himself incompetent to discuss) profess to guide themselves by common sense rather than by science; but is often countenanced by persons with greater pretensions to instruction; persons who, having sufficient acquaintance with books and with the current ideas to have heard that Bacon taught mankind to follow experience, and to ground their conclusions on facts instead of metaphysical dogmas—think that, by treating political facts in as directly experimental a method as chemical facts, they are showing themselves true Baconians, and proving their adversaries to be mere syllogizers and schoolmen. As, however, the notion of the applicability of experimental methods to political philosophy cannot coexist with any just conception of these methods themselves, the kind of arguments from experience which the chemical theory brings forth as its fruits (and which form the staple, in this country especially, of parliamentary and hustings oratory,) are such as, at no time since Bacon, would have been admitted to be valid in chemistry itself, or in any other branch of experimental science. They are such as these; that the prohibition of foreign commodities must conduce to national wealth, because England has flourished under it, or because countries in general which have adopted it have flourished; that our laws, or our internal administration, or our constitution, are excellent for a similar reason: and the eternal arguments from historical examples, from Athens or Rome, from the fires in Smithfield or the French Revolution.

I will not waste time in contending against modes of argumentation which no person, with the smallest practice in estimating evidence, could possibly be betrayed into; which draw conclusions of general application from a single unanalysed instance, or arbitrarily refer an effect to some one among its antecedents, without any process of elimination or comparison of instances. It is a rule both of justice and of good sense to grapple not with the absurdest, but with the most reasonable form of a wrong opinion. We shall suppose our inquirer acquainted with the true conditions of experimental investigation, and competent in point of acquirements for realizing them, so far as they can be realized. He shall know as much of the facts of history as mere erudition can teach— as much as can be proved by testimony, without the assistance of any theory; and if those mere facts, properly collated, can fulfil the conditions of a real induction, he shall be qualified for the task.

But, that no such attempt can have the smallest chance of success, has been abundantly shown in the tenth chapter of the Third Book.* We there examined whether effects which depend on a complication of causes can be made the subject of a true induction by observation and experiment; and concluded, on the most convincing grounds, that they cannot. Since, of all effects, none depend on so great a complication of causes as social phenomena, we might leave our case to rest in safety on that previous showing. But a logical principle as yet so little familiar to the ordinary run of thinkers, requires to be insisted on more than once, in order to make the due impression; and the present being the case which of all others exemplifies it the most strongly, there will be advantage in re-stating the grounds of the general maxim, as applied to the specialities of the class of inquiries now under consideration.

§ 2. The first difficulty which meets us in the attempt to apply experimental methods for ascertaining the laws of social phenomena, is that we are without the means of making artificial experiments. Even if we could contrive experiments at leisure, and try them without limit, we should do so under immense disadvantage; both from the impossibility of ascertaining and taking note of all the facts of each case, and because (those facts being in a perpetual state of change) before sufficient time had elapsed to ascertain the result of the experiment, some material circumstances would always have ceased to be the same. But it is unnecessary to consider the logical objections which would exist to the conclusiveness of our experiments, since we palpably never have the power of trying any. We can only watch those which nature produces, or which are produced for other reasons. We cannot adapt our logical means to our wants, by varying the circumstances as the exigencies of elimination may require. If the spontaneous instances, formed by cotemporary events and by the successions of phenomena recorded in history, afford a suffi-

* Vol. i. p. 494 to the end of the chapter.

cient variation of circumstances, an induction from specific
experience is attainable; otherwise not. The question to be
resolved is, therefore, whether the requisites for induction
respecting the causes of political effects or the properties of
political agents, are to be met with in history? including
under the term, cotemporary history. And in order to give
fixity to our conceptions, it will be advisable to suppose this
question asked in reference to some special subject of political
inquiry or controversy; such as that frequent topic of debate
in the present century, the operation of restrictive and prohibitory commercial legislation upon national wealth. Let this,
then, be the scientific question to be investigated by specific
experience.

§ 8. In order to apply to the case the most perfect of
the methods of experimental inquiry, the Method of Difference,
we require to find two instances, which tally in every particular except the one which is the subject of inquiry. If two
nations can be found which are alike in all natural advantages
and disadvantages; whose people resemble each other in every
quality, physical and moral, spontaneous and acquired; whose
habits, usages, opinions, laws and institutions are the same in
all respects, except that one of them has a more protective
tariff, or in other respects interferes more with the freedom of
industry; if one of these nations is found to be rich, and the
other poor, or one richer than the other, this will be an *experimentum crucis*: a real proof by experience, which of the
two systems is most favourable to national riches. But the
supposition that two such instances can be met with is manifestly absurd. Nor is such a concurrence even abstractedly
possible. Two nations which agreed in everything except
their commercial policy, would agree also in that. Differences
of legislation are not inherent and ultimate diversities; are
not properties of Kinds. They are effects of pre-existing
causes. If the two nations differ in this portion of their institutions, it is from some difference in their position, and
thence in their apparent interests, or in some portion or other
of their opinions, habits, [and tendencies; which opens a view

of further differences without any assignable limit, capable of
operating on their industrial prosperity, as well as on every
other feature of their condition, in more ways than can be
enumerated or imagined.   There is thus a demonstrated
impossibility of obtaining, in the investigations of the social
science, the conditions required for the most conclusive form
of inquiry by specific experience.

In the absence of the direct, we may next try, as in other
cases, the supplementary resource, called in a former place the
Indirect Method of Difference : which, instead of two instances
differing in nothing but the presence or absence of a given
circumstance, compares two *classes* of instances respectively
agreeing in nothing but the presence of a circumstance
on the one side and its absence on the other.   To choose
the most advantageous case conceivable, (a case far too advantageous to be ever obtained,) suppose that we compare
one nation which has a restrictive policy, with two or more
nations agreeing in nothing but in permitting free trade.   We
need not now suppose that either of these nations agrees with
the first in all its circumstances; one may agree with it in
some of its circumstances, and another in the remainder.
And it may be argued, that if these nations remain poorer
than the restrictive nation, it cannot be for want either of the
first or of the second set of circumstances, but it must be for
want of the protective system. If (we might say) the restrictive
nation had prospered from the one set of causes, the first
of the free-trade nations would have prospered equally; if
by reason of the other, the second would : but neither has :
therefore the prosperity was owing to the restrictions.   This
will be allowed to be a very favourable specimen of an argument from specific experience in politics, and if this be
inconclusive, it would not be easy to find another preferable
to it.

Yet, that it is inconclusive, scarcely requires to be pointed
out.   Why must the prosperous nation have prospered from
one cause exclusively ?   National prosperity is always the
collective result of a multitude of favourable circumstances ;
and of these, the restrictive nation may unite a greater number

than either of the others, though it may have all of those
circumstances in common with either one or the other of
them. Its prosperity may be partly owing to circumstances
common to it with one of those nations, and partly with the
other, while they, having each of them only half the number
of favourable circumstances, have remained inferior. So that
the closest imitation which can be made, in the social science,
of a legitimate induction from direct experience, gives but
a specious semblance of conclusiveness, without any real
value.

§ 4. The Method of Difference in either of its forms
being thus completely out of the question, there remains the
Method of Agreement. But we are already aware of how
little value this method is, in cases admitting Plurality of
Causes: and social phenomena are those in which the plurality
prevails in the utmost possible extent.

Suppose that the observer makes the luckiest hit which
could be given by any conceivable combination of chances:
that he finds two nations which agree in no circumstance
whatever, except in having a restrictive system, and in being
prosperous; or a number of nations, all prosperous, which
have no antecedent circumstances common to them all but
that of having a restrictive policy. It is unnecessary to go
into the consideration of the impossibility of ascertaining
from history, or even from cotemporary observation, that such
is really the fact: that the nations agree in no other circumstance capable of influencing the case. Let us suppose this
impossibility vanquished, and the fact ascertained that they
agree only in a restrictive system as an antecedent, and industrial prosperity as a consequent. What degree of presumption
does this raise, that the restrictive system caused the prosperity? One so trifling as to be equivalent to none at all.
That some one antecedent is the cause of a given effect,
because all other antecedents have been found capable of being
eliminated, is a just inference, only if the effect can have but
one cause. If it admits of several, nothing is more natural
than that each of these should separately admit of being

eliminated. Now, in the case of political phenomena, the supposition of unity of cause is not only wide of the truth, but at an immeasurable distance from it. The causes of every social phenomenon which we are particularly interested about, security, wealth, freedom, good government, public virtue, general intelligence, or their opposites, are infinitely numerous: especially the external or remote causes, which alone are, for the most part, accessible to direct observation. No one cause suffices of itself to produce any of these phenomena; while there are countless causes which have some influence over them, and may co-operate either in their production or in their prevention. From the mere fact, therefore, of our having been able to eliminate some circumstance, we can by no means infer that this circumstance was not instrumental to the effect in some of the very instances from which we have eliminated it. We can conclude that the effect is sometimes produced without it; but not that, when present, it does not contribute its share.

Similar objections will be found to apply to the Method of Concomitant Variations. If the causes which act upon the state of any society produced effects differing from one another in kind; if wealth depended on one cause, peace on another, a third made people virtuous, a fourth intelligent; we might, though unable to sever the causes from one another, refer to each of them that property of the effect which waxed as it waxed, and which waned as it waned. But every attribute of the social body is influenced by innumerable causes; and such is the mutual action of the coexisting elements of society, that whatever affects any one of the more important of them, will by that alone, if it does not affect the others directly, affect them indirectly. The effects, therefore, of different agents not being different in quality, while the quantity of each is the mixed result of all the agents, the variations of the aggregate cannot bear an uniform proportion to those of any one of its component parts.

§ 5. There remains the Method of Residues; which appears, on the first view, less foreign to this kind of inquiry

than the three other methods, because it only requires that
we should accurately note the circumstances of some one
country, or state of society. Making allowance, thereupon,
for the effect of all causes whose tendencies are known, the
residue which those causes are inadequate to explain may
plausibly be imputed to the remainder of the circumstances
which are known to have existed in the case. Something
similar to this is the method which Coleridge[*] describes himself as having followed in his political essays in the *Morning
Post*. "On every great occurrence I endeavoured to discover
in past history the event that most nearly resembled it. I
procured, whenever it was possible, the contemporary historians, memorialists, and pamphleteers. Then fairly subtracting the points of difference from those of likeness, as the
balance favoured the former or the latter, I conjectured that
the result would be the same or different. As, for instance, in
the series of essays entitled 'A comparison of France under
Napoleon with Rome under the first Cæsars,' and in those
which followed, 'on the probable final restoration of the
Bourbons.' The same plan I pursued at the commencement
of the Spanish Revolution, and with the same success, taking
the war of the United Provinces with Philip II. as the
groundwork of the comparison." In this inquiry he no doubt
employed the Method of Residues; for, in "subtracting the
points of difference from those of likeness," he doubtless
weighed, and did not content himself with numbering, them:
he doubtless took those points of agreement only, which he
presumed from their own nature to be capable of influencing
the effect, and, allowing for that influence, concluded that the
remainder of the result would be referable to the points of
difference.

Whatever may be the efficacy of this method, it is, as we
long ago remarked, not a method of pure observation and
experiment; it concludes, not from a comparison of instances,
but from the comparison of an instance with the result of a
previous deduction. Applied to social phenomena, it pre-

---

* *Biographia Literaria*, i. 214.

supposes that the causes from which part of the effect proceeded are already known; and as we have shown that these
cannot have been known by specific experience, they must
have been learnt by deduction from principles of human
nature; experience being called in only as a supplementary
resource, to determine the causes which produced an unexplained residue. But if the principles of human nature may
be had recourse to for the establishment of some political
truths, they may for all. If it be admissible to say, England
must have prospered by reason of the prohibitory system,
because after allowing for all the other tendencies which have
been operating, there is a portion of prosperity still to be
accounted for; it must be admissible to go to the same source
for the effect of the prohibitory system, and examine what
account the laws of human motives and actions will enable us
to give of *its* tendencies. Nor, in fact, will the experimental
argument amount to anything, except in verification of a conclusion drawn from those general laws. For we may subtract
the effect of one, two, three, or four causes, but we shall never
succeed in subtracting the effect of all causes except one:
while it would be a curious instance of the dangers of too
much caution, if, to avoid depending on *à priori* reasoning
concerning the effect of a single cause, we should oblige ourselves to depend on as many separate *à priori* reasonings as
there are causes operating concurrently with that particular
cause in some given instance.

We have now sufficiently characterized the gross misconception of the mode of investigation proper to political
phenomena, which I have termed the Chemical Method. So
lengthened a discussion would not have been necessary, if the
claim to decide authoritatively on political doctrines were confined to persons who had competently studied any one of the
higher departments of physical science. But since the generality of those who reason on political subjects, satisfactorily
to themselves and to a more or less numerous body of
admirers, know nothing whatever of the methods of physical
investigation beyond a few precepts which they continue to
rot after Bacon, being entirely unaware that Bacon's con-

ception of scientific inquiry has done its work, and that science has now advanced into a higher stage; there are probably many to whom such remarks as the foregoing may still be useful. In an age in which chemistry itself, when attempting to deal with the more complex chemical sequences, those of the animal or even the vegetable organism, has found it necessary to become, and has succeeded in becoming, a Deductive Science—it is not to be apprehended that any person of scientific habits, who has kept pace with the general progress of the knowledge of nature, can be in danger of applying the methods of elementary chemistry to explore the sequences of the most complex order of phenomena in existence.

CHAPTER VIII.

OF THE GEOMETRICAL, OR ABSTRACT METHOD.

§ 1. THE misconception discussed in the preceding chapter is, as we said, chiefly committed by persons not much accustomed to scientific investigation: practitioners in politics, who rather employ the commonplaces of philosophy to justify their practice, than seek to guide their practice by philosophic principles: or imperfectly educated persons, who, in ignorance of the careful selection and elaborate comparison of instances required for the formation of a sound theory, attempt to found one upon a few coincidences which they have casually noticed.

The erroneous method of which we are now to treat, is, on the contrary, peculiar to thinking and studious minds. It never could have suggested itself but to persons of some familiarity with the nature of scientific research; who,—being aware of the impossibility of establishing, by casual observation or direct experimentation, a true theory of sequences so complex as are those of the social phenomena,—have recourse to the simpler laws which are immediately operative in those phenomena, and which are no other than the laws of the nature of the human beings therein concerned. These thinkers perceive (what the partisans of the chemical or experimental theory do not) that the science of society must necessarily be deductive. But, from an insufficient consideration of the specific nature of the subject matter,—and often because (their own scientific education having stopped short in too early a stage) geometry stands in their minds as the type of all deductive science, it is to geometry rather than to astronomy and ʰural philosophy, that they unconsciously assimilate the ʰtive science of society.

Among the differences between geometry (a science of coexistent facts, altogether independent of the laws of the succession of phenomena), and those physical Sciences of Causation which have been rendered deductive, the following is one of the most conspicuous: That geometry affords no room for what so constantly occurs in mechanics and its applications, the case of conflicting forces; of causes which counteract or modify one another. In mechanics we continually find two or more moving forces producing, not motion, but rest; or motion in a different direction from that which would have been produced by either of the generating forces. It is true that the effect of the joint forces is the same when they act simultaneously, as if they had acted one after another, or by turns; and it is in this that the difference between mechanical and chemical laws consists. But still the effects, whether produced by successive or by simultaneous action, do, wholly or in part, cancel one another: what the one force does, the other, partly or altogether, undoes. There is no similar state of things in geometry. The result which follows from one geometrical principle has nothing that conflicts with the result which follows from another. What is proved true from one geometrical theorem, what would be true if no other geometrical principles existed, cannot be altered and made no longer true by reason of some other geometrical principle. What is once proved true is true in all cases, whatever supposition may be made in regard to any other matter.

Now a conception, similar to this last, would appear to have been formed of the social science, in the minds of the earlier of those who have attempted to cultivate it by a deductive method. Mechanics would be a science very similar to geometry, if every motion resulted from one force alone, and not from a conflict of forces. In the geometrical theory of society, it seems to be supposed that this is really the case with the social phenomena; that each of them results always from only one force, one single property of human nature.

At the point which we have now reached, it cannot be

necessary to say anything either in proof or in illustration of
the assertion that such is not the true character of the social
phenomena. There is not, among these most complex and
(for that reason) most modifiable of all phenomena, any one
over which innumerable forces do not exercise influence;
which does not depend on a conjunction of very many causes.
We have not, therefore, to prove the notion in question to be
an error, but to prove that the error has been committed;
that so mistaken a conception of the mode in which the
phenomena of society are produced, has actually been ascertained.

§ 2. One numerous division of the reasoners who have
treated social facts according to geometrical methods, not
admitting any modification of one law by another, must for
the present be left out of consideration; because in them this
error is complicated with, and is the effect of, another fundamental misconception, of which we have already taken some
notice, and which will be further treated of before we conclude. I speak of those who deduce political conclusions not
from laws of nature, not from sequences of phenomena, real
or imaginary, but from unbending practical maxims. Such,
for example, are all who found their theory of politics on what
is called abstract right, that is to say, on universal precepts;
a pretension of which we have already noticed the chimerical
nature. Such, in like manner, are those who make the assumption of a social contract, or any other kind of original obligation, and apply it to particular cases by mere interpretation.
But in this the fundamental error is the attempt to treat an
art like a science, and to have a deductive art; the irra-
tionality of which will be shown in a future chapter. It will
be proper to take our exemplification of the geometrical theory
from those thinkers who have avoided this additional error,
and who entertain, so far, a juster idea of the nature of
political inquiry.

We may cite, in the first instance, those who assume as
the principle of their political philosophy that government is

founded on fear; that the dread of each other is the one motive by which human beings were originally brought into a state of society, and are still held in it. Some of the earlier scientific inquirers into politics, in particular Hobbes, assumed this proposition, not by implication, but avowedly, as the foundation of their doctrine, and attempted to build a complete philosophy of politics thereupon. It is true that Hobbes did not find this one maxim sufficient to carry him through the whole of his subject, but was obliged to eke it out by the double sophism of an original contract. I call this a double sophism; first, as passing off a fiction for a fact, and, secondly, assuming a practical principle, or precept, as the basis of a theory; which is a *petitio principii*, since (as we noticed in treating of that Fallacy) every rule of conduct, even though it be so binding a one as the observance of a promise, must rest its own foundations on the theory of the subject, and the theory, therefore, cannot rest upon it.

§ 8. Passing over less important instances, I shall come at once to the most remarkable example afforded by our own times of the geometrical method in politics; emanating from persons who were well aware of the distinction between science and art; who knew that rules of conduct must follow, not precede, the ascertainment of laws of nature, and that the latter, not the former, is the legitimate field for the application of the deductive method. I allude to the interest-philosophy of the Bentham school.

The profound and original thinkers who are commonly known under this description, founded their general theory of government on one comprehensive premise, namely, that men's actions are always determined by their interests. There is an ambiguity in this last expression; for, as the same philosophers, especially Bentham, gave the name of an interest to anything which a person likes, the proposition may be understood to mean only this, that men's actions are always determined by their wishes. In this sense, however, it would not bear out any of the consequences which these writers drew from it; and the word, therefore, in their political reasonings, must be understood to mean (which is also the explanation they themselves, on such occasions, gave of it) what is commonly termed private, or worldly, interest.

Taking the doctrine, then, in this sense, an objection presents itself *in limine* which might be deemed a fatal one, namely, that so sweeping a proposition is far from being universally true. Human beings are not governed in all their actions by their worldly interests. This, however, is by no means so conclusive an objection as it at first appears; because in politics we are for the most part concerned with the conduct not of individual persons, but either of a series of persons (as a succession of kings) or a body or mass of persons, as a nation, an aristocracy, or a representative assembly. And whatever is true of a large majority of mankind, may without much error be taken for true of any succession of persons, considered as a whole, or of any collection of persons in which the act of the majority becomes the act of the whole body. Although, therefore, the maxim is sometimes expressed in a manner unnecessarily paradoxical, the consequences drawn from it will hold equally good if the assertion be limited as follows—Any succession of persons, or the majority of any body of persons, will be governed in the bulk of their conduct by their personal interests. We are bound to allow to this school of thinkers the benefit of this more rational statement of their fundamental maxim, which is also in strict conformity to the explanations which, when considered to be called for, have been given by themselves.

The theory goes on to infer, quite correctly, that if the actions of mankind are determined in the main by their selfish interests, the only rulers who will govern according to the interest of the governed, are those whose selfish interests are in accordance with it. And to this is added a third proposition, namely, that no rulers have their selfish interest identical with that of the governed, unless it be rendered so by accountability, that is, by dependence on the will of the governed. In other words (and as the result of the whole), that the desire of retaining or the fear of losing their power, and whatever is thereon consequent, is the sole motive which can be relied on for producing on the part of rulers a course of conduct in accordance with the general interest.

We have thus a fundamental theorem of political science, consisting of three syllogisms, and depending chiefly on two general premises, in each of which a certain effect is considered as determined only by one cause, not by a concurrence of causes. In the one, it is assumed that the actions of average rulers are determined solely by self-interest; in the other, that the sense of identity of interest with the governed, is produced and producible by no other cause than responsibility.

Neither of these propositions is by any means true; the last is extremely wide of the truth.

It is not true that the actions even of average rulers are wholly, or anything approaching to wholly, determined by their personal interest, or even by their own opinion of their personal interest. I do not speak of the influence of a sense of duty, or feelings of philanthropy, motives never to be mainly relied on, though (except in countries or during periods of great moral debasement) they influence almost all rulers in some degree, and some rulers in a very great degree. But I insist only on what is true of all rulers, viz. that the character and course of their actions is largely influenced (independently of personal calculation) by the habitual sentiments and feelings, the general modes of thinking and acting, which prevail throughout the community of which they are members; as well as by the feelings, habits, and modes of thought which characterize the particular class in that community to which they themselves belong. And no one will understand or be able to decypher their system of conduct, who does not take all these things into account. They are also much influenced by the maxims and traditions which have descended to them from other rulers, their predecessors; which maxims and traditions have been known to retain an ascendancy during long periods, even in opposition to the private interests of the rulers for the time being. I put aside the influence of other less general causes. Although, therefore, the private interest of the rulers or of the ruling class is a very powerful force, constantly in action, and exercising the most important influence upon their conduct; there is also, in what they do, a large portion which that private interest by no means affords a sufficient explanation of: and even the particulars which constitute the goodness or badness of their government, are in some, and no small degree, influenced by those among the circumstances acting upon them, which cannot, with any propriety, be included in the term self-interest.

Turning now to the other proposition, that responsibility to the governed is the only cause capable of producing in the rulers a sense of identity of interest with the community; this is still less admissible as an universal truth, than even the former. I am not speaking of perfect identity of interest, which is an impracticable chimera; which, most assuredly, responsibility to the people does not give. I speak of identity in essentials; and the essentials are different at different places and times. There are a large number of cases in which those things which it is most for the general interest that the rulers should do, are also those which they are prompted to do by their strongest personal interest, the consolidation of their power. The suppression, for instance, of anarchy and resistance to law,—the complete establishment of the authority of the central government, in a state of society like that of Europe in the middle ages,—is one of the strongest interests of the people, and also of the rulers simply because they are the rulers: and responsibility on their part could not strengthen, though in many conceivable ways it might weaken, the motives prompting them to pursue this object. During the greater part of the reign of Queen Elizabeth, and of many other monarchs who might be named, the sense of identity of interest between the sovereign and the majority of the people was probably stronger than it usually is in responsible governments: everything that the people had most at heart, the monarch had at heart too. Had Peter the Great, or the rugged savages whom he began to civilize, the truest inclination towards the things which were for the real interest of those savages?

I am not here attempting to establish a theory of government, and am not called upon to determine the proportional weight which ought to be given to the circumstances which this school of geometrical politicians left out of their system, and those which they took into it. I am only concerned to show that their method was unscientific; not to measure the amount of error which may have affected their practical conclusions.

It is but justice to them, however, to remark, that their mistake was not so much one of substance as of form; and consisted in presenting in a systematic shape, and as the scientific treatment of a great philosophical question, what should have passed for that which it really was, the mere polemics of the day. Although the actions of rulers are by no means wholly determined by their selfish interests, it is chiefly as a security against those selfish interests that constitutional checks are required; and for that purpose such checks, in England, and the other nations of modern Europe, can in no manner be dispensed with. It is likewise true, that in these same nations, and in the present age, responsibility to the governed is the only means practically available to create a feeling of identity of interest, in the cases, and on the points, where that feeling does not sufficiently exist. To all this, and to the arguments which may be founded on it in favour of measures for the correction of our representative system, I have nothing to object; but I confess my regret, that the small though highly important portion of the philosophy of government, which was wanted for the immediate purpose of serving the cause of parliamentary reform, should have been held forth by thinkers of such eminence as a complete theory.

It is not to be imagined possible, nor is it true in point of fact, that these philosophers regarded the few premises of their theory as including all that is required for explaining social phenomena, or for determining the choice of forms of govern-

ment and measures of legislation and administration. They
were too highly instructed, of too comprehensive intellect, and
some of them of too sober and practical a character, for such
an error. They would have applied and did apply their principles with innumerable allowances. But it is not allowances
that are wanted. There is little chance of making due amends
in the superstructure of a theory for the want of sufficient
breadth in its foundations. It is unphilosophical to construct
a science out of a few of the agencies by which the phenomena
are determined, and leave the rest to the routine of practice
or the sagacity of conjecture. We either ought not to pretend
to scientific forms, or we ought to study all the determining
agencies equally, and endeavour, so far as it can be done, to
include all of them within the pale of the science; else we
shall infallibly bestow a disproportionate attention upon those
which our theory takes into account, while we misestimate the
rest, and probably underrate their importance. That the deductions should be from the whole and not from a part only of
the laws of nature that are concerned, would be desirable even
if those omitted were so insignificant in comparison with the
others, that they might, for most purposes and on most occasions, be left out of the account. But this is far indeed from
being true in the social science. The phenomena of society
do not depend, in essentials, on some one agency or law of
human nature, with only inconsiderable modifications from
others. The whole of the qualities of human nature influence
those phenomena, and there is not one which influences them
in a small degree. There is not one, the removal or any great
alteration of which would not materially affect the whole aspect
of society, and change more or less the sequences of social
phenomena generally.

The theory which has been the subject of these remarks is
in this country at least, the principal cotemporary example of
what I have styled the geometrical method of philosophizing
in the social science; and our examination of it has, for this
reason, been more detailed than would otherwise have been
suitable to a work like the present. Having now sufficiently

illustrated the two erroneous methods, we shall pass without further preliminary to the true method; that which proceeds (conformably to the practice of the more complex physical sciences) deductively indeed, but by deduction from many, not from one or a very few, original premises; considering each effect as (what it really is) an aggregate result of many causes, operating sometimes through the same, sometimes through different mental agencies, or laws of human nature.

CHAPTER IX.

OF THE PHYSICAL, OR CONCRETE DEDUCTIVE METHOD.

§ 1. AFTER what has been said to illustrate the nature of the inquiry into social phenomena, the general character of the method proper to that inquiry is sufficiently evident, and needs only to be recapitulated, not proved. However complex the phenomena, all their sequences and coexistences result from the laws of the separate elements. The effect produced, in social phenomena, by any complex set of circumstances, amounts precisely to the sum of the effects of the circumstances taken singly: and the complexity does not arise from the number of the laws themselves, which is not remarkably great; but from the extraordinary number and variety of the data or elements—of the agents which, in obedience to that small number of laws, co-operate towards the effect. The Social Science, therefore (which, by a convenient barbarism, has been termed Sociology,) is a deductive science; not, indeed, after the model of geometry, but after that of the more complex physical sciences. It infers the law of each effect from the laws of causation on which that effect depends; not, however, from the law merely of one cause, as in the geometrical method; but by considering all the causes which conjunctly influence the effect, and compounding their laws with one another. Its method, in short, is the Concrete Deductive Method; that of which astronomy furnishes the most perfect, natural philosophy a somewhat less perfect example, and the employment of which, with the adaptations and precautions required by the subject, is beginning to regenerate physiology.

Nor does it admit of doubt, that similar adaptations and precautions are indispensable in sociology. In applying, to that most complex of all studies, what is demonstrably the sole method capable of throwing the light of science even upon phenomena of a far inferior degree of complication, we ought to be aware that the same superior complexity which renders the instrument of Deduction more necessary, renders it also more precarious; and we must be prepared to meet, by appropriate contrivances, this increase of difficulty.

The actions and feelings of human beings in the social state, are, no doubt, entirely governed by psychological and ethological laws: whatever influence any cause exercises upon the social phenomena, it exercises through those laws. Supposing therefore the laws of human actions and feelings to be sufficiently known, there is no extraordinary difficulty in determining from those laws, the nature of the social effects which any given cause tends to produce. But when the question is that of compounding several tendencies together, and computing the aggregate result of many coexistent causes; and especially when, by attempting to predict what will actually occur in a given case, we incur the obligation of estimating and compounding the influences of all the causes which happen to exist in that case; we attempt a task, to proceed far in which, surpasses the compass of the human faculties.

If all the resources of science are not sufficient to enable us to calculate à priori, with complete precision, the mutual action of three bodies gravitating towards one another; it may be judged with what prospect of success we should endeavour to calculate the result of the conflicting tendencies which are acting in a thousand different directions and promoting a thousand different changes at a given instant in a given society: although we might and ought to be able, from the laws of human nature, to distinguish correctly enough the tendencies themselves, so far as they depend on causes accessible to our observation; and to determine the direction which each of them, if acting alone, would impress upon society, as well as, in a general way at least, to pronounce that some of these tendencies are more powerful than others.

But, without dissembling the necessary imperfections of the à priori method when applied to such a subject, neither ought we, on the other hand, to exaggerate them. The same objections, which apply to the Method of Deduction in this its most difficult employment, apply to it, as we formerly showed,* in its easiest; and would even there have been insuperable, if there had not existed, as was then fully explained, an appropriate remedy. This remedy consists in the process which, under the name of Verification, we have characterized as the third essential constituent part of the Deductive Method; that of collating the conclusions of the ratiocination either with the concrete phenomena themselves, or, when such are obtainable, with their empirical laws. The ground of confidence in any concrete deductive science is not the *à priori* reasoning itself, but the accordance between its results and those of observation *à posteriori*. Either of these processes, apart from the other, diminishes in value as the subject increases in complication, and this in so rapid a ratio as soon to become entirely worthless; but the reliance to be placed in the concurrence of the two sorts of evidence, not only does not diminish in anything like the same proportion, but is not necessarily much diminished at all. Nothing more results than a disturbance in the order of precedency of the two processes, sometimes amounting to its actual inversion: insomuch that instead of deducing our conclusions by reasoning, and verifying them by observation, we in some cases begin by obtaining them conjecturally from specific experience, and afterwards connect them with the principles of human nature by *à priori* reasonings, which reasonings are thus a real Verification.

The only thinker who, with a competent knowledge of scientific methods in general, has attempted to characterize the Method of Sociology, M. Comte, considers this inverse order as inseparably inherent in the nature of sociological speculation. He looks upon the social science as essentially consisting of generalizations from history, verified, not originally suggested, by deduction from the laws of human nature. Though there is a truth contained in this opinion, of which I

---

* Supra, vol. i. p. 500.

shall presently endeavour to show the importance, I cannot but think that this truth is enunciated in too unlimited a manner, and that there is considerable scope in sociological inquiry for the direct, as well as for the inverse, Deductive Method.

It will, in fact, be shown in the next chapter, that there is a kind of sociological inquiries to which, from their prodigious complication, the method of direct deduction is altogether inapplicable, while by a happy compensation it is precisely in these cases that we are able to obtain the best empirical laws: to these inquiries, therefore, the Inverse Method is exclusively adapted. But there are also, as will presently appear, other cases in which it is impossible to obtain from direct observation anything worthy the name of an empirical law; and it fortunately happens that these are the very cases in which the Direct Method is least affected by the objection which undoubtedly must always affect it in a certain degree.

We shall begin, then, by looking at the Social Science as a science of direct Deduction, and considering what can be accomplished in it, and under what limitations, by that mode of investigation. We shall, then, in a separate chapter, examine and endeavour to characterize the inverse process.

§ 2. It is evident, in the first place, that Sociology, considered as a system of deductions à priori, cannot be a science of positive predictions, but only of tendencies. We may be able to conclude, from the laws of human nature applied to the circumstances of a given state of society, that a particular cause will operate in a certain manner unless counteracted; but we can never be assured to what extent or amount it will so operate, or affirm with certainty that it will not be counteracted; because we can seldom know, even approximately, all the agencies which may coexist with it, and still less calculate the collective result of so many combined elements. The remark, however, must here be once more repeated, that knowledge insufficient for prediction may be most valuable for guidance. It is not necessary for the

wise conduct of the affairs of society, no more than of any
one's private concerns, that we should be able to foresee infallibly the results of what we do. We must seek our objects by
means which may perhaps be defeated, and take precautions
against dangers which possibly may never be realized. The
aim of practical politics is to surround any given society with
the greatest possible number of circumstances of which the
tendencies are beneficial, and to remove or counteract, as far
as practicable, those of which the tendencies are injurious. A
knowledge of the tendencies only, though without the power
of accurately predicting their conjunct result, gives us to a
certain extent this power.

It would, however, be an error to suppose that even with
respect to tendencies, we could arrive in this manner at any
great number of propositions which will be true in all societies
without exception. Such a supposition would be inconsistent
with the eminently modifiable nature of the social phenomena,
and the multitude and variety of the circumstances by which
they are modified; circumstances never the same, or even
nearly the same, in two different societies, or in two different
periods of the same society. This would not be so serious an
obstacle if, though the causes acting upon society in general
are numerous, those which influence any one feature of society
were limited in number; for we might then insulate any particular social phenomenon, and investigate its laws without
disturbance from the rest. But the truth is the very opposite
of this. Whatever affects, in an appreciable degree, any one
element of the social state, affects through it all the other
elements. The mode of production of all social phenomena is
one great case of Intermixture of Laws. We can never either
understand in theory or command in practice the condition of
a society in any one respect, without taking into consideration its condition in all other respects. There is no social
phenomenon which is not more or less influenced by every
other part of the condition of the same society, and therefore
by every cause which is influencing any other of the contemporaneous social phenomena. There is, in short, what physiologists term a *consensus*, similar to that existing among the

various organs and functions of the physical frame of man and the more perfect animals; and constituting one of the many analogies which have rendered universal such expressions as the "body politic" and "body natural." It follows from this *consensus*, that unless two societies could be alike in all the circumstances which surround and influence them, (which would imply their being alike in their previous history,) no portion whatever of the phenomena will, unless by accident, precisely correspond; no one cause will produce exactly the same effects in both. Every cause, as its effect spreads through society, comes somewhere in contact with different sets of agencies, and thus has its effects on some of the social phenomena differently modified; and these differences, by their reaction, produce a difference even in those of the effects which would otherwise have been the same. We can never, therefore, affirm with certainty that a cause which has a particular tendency in one people or in one age will have exactly the same tendency in another, without referring back to our premises, and performing over again for the second age or nation, that analysis of the whole of its influencing circumstances which we had already performed for the first. The deductive science of society will not lay down a theorem, asserting in an universal manner the effect of any cause; but will rather teach us how to frame the proper theorem for the circumstances of any given case. It will not give the laws of society in general, but the means of determining the phenomena of any given society from the particular elements or data of that society.

All the general propositions which can be framed by the deductive science, are therefore, in the strictest sense of the word, hypothetical. They are grounded on some supposititious set of circumstances, and declare how some given cause would operate in those circumstances, supposing that no others were combined with them. If the set of circumstances supposed have been copied from those of any existing society, the conclusions will be true of that society, provided, and in as far as, the effect of those circumstances shall not be modified by others which have not been taken into the account.

If we desire a nearer approach to concrete truth, we can only aim at it by taking, or endeavouring to take, a greater number of individualizing circumstances into the computation.

Considering, however, in how accelerating a ratio the uncertainty of our conclusions increases, as we attempt to take the effect of a greater number of concurrent causes into our calculations; the hypothetical combinations of circumstances on which we construct the general theorems of the science, cannot be made very complex, without so rapidly-accumulating a liability to error as must soon deprive our conclusions of all value. This mode of inquiry, considered as a means of obtaining general propositions, must, therefore, on pain of frivolity, be limited to those classes of social facts which, though influenced like the rest by all sociological agents, are under the *immediate* influence, principally at least, of a few only.

§ 3. Notwithstanding the universal *consensus* of the social phenomena, whereby nothing which takes place in any part of the operations of society is without its share of influence on every other part; and notwithstanding the paramount ascendancy which the general state of civilization and social progress in any given society must hence exercise over all the partial and subordinate phenomena; it is not the less true that different species of social facts are in the main dependent, immediately and in the first resort, on different kinds of causes; and therefore not only may with advantage, but must, be studied apart: just as in the natural body we study separately the physiology and pathology of each of the principal organs and tissues, though every one is acted upon by the state of all the others: and though the peculiar constitution and general state of health of the organism co-operates with, and often preponderates over, the local causes, in determining the state of any particular organ.

On these considerations is grounded the existence of distinct and separate, though not independent, branches or departments of sociological speculation.

There is, for example, one large class of social phenomena,

in which the immediately determining causes are principally those which act through the desire of wealth; and in which the psychological law mainly concerned is the familiar one, that a greater gain is preferred to a smaller. I mean, of course, that portion of the phenomena of society which emanate from the industrial, or productive, operations of mankind; and from those of their acts through which the distribution of the products of those industrial operations takes place, in so far as not effected by force, or modified by voluntary gift. By reasoning from that one law of human nature, and from the principal outward circumstances (whether universal or confined to particular states of society) which operate upon the human mind through that law, we may be enabled to explain and predict this portion of the phenomena of society, so far as they depend on that class of circumstances only; overlooking the influence of any other of the circumstances of society; and therefore neither tracing back the circumstances which we do take into account, to their possible origin in some other facts in the social state, nor making allowance for the manner in which any of those other circumstances may interfere with, and counteract or modify, the effect of the former. A science may thus be constructed, which has received the name of Political Economy.

The motive which suggests the separation of this portion of the social phenomena from the rest, and the creation of a distinct science relating to them is,—that they do *mainly* depend, at least in the first resort, on one class of circumstances only; and that even when other circumstances interfere, the ascertainment of the effect due to the one class of circumstances alone, is a sufficiently intricate and difficult business to make it expedient to perform it once for all, and then allow for the effect of the modifying circumstances; especially as certain fixed combinations of the former are apt to recur often, in conjunction with ever-varying circumstances of the latter class.

Political Economy, as I have said on another occasion, concerns itself only with " such of the phenomena of the social state as take place in consequence of the pursuit of wealth.

It makes entire abstraction of every other human passion or motive; except those which may be regarded as perpetually antagonizing principles to the desire of wealth, namely, aversion to labour, and desire of the present enjoyment of costly indulgences.   These it takes, to a certain extent, into its calculations, because these do not merely, like our other desires, occasionally conflict with the pursuit of wealth, but accompany it always as a drag or impediment, and are therefore inseparably mixed up in the consideration of it.   Political Economy considers mankind as occupied solely in acquiring and consuming wealth; and aims at showing what is the course of action into which mankind, living in a state of society, would be impelled, if that motive, except in the degree in which it is checked by the two perpetual counter-motives above adverted to, were absolute ruler of all their actions. Under the influence of this desire, it shows mankind accumulating wealth, and employing that wealth in the production of other wealth; sanctioning by mutual agreement the institution of property; establishing laws to prevent individuals from encroaching upon the property of others by force or fraud; adopting various contrivances for increasing the productiveness of their labour; settling the division of the produce by agreement, under the influence of competition (competition itself being governed by certain laws, which laws are therefore the ultimate regulators of the division of the produce); and employing certain expedients (as money, credit, &c.) to facilitate the distribution.   All these operations, though many of them are really the result of a plurality of motives, are considered by political economy as flowing solely from the desire of wealth.   The science then proceeds to investigate the laws which govern these several operations, under the supposition that man is a being who is determined, by the necessity of his nature, to prefer a greater portion of wealth to a smaller, in all cases, without any other exception than that constituted by the two counter-motives already specified.   Not that any political economist was ever so absurd as to suppose that mankind are really thus constituted, but because this is the mode in which science must necessarily proceed.   When an effect depends on a concurrence of causes, these causes must be studied one at a time, and their laws separately investigated, if we wish, through the causes, to obtain the power of either predicting or controlling the effect; since the law of the effect is compounded of the laws of all the causes which determine it. The law of the centripetal and that of the projectile force must have been known, before the motions of the earth and planets could be explained, or many of them predicted. The same is the case with the conduct of man in society. In order to judge how he will act under the variety of desires and aversions which are concurrently operating upon him, we must know how he would act under the exclusive influence of each one in particular. There is, perhaps, no action of a man's life in which he is neither under the immediate nor under the remote influence of any impulse but the mere desire of wealth. With respect to those parts of human conduct of which wealth is not even the principal object, to these political economy does not pretend that its conclusions are applicable. But there are also certain departments of human affairs, in which the acquisition of wealth is the main and acknowledged end. It is only of these that political economy takes notice. The manner in which it necessarily proceeds is that of treating the main and acknowledged end as if it were the sole end; which, of all hypotheses equally simple, is the nearest to the truth. The political economist inquires, what are the actions which would be produced by this desire, if within the departments in question it were unimpeded by any other. In this way a nearer approximation is obtained than would otherwise be practicable to the real order of human affairs in those departments. This approximation has then to be corrected by making proper allowance for the effects of any impulses of a different description, which can be shown to interfere with the result in any particular case. Only in a few of the most striking cases (such as the important one of the principle of population) are these corrections interpolated into the expositions of political economy itself; the strictness of purely scientific arrangement being thereby somewhat departed from, for the sake of practical utility. So far as it is known or may be presumed, that the conduct of mankind in the pursuit of wealth is under the collateral influence of any other of the properties of our nature, than the desire of obtaining the greatest quantity of wealth with the least labour and self-denial, the conclusions of political economy will so far fail of being applicable to the explanation or prediction of real events, until they are modified by a correct allowance for the degree of influence exercised by the other cause."*

Extensive and important practical guidance may be derived, in any given state of society, from general propositions such as those above indicated; even though the modifying influence of the miscellaneous causes which the theory does not take into account, as well as the effect of the general social changes in progress, be provisionally overlooked. And though it has been a very common error of political economists to draw conclusions from the elements of one state of society, and apply them to other states in which many of the elements are not the same; it is even then not difficult, by tracing back the demonstrations, and introducing the new premises in their proper places, to make the same general course of argument which served for the one case, serve for the others too.

For example, it has been greatly the custom of English political economists to discuss the laws of the distribution of the produce of industry, on a supposition which is scarcely realized anywhere out of England and Scotland, namely, that the produce is "shared among three classes, altogether distinct from one another, labourers, capitalists, and landlords; and that all these are free agents, permitted in law and in fact to set upon their labour, their capital, and their land, whatever price they are able to get for it. The conclusions of the science, being all adapted to a society thus constituted, require to be revised whenever they are applied to any other. They are inapplicable where the only capitalists are the landlords, and the labourers are their property, as in slave countries. They are inapplicable where the almost universal landlord is the

---

* *Essays on some Unsettled Questions of Political Economy*, pp. 137-140.

state as in India. They are inapplicable where the agricultural labourer is generally the owner both of the land itself and of the capital, as frequently in France, or of the capital only, as in Ireland." But though it may often be very justly objected to the existing race of political economists " that they attempt to construct a permanent fabric out of transitory materials; that they take for granted the immutability of arrangements of society, many of which are in their nature fluctuating or progressive, and enunciate with as little qualification as if they were universal and absolute truths, propositions which are perhaps applicable to no state of society except the particular one in which the writer happened to live ;" this does not take away the value of the propositions, considered with reference to the state of society from which they were drawn. And even as applicable to other states of society, "it must not be supposed that the science is so incomplete and unsatisfactory as this might seem to prove. Though many of its conclusions are only locally true, its method of investigation is applicable universally ; and as whoever has solved a certain number of algebraic equations, can without difficulty solve all others of the same kind, so whoever knows the political economy of England, or even of Yorkshire, knows that of all nations, actual or possible, provided he have good sense enough not to expect the same conclusion to issue from varying premises." Whoever is thoroughly master of the laws which, under free competition, determine the rent, profits, and wages, received by landlords, capitalists, and labourers, in a state of society in which the three classes are completely separate, will have no difficulty in determining the very different laws which regulate the distribution of the produce among the classes interested in it, in any of the states of cultivation and landed property set forth in the foregoing extract.*

§ 4. I would not here undertake to decide what ether hypothetical or abstract sciences similar to Political Economy,

---

* The quotations in this paragraph are from a paper written by the author, and published in a periodical in 1884.

may admit of being carved out of the general body of the
social science; what other portions of the social phenomena
are in a sufficiently close and complete dependence, in the first
resort, on a peculiar class of causes, to make it convenient to
create a preliminary science of those causes; postponing the
consideration of the causes which act through them, or in concurrence with them, to a later period of the inquiry.   There is
however among these separate departments one which cannot
be passed over in silence, being of a more comprehensive and
commanding character than any of the other branches into
which the social science may admit of being divided. Like
them, it is directly conversant with the causes of only one
class of social facts, but a class which exercises, immediately
or remotely, a paramount influence over the rest. I allude to
what may be termed Political Ethology, or the theory of the
causes which determine the type of character belonging to a
people or to an age. Of all the subordinate branches of the
social science, this is the most completely in its infancy. The
causes of national character are scarcely at all understood, and
the effect of institutions or social arrangements upon the
character of the people is generally that portion of their effects
which is least attended to, and least comprehended. Nor is
this wonderful, when we consider the infant state of the
Science of Ethology itself, from whence the laws must be
drawn, of which the truths of political ethology can be but
results and exemplifications.

Yet to whoever well considers the matter, it must appear
that the laws of national (or collective) character are by far
the most important class of sociological laws. In the first
place, the character which is formed by any state of social
circumstances is in itself the most interesting phenomenon
which that state of society can possibly present. Secondly, it
is also a fact which enters largely into the production of all
the other phenomena. And above all, the character, that is,
the opinions, feelings, and habits, of the people, though greatly
the results of the state of society which precedes them, are also
greatly the causes of the state of society which follows them;
and are the power by which all those of the circumstances of

society which are artificial, laws and customs for instance, are altogether moulded : customs evidently, laws no less really, either by the direct influence of public sentiment upon the ruling powers, or by the effect which the state of national opinion and feeling has in determining the form of government and shaping the character of the governors.

As might be expected, the most imperfect part of those branches of social inquiry which have been cultivated as separate sciences, is the theory of the manner in which their conclusions are affected by ethological considerations. The omission is no defect in them as abstract or hypothetical sciences, but it vitiates them in their practical application as branches of a comprehensive social science. In political economy for instance, empirical laws of human nature are tacitly assumed by English thinkers, which are calculated only for Great Britain and the United States. Among other things, an intensity of competition is constantly supposed, which, as a general mercantile fact, exists in no country in the world except those two. An English political economist, like his countrymen in general, has seldom learned that it is possible that men, in conducting the business of selling their goods over a counter, should care more about their ease or their vanity than about their pecuniary gain. Yet those who know the habits of the Continent of Europe are aware how apparently small a motive often outweighs the desire of money-getting, even in the operations which have money-getting for their direct object. The more highly the science of ethology is cultivated, and the better the diversities of individual and national character are understood, the smaller, probably, will the number of propositions become, which it will be considered safe to build on as universal principles of human nature.

These considerations show that the process of dividing off the social science into compartments, in order that each may be studied separately, and its conclusions afterwards corrected for practice by the modifications supplied by the others, must be subject to at least one important limitation. Those portions alone of the social phenomena can with advantage be made the subjects, even provisionally, of distinct branches of science, into which the diversities of character between different nations or different times enter as influencing causes only in a secondary degree. Those phenomena, on the contrary, with which the influences of the ethological state of the people are mixed up at every step (so that the connexion of effects and causes cannot be even rudely marked out without taking those influences into consideration) could not with any advantage, nor without great disadvantage, be treated independently of political ethology, nor, therefore, of all the circumstances by which the qualities of a people are influenced. For this reason (as well as for others which will hereafter appear) there can be no separate Science of Government; that being the fact which, of all others, is most mixed up, both as cause and effect, with the qualities of the particular people or of the particular age. All questions respecting the tendencies of forms of government must stand part of the general science of society, not of any separate branch of it.

This general Science of Society, as distinguished from the separate departments of the science (each of which asserts its conclusions only conditionally, subject to the paramount control of the laws of the general science) now remains to be characterized. And as will be shown presently, nothing of a really scientific character is here possible, except by the inverse deductive method. But before we quit the subject of those sociological speculations which proceed by way of direct deduction, we must examine in what relation they stand to that indispensable element in all deductive sciences, Verification by Specific Experience—comparison between the conclusions of reasoning and the results of observation.

§ 5. We have seen that, in most deductive sciences, and among the rest in Ethology itself, which is the immediate foundation of the Social Science, a preliminary work of preparation is performed on the observed facts, to fit them for being rapidly and accurately collated (sometimes even for being collated at all) with the conclusions of theory. This preparatory treatment consists in finding general propositions which express concisely what is common to large classes of observed facts: and these are called the empirical laws of the phenomena. We have, therefore, to inquire, whether any similar preparatory process can be performed on the facts of the social science; whether there are any empirical laws in history or statistics.

In statistics, it is evident that empirical laws may sometimes be traced; and the tracing them forms an important part of that system of indirect observation on which we must often rely for the data of the Deductive Science. The process of the science consists in inferring effects from their causes; but we have often no means of observing the causes, except through the medium of their effects. In such cases the deductive science is unable to predict the effects, for want of the necessary data; it can determine what causes are capable of producing any given effect, but not with what frequency and in what quantities those causes exist. An instance in point is afforded by a newspaper now lying before me. A statement was furnished by one of the official assignees in bankruptcy, showing among the various bankruptcies which it had been his duty to investigate, in how many cases the losses had been caused by misconduct of different kinds, and in how many by unavoidable misfortunes. The result was, that the number of failures caused by misconduct greatly preponderated over those arising from all other causes whatever. Nothing but specific experience could have given sufficient ground for a conclusion to this purport. To collect, therefore, such empirical laws (which are never more than approximate generalizations) from direct observation, is an important part of the process of sociological inquiry.

The experimental process is not here to be regarded as a distinct road to the truth, but as a means (happening accidentally to be the only, or the best, available) for obtaining the necessary data for the deductive science. When the immediate causes of social facts are not open to direct observation, the empirical law of the effects gives us the empirical law (which in that case is all that we can obtain) of the causes likewise. But those immediate causes depend on remote causes; and the empirical law, obtained by this indirect mode of observation, can only be relied on as applicable to unobserved cases, so long as there is reason to think that no change has taken place in any of the remote causes on which the immediate causes depend. In making use, therefore, of even the best statistical generalizations for the purpose of inferring (though it be only conjecturally) that the same empirical laws will hold in any new case, it is necessary that we be well acquainted with the remoter causes, in order that we may avoid applying the empirical law to cases which differ in any of the circumstances on which the truth of the law ultimately depends. And thus, even where conclusions derived from specific observation are available for practical inferences in new cases, it is necessary that the deductive science should stand sentinel over the whole process; that it should be constantly referred to, and its sanction obtained to every inference.

The same thing holds true of all generalizations which can be grounded on history. Not only there are such generalizations, but it will presently be shown that the general science of society, which inquires into the laws of succession and co-existence of the great facts constituting the state of society and civilization at any time, can proceed in no other manner than by making such generalizations—afterwards to be confirmed by connecting them with the psychological and ethological laws on which they must really depend.

§ 6. But (reserving this question for its proper place) in those more special inquiries which form the subject of the separate branches of the social science, this twofold logical process and reciprocal verification is not possible: specific experience affords nothing amounting to empirical laws. This is particularly the case where the object is to determine the effect of any one social cause among a great number acting simultaneously; the effect, for example, of corn laws, or of a prohibitive commercial system generally. Though it may be perfectly certain, from theory, what *kind* of effects corn laws must produce, and in what general direction their influence must tell upon industrial prosperity; their effect is

yet of necessity so much disguised by the similar or contrary
effects of other influencing agents, that specific experience
can at most only show that on the average of some great
number of instances, the cases where there were corn laws
exhibited the effect in a greater degree than those where there
were not. Now the number of instances necessary to exhaust
the whole round of combinations of the various influential
circumstances, and thus afford a fair average, never can be
obtained. Not only we can never learn with sufficient authenticity the facts of so many instances, but the world itself
does not afford them in sufficient numbers, within the limits
of the given state of society and civilization which such inquiries always presuppose. Having thus no previous empirical
generalizations with which to collate the conclusions of theory,
the only mode of direct verification which remains is to compare those conclusions with the result of an individual experiment or instance. But here the difficulty is equally great.
For in order to verify a theory by an experiment, the circumstances of the experiment must be exactly the same with
those contemplated in the theory. But in social phenomena
the circumstances of no two cases are exactly alike. A trial
of corn laws in another country or in a former generation,
would go a very little way towards verifying a conclusion
drawn respecting their effect in this generation and in this
country. It thus happens, in most cases, that the only individual instance really fitted to verify the predictions of theory
is the very instance for which the predictions were made; and
the verification comes too late to be of any avail for practical
guidance.

Although, however, direct verification is impossible, there
is an indirect verification, which is scarcely of less value, and
which is always practicable. The conclusion drawn as to the
individual case, can only be directly verified in that case; but
it is verified indirectly, by the verification of other conclusions,
drawn in other individual cases from the same laws. The
experience which comes too late to verify the particular proposition to which it refers, is not too late to help towards veri-

fying the general sufficiency of the theory. The test of the degree in which the science affords safe ground for predicting (and consequently for practically dealing with) what has not yet happened, is the degree in which it would have enabled us to predict what has actually occurred. Before our theory of the influence of a particular cause, in a given state of circumstances, can be entirely trusted, we must be able to explain and account for the existing state of all that portion of the social phenomena which that cause has a tendency to influence. If, for instance, we would apply our speculations in political economy to the prediction or guidance of the phenomena of any country, we must be able to explain all the mercantile or industrial facts of a general character, appertaining to the present state of that country: to point out causes sufficient to account for all of them, and prove, or show good ground for supposing, that these causes have really existed. If we cannot do this, it is a proof either that the facts which ought to be taken into account are not yet completely known to us, or that although we know the facts, we are not masters of a sufficiently perfect theory to enable us to assign their consequences. In either case we are not, in the present state of our knowledge, fully competent to draw conclusions, speculative or practical, for that country. In like manner if we would attempt to judge of the effect which any political institution would have, supposing that it could be introduced into any given country; we must be able to show that the existing state of the practical government of that country, and of whatever else depends thereon, together with the particular character and tendencies of the people, and their state in respect to the various elements of social well-being, are such as the institutions they have lived under, in conjunction with the other circumstances of their nature or of their position, were calculated to produce.

To prove (in short) that our science, and our knowledge of the particular case, render us competent to predict the future, we must show that they would have enabled us to predict the present and the past. If there be anything which we could not have predicted, this constitutes a residual phenomenon, requiring further study for the purpose of explanation; and we must either search among the circumstances of the particular case until we find one which, on the principles of our existing theory, accounts for the unexplained phenomenon, or we must turn back, and seek the explanation by an extension and improvement of the theory itself.

CHAPTER X.

OF THE INVERSE DEDUCTIVE, OR HISTORICAL METHOD.

§ 1. THERE are two kinds of sociological inquiry. In the first kind, the question proposed is, what effect will follow from a given cause, a certain general condition of social circumstances being presupposed. As, for example, what would be the effect of imposing or of repealing corn laws, of abolishing monarchy or introducing universal suffrage, in the present condition of society and civilization in any European country, or under any other given supposition with regard to the circumstances of society in general: without reference to the changes which might take place, or which may already be in progress, in those circumstances. But there is also a second inquiry, namely, what are the laws which determine those general circumstances themselves. In this last the question is, not what will be the effect of a given cause in a certain state of society, but what are the causes which produce, and the phenomena which characterize, States of Society generally. In the solution of this question consists the general Science of Society; by which the conclusions of the other and more special kind of inquiry must be limited and controlled.

§ 2. In order to conceive correctly the scope of this general science, and distinguish it from the subordinate departments of sociological speculation, it is necessary to fix the ideas attached to the phrase, "a State of Society." What is called a state of society, is the simultaneous state of all the greater social facts or phenomena. Such are, the degree of knowledge, and of intellectual and moral culture, existing in the community, and in every class of it; the state of industry, of wealth and its distribution; the habitual occupations of the community; their division into classes, and the relations of those classes to one another; the common beliefs which they entertain on all the subjects most important to mankind, and the degree of assurance with which those beliefs are held; their tastes, and the character and degree of their æsthetic development; their form of government, and the more important of their laws and customs. The condition of all these things, and of many more which will readily suggest themselves, constitute the state of society or the state of civilization at any given time.

When states of society, and the causes which produce them, are spoken of as a subject of science, it is implied that there exists a natural correlation among these different elements; that not every variety of combination of these general social facts is possible, but only certain combinations; that, in short, there exist Uniformities of Coexistence between the states of the various social phenomena. And such is the truth; as is indeed a necessary consequence of the influence exercised by every one of those phenomena over every other. It is a fact implied in the *consensus* of the various parts of the social body.

States of society are like different constitutions or different ages in the physical frame; they are conditions not of one or a few organs or functions, but of the whole organism. Accordingly, the information which we possess respecting past ages, and respecting the various states of society now existing in different regions of the earth, does, when duly analysed, exhibit uniformities. It is found that when one of the features of society is in a particular state, a state of many other features, more or less precisely determinate, always or usually coexists with it.

But the uniformities of coexistence obtaining among phenomena which are effects of causes, must (as we have so often observed) be corollaries from the laws of causation by which these phenomena are really determined. The mutual correlation between the different elements of each state of society, is therefore a derivative law, resulting from the laws which regu-

late the succession between one state of society and another:
for the proximate cause of every state of society is the state
of society immediately preceding it. The fundamental problem, therefore, of the social science, is to find the laws according to which any state of society produces the state which
succeeds it and takes its place. This opens the great and
vexed question of the progressiveness of man and society ; an
idea involved in every just conception of social phenomena as
the subject of a science.

§ 3. It is one of the characters, not absolutely peculiar
to the sciences of human nature and society, but belonging to
them in a peculiar degree, to be conversant with a subject-
matter whose properties are changeable. I do not mean
changeable from day to day, but from age to age ; so that not
only the qualities of individuals vary, but those of the majority
are not the same in one age as in another.

The principal cause of this peculiarity is the extensive and
constant reaction of the effects upon their causes. The circumstances in which mankind are placed, operating according to
their own laws and to the laws of human nature, form the
characters of the human beings ; but the human beings, in
their turn, mould and shape the circumstances, for themselves
and for those who come after them.   From this reciprocal
action there must necessarily result either a cycle or a progress.
In astronomy also, every fact is at once effect and cause ; the
successive positions of the various heavenly bodies produce
changes both in the direction and in the intensity of the forces
by which those positions are determined.   But in the case of
the solar system, these mutual actions bring round again, after
a certain number of changes, the former state of circumstances ; which of course leads to the perpetual recurrence of
the same series in an unvarying order. Those bodies, in
short, revolve in orbits : but there are (or, conformably to the
laws of astronomy, there might be) others which, instead of
an orbit, describe a trajectory—a course not returning into
itself. One or other of these must be the type to which human
affairs must conform.

One of the thinkers who earliest conceived the succession of historical events as subject to fixed laws, and endeavoured to discover these laws by an analytical survey of history, Vico, the celebrated author of the *Scienza Nuova*, adopted the former of these opinions. He conceived the phenomena of human society as revolving in an orbit; as going through periodically the same series of changes. Though there were not wanting circumstances tending to give some plausibility to this view, it would not bear a close scrutiny: and those who have succeeded Vico in this kind of speculations have universally adopted the idea of a trajectory or progress, in lieu of an orbit or cycle.

The words Progress and Progressiveness are not here to be understood as synonymous with improvement and tendency to improvement. It is conceivable that the laws of human nature might determine, and even necessitate, a certain series of changes in man and society, which might not in every case, or which might not on the whole, be improvements. It is my belief indeed that the general tendency is, and will continue to be, saving occasional and temporary exceptions, one of improvement; a tendency towards a better and happier state. This, however, is not a question of the method of the social science, but a theorem of the science itself. For our purpose it is sufficient, that there is a progressive change both in the character of the human race, and in their outward circumstances so far as moulded by themselves: that in each successive age the principal phenomena of society are different from what they were in the age preceding, and still more different from any previous age: the periods which most distinctly mark these successive changes being intervals of one generation, during which a new set of human beings have been educated, have grown up from childhood, and taken possession of society.

The progressiveness of the human race is the foundation on which a method of philosophizing in the social science has been of late years erected, far superior to either of the two modes which had previously been prevalent, the chemical or experimental, and the geometrical modes. This method, which is now generally adopted by the most advanced thinkers on the Continent, consists in attempting, by a study and analysis of the general facts of history, to discover (what these philosophers term) the law of progress : which law, once ascertained, must according to them enable us to predict future events, just as after a few terms of an infinite series in algebra we are able to detect the principle of regularity in their formation, and to predict the rest of the series to any number of terms we please. The principal aim of historical speculation in France, of late years, has been to ascertain this law. But while I gladly acknowledge the great services which have been rendered to historical knowledge by this school, I cannot but deem them to be mostly chargeable with a fundamental misconception of the true method of social philosophy. The misconception consists in supposing that the order of succession which we may be able to trace among the different states of society and civilization which history presents to us, even if that order were more rigidly uniform than it has yet been proved to be, could ever amount to a law of nature. It can only be an empirical law. The succession of states of the human mind and of human society cannot have an independent law of its own; it must depend on the psychological and ethological laws which govern the action of circumstances on men and of men on circumstances. It is conceivable that those laws might be such, and the general circumstances of the human race such, as to determine the successive transformations of man and society to one given and unvarying order. But even if the case were so, it cannot be the ultimate aim of science to discover an empirical law. Until that law could be connected with the psychological and ethological laws on which it must depend, and, by the consilience of deduction à *priori* with historical evidence, could be converted from an empirical law into a scientific one, it could not be relied on for the prediction of future events, beyond, at most, strictly adjacent cases. M. Comte alone, among the new historical school, has seen the necessity of thus connecting all our generalizations from history with the laws of human nature.

§ 4. But, while it is an imperative rule never to introduce any generalization from history into the social science unless sufficient grounds can be pointed out for it in human nature, I do not think any one will contend that it would have been possible, setting out from the principles of human nature and from the general circumstances of the position of our species, to determine *à priori* the order in which human development must take place, and to predict, consequently, the general facts of history up to the present time. After the first few terms of the series, the influence exercised over each generation by the generations which preceded it, becomes (as is well observed by the writer last referred to) more and more preponderant over all other influences; until at length what we now are and do, is in a very small degree the result of the universal circumstances of the human race, or even of our own circumstances acting through the original qualities of our species, but mainly of the qualities produced in us by the whole previous history of humanity. So long a series of actions and reactions between Circumstances and Man, each successive term being composed of an ever greater number and variety of parts, could not possibly be computed by human faculties from the elementary laws which produce it. The mere length of the series would be a sufficient obstacle, since a slight error in any one of the terms would augment in rapid progression at every subsequent step.

If, therefore, the series of the effects themselves did not, when examined as a whole, manifest any regularity, we should in vain attempt to construct a general science of society. We must in that case have contented ourselves with that subordinate order of sociological speculation formerly noticed, namely, with endeavouring to ascertain what would be the effect of the introduction of any new cause, in a state of society supposed to be fixed; a knowledge sufficient for the more common exigencies of daily political practice, but liable to fail in all cases in which the progressive movement of society is one of the influencing elements; and therefore more precarious in proportion as the case is more important. But since both the natural varieties of mankind, and the original diversities of local circumstances, are much less considerable than the points of agreement, there will naturally be a certain degree of uniformity in the progressive development of the species and of its works. And this uniformity tends to become greater, not less, as society advances; since the evolution of each people, which is at first determined exclusively by the nature and circumstances of that people, is gradually brought under the influence (which becomes stronger as civilization advances) of the other nations of the earth, and of the circumstances by which they have been influenced. History accordingly does, when judiciously examined, afford Empirical Laws of Society. And the problem of general sociology is to ascertain these, and connect them with the laws of human nature, by deductions showing that such were the derivative laws naturally to be expected as the consequences of those ultimate ones.

It is, indeed, hardly ever possible, even after history has suggested the derivative law, to demonstrate *à priori* that such was the only order of succession or of coexistence in which the effects could, consistently with the laws of human nature, have been produced. We can at most make out that there were strong *à priori* reasons for expecting it, and that no other order of succession or coexistence would have been so likely to result from the nature of man and the general circumstances of his position. Often we cannot do even this; we cannot even show that what did take place was probable *à priori*, but only that it was possible. This, however,— which, in the Inverse Deductive Method that we are now characterizing, is a real process of verification,—is as indispensable, as verification by specific experience has been shown to be, where the conclusion is originally obtained by the direct way of deduction. The empirical laws must be the result of but a few instances, since few nations have ever attained at all, and still fewer by their own independent development, a high stage of social progress. If, therefore, even one or two of these few instances be insufficiently known, or imperfectly analysed into their elements, and therefore not adequately compared with other instances, nothing is more probable than that a wrong empirical law will emerge instead of the right one. Accordingly, the most erroneous generalizations are continually made from the course of history: not only in this country, where history cannot yet be said to be at all cultivated as a science, but in other countries, where it is so cultivated, and by persons well versed in it. The only check or corrective is, constant verification by psychological and ethological laws. We may add to this, that no one but a person competently skilled in those laws is capable of preparing the materials for ·historical generalization, by analysing the facts of history, or even by observing the social phenomena of his own time. No other will be aware of the comparative importance of different facts, nor consequently know what facts to look for, or to observe; still less will he be capable of estimating the evidence of facts which, as is the case with most, cannot be ascertained by direct observation or learnt from testimony, but must be inferred from marks.

§ 5. The Empirical Laws of Society are of two kinds; some are uniformities of coexistence, some of succession. According as the science is occupied in ascertaining and verifying the former sort of uniformities or the latter, M. Comte gives it the title of Social Statics, or of Social Dynamics; conformably to the distinction in mechanics between the conditions of equilibrium and those of movement; or in biology, between the laws of organization and those of life. The first branch of the science ascertains the conditions of stability in the social union: the second, the laws of progress. Social Dynamics is the theory of Society considered in a state of progressive movement; while Social Statics is the theory of the *consensus* already spoken of as existing among the different parts of the social organism; in other words, the theory of the mutual actions and reactions of contemporaneous social phenomena; "making* provisionally, as far as possible, abstraction, for scientific purposes, of the fundamental move-

* *Cours de Philosophie Positive*, iv. 325-9.

ment which is at all times gradually modifying the whole of them.

"In this first point of view, the provisions of sociology will enable us to infer one from another (subject to ulterior verification by direct observation) the various characteristic marks of each distinct mode of social existence; in a manner essentially analogous to what is now habitually practised in the anatomy of the physical body. This preliminary aspect, therefore, of political science, of necessity supposes that (contrary to the existing habits of philosophers) each of the numerous elements of the social state, ceasing to be looked at independently and absolutely, shall be always and exclusively considered relatively to all the other elements, with the whole of which it is united by mutual interdependence. It would be superfluous to insist here upon the great and constant utility of this branch of sociological speculation. It is, in the first place, the indispensable basis of the theory of social progress. It may, moreover, be employed, immediately, and of itself, to supply the place, provisionally at least, of direct observation, which in many cases is not always practicable for some of the elements of society, the real condition of which may however be sufficiently judged of by means of the relations which connect them with others previously known. The history of the sciences may give us some notion of the habitual importance of this auxiliary resource, by reminding us, for example, how the vulgar errors of mere erudition concerning the pretended acquirements of the ancient Egyptians in the higher astronomy, were irrevocably dissipated (even before sentence had been passed on them by a sounder erudition) from the single consideration of the inevitable connexion between the general state of astronomy and that of abstract geometry, then evidently in its infancy. It would be easy to cite a multitude of analogous cases, the character of which could admit of no dispute. In order to avoid exaggeration, however, it should be remarked, that these necessary relations among the different aspects of society cannot, from their very nature, be so simple and precise that the results observed could only have arisen from some one mode of mutual co-ordination.

Such a notion, already too narrow in the science of life, would be completely at variance with the still more complex nature of sociological speculations. But the exact estimation of these limits of variation, both in the healthy and in the morbid state, constitutes, at least as much as in the anatomy of the natural body, an indispensable complement to every theory of Sociological Statics; without which the indirect exploration above spoken of would often lead into error.

"This is not the place for methodically demonstrating the existence of a necessary relation among all the possible aspects of the same social organism; a point on which, in principle at least, there is now little difference of opinion among sound thinkers. From whichever of the social elements we choose to set out, we may easily recognise that it has always a connexion, more or less immediate, with all the other elements, even with those which at first sight appear the most independent of it. The dynamical consideration of the progressive development of civilized humanity, affords, no doubt, a still more efficacious means of effecting this interesting verification of the *consensus* of the social phenomena, by displaying the manner in which every change in any one part, operates immediately, or very speedily, upon all the rest. But this indication may be preceded, or at all events followed, by a confirmation of a purely statical kind; for, in politics as in mechanics, the communication of motion from one object to another proves a connexion between them. Without descending to the minute interdependence of the different branches of any one science or art, is it not evident that among the different sciences, as well as among most of the arts, there exists such a connexion, that if the state of any one well marked division of them is sufficiently known to us, we can with real scientific assurance infer, from their necessary correlation, the contemporaneous state of every one of the others? By a further extension of this consideration, we may conceive the necessary relation which exists between the condition of the sciences in general and that of the arts in general, except that the mutual dependence is less intense in proportion as it is more indirect. The same is the

33—2

case when, instead of considering the aggregate of the social phenomena in some one people, we examine it simultaneously in different contemporaneous nations; between which the perpetual reciprocity of influence, especially in modern times, cannot be contested, though the *consensus* must in this case be ordinarily of a less decided character, and must decrease gradually with the affinity of the cases and the multiplicity of the points of contact, so as at last, in some cases, to disappear almost entirely; as for example between Western Europe and Eastern Asia, of which the various general states of society appear to have been hitherto almost independent of one another."

These remarks are followed by illustrations of one of the most important, and until lately, most neglected, of the general principles which, in this division of the social science, may be considered as established; namely, the necessary correlation between the form of government existing in any society and the contemporaneous state of civilization: a natural law, which stamps the endless discussions and innumerable theories respecting forms of government in the abstract, as fruitless and worthless, for any other purpose than as a preparatory treatment of materials to be afterwards used for the construction of a better philosophy.

As already remarked, one of the main results of the science of social statics would be to ascertain the requisites of stable political union. There are some circumstances which, being found in all societies without exception, and in the greatest degree where the social union is most complete, may be considered (when psychological and ethological laws confirm the indication) as conditions of the existence of the complex phenomenon called a State. For example, no numerous society has ever been held together without laws, or usages equivalent to them; without tribunals, and an organized force of some sort to execute their decisions. There have always been public authorities whom, with more or less strictness and in cases more or less accurately defined, the rest of the community obeyed, or according to general opinion were bound to obey. By following out this course of inquiry we shall find

a number of requisites, which have been present in every
society that has maintained a collective existence, and on the
cessation of which it has either merged in some other society,
or reconstructed itself on some new basis, in which the conditions were conformed to. Although these results, obtained by
comparing different forms and states of society, amount in
themselves only to empirical laws; some of them, when once
suggested, are found to follow with so much probability from
general laws of human nature, that the consilience of the two
processes raises the evidence to proof, and the generalizations
to the rank of scientific truths.

This seems to be affirmable (for instance) of the conclusions
arrived at in the following passage; extracted, with some
alterations, from a criticism on the negative philosophy of the
eighteenth century,* and which I quote, though (as in some
former instances) from myself, because I have no better way
of illustrating the conception I have formed of the kind of
theorems of which sociological statics would consist.

"The very first element of the social union, obedience to a
government of some sort, has not been found so easy a thing
to establish in the world. Among a timid and spiritless race
like the inhabitants of the vast plains of tropical countries,
passive obedience may be of natural growth; though even
there we doubt whether it has ever been found among any
people with whom fatalism, or in other words, submission to
the pressure of circumstances as a divine decree, did not prevail as a religious doctrine. But the difficulty of inducing a
brave and warlike race to submit their individual *arbitrium* to
any common umpire, has always been felt to be so great, that
nothing short of supernatural power has been deemed adequate
to overcome it; and such tribes have always assigned to the
first institution of civil society a divine origin. So differently
did those judge who knew savage men by actual experience,
from those who had no acquaintance with them except in
the civilized state. In modern Europe itself, after the fall of

---

* Since reprinted entire in *Dissertations and Discussions*, as the concluding
paper of the first volume.

the Roman empire, to subdue the feudal anarchy and bring the whole people of any European nation into subjection to government (though Christianity in the most concentrated form of its influence was co-operating in the work) required thrice as many centuries as have elapsed since that time.

"Now if these philosophers had known human nature under any other type than that of their own age, and of the particular classes of society among whom they lived, it would have occurred to them, that wherever this habitual submission to law and government has been firmly and durably established, and yet the vigour and manliness of character which resisted its establishment have been in any degree preserved, certain requisites have existed, certain conditions have been fulfilled, of which the following may be regarded as the principal.

"First: there has existed, for all who were accounted citizens,—for all who were not slaves, kept down by brute force,—a system of *education*, beginning with infancy and continued through life, of which whatever else it might include, one main and incessant ingredient was *restraining discipline*. To train the human being in the habit, and thence the power, of subordinating his personal impulses and aims, to what were considered the ends of society; of adhering, against all temptation, to the course of conduct which those ends prescribed; of controlling in himself all feelings which were liable to militate against those ends, and encouraging all such as tended towards them; this was the purpose, to which every outward motive that the authority directing the system could command, and every inward power or principle which its knowledge of human nature enabled it to evoke, were endeavoured to be rendered instrumental. The entire civil and military policy of the ancient commonwealths was such a system of training; in modern nations its place has been attempted to be supplied, principally, by religious teaching. And whenever and in proportion as the strictness of the restraining discipline was relaxed, the natural tendency of mankind to anarchy re-asserted itself; the state became disorganized from within; mutual conflict for selfish ends, neutralized the energies which were required to keep up the contest against natural causes of evil; and the nation, after a longer or briefer interval of progressive decline, became either the slave of a despotism, or the prey of a foreign invader.

"The second condition of permanent political society has been found to be, the existence, in some form or other, of the feeling of allegiance or loyalty. This feeling may vary in its objects, and is not confined to any particular form of government; but whether in a democracy or in a monarchy, its essence is always the same; viz. that there be in the constitution of the state *something* which is settled, something permanent, and not to be called in question; something which, by general agreement, has a right to be where it is, and to be secure against disturbance, whatever else may change. This feeling may attach itself, as among the Jews (and in most of the commonwealths of antiquity), to a common God or gods, the protectors and guardians of their state. Or it may attach itself to certain persons, who are deemed to be, whether by divine appointment, by long prescription, or by the general recognition of their superior capacity and worthiness, the rightful guides and guardians of the rest. Or it may connect itself with laws; with ancient liberties or ordinances. Or, finally, (and this is the only shape in which the feeling is likely to exist hereafter,) it may attach itself to the principles of individual freedom and political and social equality, as realized in institutions which as yet exist nowhere, or exist only in a rudimentary state. But in all political societies which have had a durable existence, there has been some fixed point: something which people agreed in holding sacred; which, wherever freedom of discussion was a recognised principle, it was of course lawful to contest in theory, but which no one could either fear or hope to see shaken in practice; which, in short (except perhaps during some temporary crisis) was in the common estimation placed beyond discussion. And the necessity of this may easily be made evident. A state never is, nor until mankind are vastly improved, can

hope to be, for any long time exempt from internal dissension;
for there neither is nor has ever been any state of society in
which collisions did not occur between the immediate interests
and passions of powerful sections of the people.   What, then,
enables nations to weather these storms, and pass through
turbulent times without any permanent weakening of the
securities for peaceable existence ?   Precisely this—that however important the interests about which men fell out, the
conflict did not affect the fundamental principle of the system
of social union which happened to exist ; nor threaten large
portions of the community with the subversion of that on
which they had built their calculations, and with which their
hopes and aims had become identified.   But when the questioning of these fundamental principles is (not the occasional
disease, or salutary medicine, but) the habitual condition of
the body politic ; and when all the violent animosities are
called forth, which spring naturally from such a situation, the
state is virtually in a position of civil war ; and can never long
remain free from it in act and fact.

"The third essential condition of stability in political
society, is a strong and active principle of cohesion among
the members of the same community or state.   We need
scarcely say that we do not mean nationality, in the vulgar
sense of the term ; a senseless antipathy to foreigners ;
indifference to the general welfare of the human race, or an
unjust preference of the supposed interests of our own country ;
a cherishing of bad peculiarities because they are national, or
a refusal to adopt what has been found good by other countries.   We mean a principle of sympathy, not of hostility ;
of union, not of separation.   We mean a feeling of common
interest among those who live under the same government,
and are contained within the same natural or historical boundaries.   We mean, that one part of the community do not
consider themselves as foreigners with regard to another part ;
that they set a value on their connexion—feel that they are
one people, that their lot is cast together, that evil to any of
their fellow-countrymen is evil to themselves, and do not
desire selfishly to free themselves from their share of any

common inconvenience by severing the connexion. How strong this feeling was in those ancient commonwealths which attained any durable greatness, every one knows. How happily Rome, in spite of all her tyranny, succeeded in establishing the feeling of a common country among the provinces of her vast and divided empire, will appear when any one who has given due attention to the subject shall take the trouble to point it out. In modern times the countries which have had that feeling in the strongest degree have been the most powerful countries; England, France, and, in proportion to their territory and resources, Holland and Switzerland; while England in her connexion with Ireland, is one of the most signal examples of the consequences of its absence. Every Italian knows why Italy is under a foreign yoke; every German knows what maintains despotism in the Austrian empire; the evils of Spain flow as much from the absence of nationality among the Spaniards themselves, as from the presence of it in their relations with foreigners: while the completest illustration of all is afforded by the republics of South America, where the parts of one and the same state adhere so slightly together, that no sooner does any province think itself aggrieved by the general government than it proclaims itself a separate nation."

§ 6. While the derivative laws of social statics are ascertained by analysing different states of society, and comparing them with one another, without regard to the order of their succession; the consideration of the successive order is, on the contrary, predominant in the study of social dynamics, of which the aim is to observe and explain the sequences of social conditions. This branch of the social science would be as complete as it can be made, if every one of the leading general circumstances of each generation were traced to its causes in the generation immediately preceding. But the *consensus* is so complete, (especially in modern history,) that in the filiation of one generation and another, it is the whole which produces the whole, rather than any part a part. Little progress, therefore, can be made in establishing the filiation, directly from laws of human nature, without having first ascertained the immediate or derivative laws according to which social states generate one another as society advances; the *axiomata media* of General Sociology.

The empirical laws which are most readily obtained by generalization from history do not amount to this. They are not the "middle principles" themselves, but only evidence towards the establishment of such principles. They consist of certain general tendencies which may be perceived in society; a progressive increase of some social elements and diminution of others, or a gradual change in the general character of certain elements. It is easily seen, for instance, that as society advances, mental tend more and more to prevail over bodily qualities, and masses over individuals: that the occupation of all that portion of mankind who are not under external restraint is at first chiefly military, but society becomes progressively more and more engrossed with productive pursuits, and the military spirit gradually gives way to the industrial; to which many similar truths might be added. And with generalizations of this description, ordinary inquirers, even of the historical school now predominant on the Continent, are satisfied. But these and all such results are still at too great a distance from the elementary laws of human nature on which they depend,—too many links intervene, and the concurrence of causes at each link is far too complicated,—to enable these propositions to be presented as direct corollaries from those elementary principles. They have, therefore, in the minds of most inquirers, remained in the state of empirical laws, applicable only within the bounds of actual observation; without any means of determining their real limits, and of judging whether the changes which have hitherto been in progress are destined to continue indefinitely, or to terminate, or even to be reversed.

§ 7. In order to obtain better empirical laws, we must not rest satisfied with noting the progressive changes which manifest themselves in the separate elements of society, and in which nothing is indicated but the relation of fragments of the effect to corresponding fragments of the cause. It is necessary to combine the statical view of social phenomena with the dynamical, considering not only the progressive changes of the different elements, but the contemporaneous condition of each; and thus obtain empirically the law of correspondence not only between the simultaneous states, but between the simultaneous changes, of those elements. This law of correspondence it is, which, duly verified *à priori*, would become the real scientific derivative law of the development of humanity and human affairs.

In the difficult process of observation and comparison which is here required, it would evidently be a great assistance if it should happen to be the fact, that some one element in the complex existence of social man is pre-eminent over all others as the prime agent of the social movement. For we could then take the progress of that one element as the central chain, to each successive link of which, the corresponding links of all the other progressions being appended, the succession of the facts would by this alone be presented in a kind of spontaneous order, far more nearly approaching to the real order of their filiation than could be obtained by any other merely empirical process.

Now, the evidence of history and that of human nature combine, by a striking instance of consilience, to show that there really is one social element which is thus predominant, and almost paramount, among the agents of the social progression. This is, the state of the speculative faculties of mankind; including the nature of the beliefs which by any means they have arrived at, concerning themselves and the world by which they are surrounded.

It would be a great error, and one very little likely to be committed, to assert that speculation, intellectual activity, the pursuit of truth, is among the more powerful propensities of human nature, or holds a predominating place in the lives of any, save decidedly exceptional, individuals. But, notwithstanding the relative weakness of this principle among other sociological agents, its influence is the main determining cause of the social progress; all the other dispositions of our nature which contribute to that progress, being dependent on it for the means of accomplishing their share of the work. Thus (to take the most obvious case first,) the impelling force to most of the improvements effected in the arts of life, is the desire of increased material comfort; but as we can only act upon external objects in proportion to our knowledge of them, the state of knowledge at any time is the limit of the industrial improvements possible at that time; and the progress of industry must follow, and depend on, the progress of knowledge. The same thing may be shown to be true, though it is not quite so obvious, of the progress of the fine arts. Further, as the strongest propensities of uncultivated or half-cultivated human nature (being the purely selfish ones, and those of a sympathetic character which partake most of the nature of selfishness) evidently tend in themselves to disunite mankind, not to unite them,—to make them rivals, not confederates: social existence is only possible by a disciplining of those more powerful propensities, which consists in subordinating them to a common system of opinions. The degree of this subordination is the measure of the completeness of the social union, and the nature of the common opinions determines its kind. But in order that mankind should conform their actions to any set of opinions, these opinions must exist, must be believed by them. And thus, the state of the speculative faculties, the character of the propositions assented to by the intellect, essentially determines the moral and political state of the community, as we have already seen that it determines the physical.

These conclusions, deduced from the laws of human nature, are in entire accordance with the general facts of history. Every considerable change historically known to us in the condition of any portion of mankind, when not brought about by external force, has been preceded by a change, of proportional extent, in the state of their knowledge, or in their prevalent beliefs. As between any given state of speculation, and the correlative state of everything else, it was almost always the former which first showed itself; though the effects, no doubt, reacted potently upon the cause. Every considerable advance in material civilization has been preceded by an advance in knowledge: and when any great social change has come to pass, either in the way of gradual development or of sudden conflict, it has had for its precursor a great change in the opinions and modes of thinking of society. Polytheism, Judaism, Christianity, Protestantism, the critical philosophy of modern Europe, and its positive science—each of these has been a primary agent 'in making society what it was at each successive period, while society was but secondarily instrumental in making *them*, each of them (so far as causes can be assigned for its existence) being mainly an emanation not from the practical life of the period, but from the previous state of belief and thought. The weakness of the speculative propensity in mankind generally, has not, therefore, prevented the progress of speculation from governing that of society at large; it has only, and too often, prevented progress altogether, where the intellectual progression has come to an early stand for want of sufficiently favourable circumstances.

From this accumulated evidence, we are justified in concluding, that the order of human progression in all respects will mainly depend on the order of progression in the intellectual convictions of mankind, that is, on the law of the successive transformations of human opinions. The question remains, whether this law can be determined; at first from history as an empirical law, then converted into a scientific theorem by deducing it *à priori* from the principles of human nature. As the progress of knowledge and the changes in the opinions of mankind are very slow, and manifest themselves in a well-defined manner only at long intervals; it cannot be expected that the general order of sequence should be discoverable from the examination of less than a very considerable part of the duration of the social progress. It is necessary to take into consideration the whole of past time, from the first recorded condition of the human race, to the memorable phenomena of the last and present generations.

§ 8. The investigation which I have thus endeavoured to

characterize, has been systematically attempted, up to the present time, by M. Comte alone. His work is hitherto the only known example of the study of social phenomena according to this conception of the Historical Method. Without discussing here the worth of his conclusions, and especially of his predictions and recommendations with respect to the Future of society, which appear to me greatly inferior in value to his appreciation of the Past, I shall confine myself to mentioning one important generalization, which M. Comte regards as the fundamental law of the progress of human knowledge. Speculation he conceives to have, on every subject of human inquiry, three successive stages; in the first of which it tends to explain the phenomena by supernatural agencies, in the second by metaphysical abstractions, and in the third or final state confines itself to ascertaining their laws of succession and similitude. This generalization appears to me to have that high degree of scientific evidence, which is derived from the concurrence of the indications of history with the probabilities derived from the constitution of the human mind. Nor could it be easily conceived, from the mere enunciation of such a proposition, what a flood of light it lets in upon the whole course of history; when its consequences are traced, by connecting with each of the three states of human intellect which it distinguishes, and with each successive modification of those three states, the correlative condition of other social phenomena.*

---

* This great generalization is often unfavourably criticised (as by Dr. Whewell for instance) under a misapprehension of its real import. The doctrine, that the theological explanation of phenomena belongs only to the infancy of our knowledge of them, ought not to be construed as if it was equivalent to the assertion, that mankind, as their knowledge advances, will necessarily cease to believe in any kind of theology. This was M. Comte's opinion; but it is by no means implied in his fundamental theorem. All that is implied is, that in an advanced state of human knowledge, no other Ruler of the World will be acknowledged than one who rules by universal laws, and does not at all, or does not unless in very peculiar cases, produce events by special interpositions. Originally all natural events were ascribed to such interpositions. At present every educated person rejects this explanation in regard to all classes of phenomena of which the laws have been fully ascertained; though some have not yet reached the point of referring all phenomena to the idea of Law, but

But whatever decision competent judges may pronounce on the results arrived at by any individual inquirer, the method now characterized is that by which the derivative laws of social order and of social progress must be sought. By its aid we may hereafter succeed not only in looking far forward into the future history of the human race, but in determining what artificial means may be used, and to what extent, to accelerate the natural progress in so far as it is beneficial; to compensate for whatever may be its inherent inconveniences or disadvantages; and to guard against the dangers or accidents to which our species is exposed from the necessary incidents of its progression. Such practical instructions, founded on the highest branch of speculative sociology, will form the noblest and most beneficial portion of the Political Art.

That of this science and art even the foundations are but beginning to be laid, is sufficiently evident. But the superior

---

believe that rain and sunshine, famine and pestilence, victory and defeat, death and life, are issues which the Creator does not leave to the operation of his general laws, but reserves to be decided by express acts of volition. M. Comte's theory is the negation of this doctrine.

Dr. Whewell equally misunderstands M. Comte's doctrine respecting the second, or metaphysical stage of speculation. M. Comte did not mean that "discussions concerning ideas" are limited to an early stage of inquiry, and cease when science enters into the positive stage. (*Philosophy of Discovery*, pp. 226 et seq.) In all M. Comte's speculations as much stress is laid on the process of clearing up our conceptions, as on the ascertainment of facts. When M. Comte speaks of the metaphysical stage of speculation, he means the stage in which men speak of "Nature" and other abstractions as if they were active forces, producing effects; when Nature is said to do this, or forbid that; when Nature's horror of a vacuum, Nature's non-admission of a break, Nature's *vis medicatrix*, were offered as explanations of phenomena; when the qualities of things were mistaken for real entities dwelling in the things; when the phenomena of living bodies were thought to be accounted for by being referred to a "vital force;" when, in short, the abstract names of phenomena were mistaken for the causes of their existence. In this sense of the word it cannot be reasonably denied that the metaphysical explanation of phenomena, equally with the theological, gives way before the advance of real science.

That the final, or positive stage, as conceived by M. Comte, has been equally misunderstood, and that, notwithstanding some expressions open to just criticism, M. Comte never dreamed of denying the legitimacy of inquiry into all causes which are accessible to human investigation, I have pointed out in a former place.

minds are fairly turning themselves towards that object. It
has become the aim of really scientific thinkers to connect by
theories the facts of universal history: it is acknowledged to
be one of the requisites of a general system of social doctrine,
that it should explain, so far as the data exist, the main facts
of history; and a Philosophy of History is generally admitted
to be at once the verification, and the initial form, of the Philosophy of the Progress of Society.

If the endeavours now making in all the more cultivated
nations, and beginning to be made even in England (usually
the last to enter into the general movement of the European
mind) for the construction of a Philosophy of History, shall
be directed and controlled by those views of the nature of
sociological evidence which I have (very briefly and imperfectly) attempted to characterize; they cannot fail to give
birth to a sociological system widely removed from the vague
and conjectural character of all former attempts, and worthy
to take its place, at last, among the sciences. When this time
shall come, no important branch of human affairs will be any
longer abandoned to empiricism and unscientific surmise: the
circle of human knowledge will be complete, and it can only
thereafter receive further enlargement by perpetual expansion
from within.

CHAPTER XI.

§ 1. THE doctrine which the preceding chapters were
intended to enforce and elucidate—that the collective series
of social phenomena, in other words the course of history, is
subject to general laws, which philosophy may possibly detect
—has been familiar for generations to the scientific thinkers of
the Continent, and has for the last quarter of a century passed
out of their peculiar domain, into that of newspapers and
ordinary political discussion. In our own country, however,
at the time of the first publication of this Treatise, it was
almost a novelty, and the prevailing habits of thought on historical subjects were the very reverse of a preparation for it.
Since then a great change has taken place, and has been
eminently promoted by the important work of Mr. Buckle;
who, with characteristic energy, flung down this great principle, together with many striking exemplifications of it, into
the arena of popular discussion, to be fought over by a sort
of combatants, in the presence of a sort of spectators, who
would never even have been aware that there existed such
a principle if they had been left to learn its existence from the
speculations of pure science. And hence has arisen a considerable amount of controversy, tending not only to make the
principle rapidly familiar to the majority of cultivated minds,
but also to clear it from the confusions and misunderstandings
by which it was but natural that it should for a time be
clouded, and which impair the worth of the doctrine to those
who accept it, and are the stumbling-block of many who
do not.

Among the impediments to the general acknowledgment,
by thoughtful minds, of the subjection of historical facts to
scientific laws, the most fundamental continues to be that

which is grounded on the doctrine of Free Will, or in other
words, on the denial that the law of invariable Causation holds
true of human volitions: for if it does not, the course of
history, being the result of human volitions, cannot be a subject of scientific laws, since the volitions on which it depends
can neither be foreseen, nor reduced to any canon of regularity
even after they have occurred.   I have discussed this question,
as far as seemed suitable to the occasion, in a former chapter: and I only think it necessary to repeat, that the doctrine
of the Causation of human actions, improperly called the
doctrine of Necessity, affirms no mysterious *nexus*, or over-
ruling fatality: it asserts only that men's actions are the
joint result of the general laws and circumstances of human
nature, and of their own particular characters; those characters
again being the consequence of the natural and artificial circumstances that constituted their education, among which
circumstances must be reckoned their own conscious efforts.
Any one who is willing to take (if the expression may be permitted) the trouble of thinking himself into the doctrine as
thus stated, will find it, I believe, not only a faithful interpretation of the universal experience of human conduct, but a
correct representation of the mode in which he himself, in every
particular case, spontaneously interprets his own experience of
that conduct.

But if this principle is true of individual man, it must be
true of collective man.   If it is the law of human life, the law
must be realized in history.   The experience of human affairs
when looked at *en masse*, must be in accordance with it if true,
or repugnant to it if false.   The support which this *à posteriori*
verification affords to the law, is the part of the case which
has been most clearly and triumphantly brought out by Mr.
Buckle.

The facts of statistics, since they have been made a
subject of careful recordation and study, have yielded conclusions, some of which have been very startling to persons not
accustomed to regard moral actions as subject to uniform laws.
The very events which in their own nature appear most
capricious and uncertain, and which in any individual case no

attainable degree of knowledge would enable us to foresee, occur, when considerable numbers are taken into the account, with a degree of regularity approaching to mathematical. What act is there which all would consider as more completely dependent on individual character, and on the exercise of individual free will, than that of slaying a fellow creature? Yet in any large country, the number of murders, in proportion to the population, varies (it has been found) very little from one year to another, and in its variations never deviates widely from a certain average. What is still more remarkable, there is a similar approach to constancy in the proportion of these murders annually committed with every particular kind of instrument. There is a like approximation to identity, as between one year and another, in the comparative number of legitimate and of illegitimate births. The same thing is found true of suicides, accidents, and all other social phenomena of which the registration is sufficiently perfect; one of the most curiously illustrative examples being the fact, ascertained by the registers of the London and Paris post-offices, that the number of letters posted which the writers have forgotten to direct, is nearly the same, in proportion to the whole number of letters posted, in one year as in another. "Year after year," says Mr. Buckle, "the same proportion of letter-writers forget this simple act; so that for each successive period we can actually foretell the number of persons whose memory will fail them in regard to this trifling, and as it might appear, accidental occurrence."*

This singular degree of regularity *en masse*, combined with the extreme of irregularity in the cases composing the mass, is a felicitous verification *à posteriori* of the law of causation in its application to human conduct. Assuming the truth of that law, every human action, every murder for instance, is the concurrent result of two sets of causes. On the one part, the general circumstances of the country and its inhabitants; the moral, educational, economical, and other influences operating on the whole people, and constituting what we term the state

* Buckle's *History of Civilisation*, i. 30.

of civilization. On the other part, the great variety of influences special to the individual: his temperament, and other peculiarities of organization, his parentage, habitual associates, temptations, and so forth. If we now take the whole of the instances which occur within a sufficiently large field to exhaust all the combinations of these special influences, or in other words, to eliminate chance; and if all these instances have occurred within such narrow limits of time, that no material change can have taken place in the general influences constituting the state of civilization of the country; we may be certain, that if human actions are governed by invariable laws, the aggregate result will be something like a constant quantity. The number of murders committed within that space and time, being the effect partly of general causes which have not varied, and partly of partial causes the whole round of whose variations has been included, will be, practically speaking, invariable.

Literally and mathematically invariable it is not, and could not be expected to be: because the period of a year is too short to include *all* the possible combinations of partial causes, while it is, at the same time, sufficiently long to make it probable that in some years at least, of every series, there will have been introduced new influences of a more or less general character; such as a more vigorous or a more relaxed police; some temporary excitement from political or religious causes; or some incident generally notorious, of a nature to act morbidly on the imagination. That in spite of these unavoidable imperfections in the data, there should be so very trifling a margin of variation in the annual results, is a brilliant confirmation of the general theory.

§ 2. The same considerations which thus strikingly corroborate the evidence of the doctrine, that historical facts are the invariable effects of causes, tend equally to clear that doctrine from various misapprehensions, the existence of which has been put in evidence by the recent discussions. Some persons, for instance, seemingly imagine the doctrine to imply, not merely that the total number of murders committed in a given space and time, is entirely the effect of the general circumstances of society, but that every particular murder is so too: that the individual murderer is, so to speak, a mere instrument in the hands of general causes; that he himself has no option, or if he has, and chose to exercise it, some one else would be necessitated to take his place: that if any one of the actual murderers had abstained from the crime, some person who would otherwise have remained innocent, would have committed an extra murder to make up the average. Such a corollary would certainly convict any theory which necessarily led to it of absurdity. It is obvious, however, that each particular murder depends, not on the general state of society only, but on that combined with causes special to the case, which are generally much more powerful: and if these special causes, which have greater influence than the general ones in causing every particular murder, have no influence on the number of murders in a given period, it is because the field of observation is so extensive as to include all possible combinations of the special causes—all varieties of individual character and individual temptation compatible with the general state of society. The collective experiment, as it may be termed, exactly separates the effect of the general from that of the special causes, and shows the net result of the former: but it declares nothing at all respecting the amount of influence of the special causes, be it greater or smaller, since the scale of the experiment extends to the number of cases within which the effects of the special causes balance one another, and disappear in that of the general causes.

I will not pretend that all the defenders of the theory have always kept their language free from this same confusion, and have shown no tendency to exalt the influence of general causes at the expense of special. I am of opinion, on the contrary, that they have done so in a very great degree, and by so doing have encumbered their theory with difficulties, and laid it open to objections, which do not necessarily affect it. Some, for example (among whom is Mr. Buckle himself), have inferred, or allowed it to be supposed that they inferred, from the regularity in the recurrence of events which depend on moral qualities, that the moral qualities of mankind are little capable of being improved, or are of little importance in the general progress of society, compared with intellectual or economic causes. But to draw this inference is to forget that the statistical tables, from which the invariable averages are deduced, were compiled from facts occurring within narrow geographical limits and in a small number of successive years; that is, from a field the whole of which was under the operation of the same general causes, and during too short a time to allow of much change therein. All moral causes but those common to the country generally, have been eliminated by the great number of instances taken; and those which are common to the whole country have not varied considerably, in the short space of time comprised in the observations. If we admit the supposition that they have varied; if we compare one age with another, or one country with another, or even one part of a country with another, differing in position and character as to the moral elements, the crimes committed within a year give no longer the same, but a widely different numerical aggregate. And this cannot but be the case: for inasmuch as every single crime committed by an individual mainly depends on his moral qualities, the crimes committed by the entire population of the country must depend in an equal degree on their collective moral qualities. To render this element inoperative upon the large scale, it would be necessary to suppose that the general moral average of mankind does not vary from country to country or from age to age; which is not true, and even if it were true, could not possibly be proved by any existing statistics. I do not on this account the less agree in the opinion of Mr. Buckle, that the intellectual element in mankind, including in that expression the nature of their beliefs, the amount of their knowledge, and the development of their intelligence, is the predominant circumstance in determining their progress. But I am of this opinion, not because I regard their moral or economical condition either as less powerful or less variable agencies, but because these are in a great degree the consequences of the intellectual condition, and are, in all cases, limited by it; as was observed in the preceding chapter. The intellectual changes are the most conspicuous agents in history, not from their superior force, considered in themselves, but because practically they work with the united power belonging to all three.*

§ 3. There is another distinction often neglected in the discussion of this subject, which it is extremely important to observe. The theory of the subjection of social progress to invariable laws, is often held in conjunction with the doctrine, that social progress cannot be materially influenced by the exertions of individual persons, or by the acts of governments. But though these opinions are often held by the same persons, they are two very different opinions, and the confusion between them is the eternally recurring error of confounding Causation with Fatalism. Because whatever happens will be the effect of causes, human volitions among the rest, it does not follow that volitions, even those of peculiar individuals, are not of great efficacy as causes. If any one in a storm at sea, because about the same number of persons in every year perish by shipwreck, should conclude that it was useless for him to attempt to save

---

* I have been assured by an intimate friend of Mr. Buckle that he would not have withheld his assent from these remarks, and that he never intended to affirm or imply that mankind are not progressive in their moral as well as in their intellectual qualities. "In dealing with his problem, he availed himself of the artifice resorted to by the Political Economist, who leaves out of consideration the generous and benevolent sentiments, and founds his science on the proposition that mankind are actuated by acquisitive propensities alone," not because such is the fact, but because it is necessary to begin by treating the principal influence as if it was the sole one, and make the due corrections afterwards. "He desired to make abstraction of the intellect as the determining and dynamical element of the progression, eliminating the more dependent set of conditions, and treating the more active one as if it were an entirely independent variable."

The same friend of Mr. Buckle states that when he used expressions which seemed to exaggerate the influence of general at the expense of special causes, and especially at the expense of the influence of individual minds, Mr. Buckle really intended no more than to affirm emphatically that the greatest men cannot effect great changes in human affairs unless the general mind has been in some considerable degree prepared for them by the general circumstances of the age ; a truth which, of course, no one thinks of denying. And there certainly are passages in Mr. Buckle's writings which speak of the influence exercised by great individual intellects in as strong terms as could be desired.

his own life, we should call him a Fatalist; and should remind him that the efforts of shipwrecked persons to save their lives are so far from being immaterial, that the average amount of those efforts is one of the causes on which the ascertained annual number of deaths by shipwreck depend. However universal the laws of social development may be, they cannot be more universal or more rigorous than those of the physical agencies of nature; yet human will can convert these into instruments of its designs, and the extent to which it does so makes the chief difference between savages and the most highly civilized people. Human and social facts, from their more complicated nature, are not less, but more, modifiable, than mechanical and chemical facts; human agency, therefore, has still greater power over them. And accordingly, those who maintain that the evolution of society depends exclusively, or almost exclusively, on general causes, always include among these the collective knowledge and intellectual development of the race. But if of the race, why not also of some powerful monarch or thinker, or of the ruling portion of some political society, acting through its government? Though the varieties of character among ordinary individuals neutralize one another on any large scale, exceptional individuals in important positions do not in any given age neutralize one another; there was not another Themistocles, or Luther, or Julius Cæsar, of equal powers and contrary dispositions, who exactly balanced the given Themistocles, Luther, and Cæsar, and prevented them from having any permanent effect. Moreover, for aught that appears, the volitions of exceptional persons, or the opinions and purposes of the individuals who at some particular time compose a government, may be indispensable links in the chain of causation by which even the general causes produce their effects; and I believe this to be the only tenable form of the theory.

Lord Macaulay, in a celebrated passage of one of his early essays (let me add that it was one which he did not himself choose to reprint), gives expression to the doctrine of the absolute inoperativeness of great men, more unqualified, I should think, than has been given to it by any writer of equal abilities.

He compares them to persons who merely stand on a loftier height, and thence receive the sun's rays a little earlier, than the rest of the human race. "The sun illuminates the hills while it is still below the horizon, and truth is discovered by the highest minds a little before it becomes manifest to the multitude. This is the extent of their superiority. They are the first to catch and reflect a light which, without their assistance, must in a short time be visible to those who lie far beneath them."* If this metaphor is to be carried out, it follows that if there had been no Newton, the world would not only have had the Newtonian system, but would have had it equally soon; as the sun would have risen just as early to spectators in the plain if there had been no mountain at hand to catch still earlier rays. And so it would be, if truths, like the sun, rose by their own proper motion, without human effort; but not otherwise. I believe that if Newton had not lived, the world must have waited for the Newtonian philosophy until there had been another Newton, or his equivalent. No ordinary man, and no succession of ordinary men, could have achieved it. I will not go the length of saying that what Newton did in a single life, might not have been done in successive steps by some of those who followed him, each singly inferior to him in genius. But even the least of those steps required a man of great intellectual superiority. Eminent men do not merely see the coming light from the hill-top, they mount on the hill-top and evoke it; and if no one had ever ascended thither, the light, in many cases, might never have risen upon the plain at all. Philosophy and religion are abundantly amenable to general causes; yet few will doubt, that had there been no Socrates, no Plato, and no Aristotle, there would have been no philosophy for the next two thousand years, nor in all probability then; and that if there had been no Christ, and no St. Paul, there would have been no Christianity.

The point in which, above all, the influence of remarkable individuals is decisive, is in determining the celerity of the movement. In most states of society it is the existence of

* Essay on Dryden, in Miscellaneous Writings, i. 186.

great men which decides even whether there shall be any progress. It is conceivable that Greece, or that Christian Europe, might have been progressive in certain periods of their history through general causes only: but if there had been no Mahomet, would Arabia have produced Avicenna or Averroes, or Caliphs of Bagdad or of Cordova? In determining, however, in what manner and order the progress of mankind shall take place if it take place at all, much less depends on the character of individuals. There is a sort of necessity established in this respect by the general laws of human nature; by the constitution of the human mind. Certain truths cannot be discovered, or inventions made, unless certain others have been made first; certain social improvements, from the nature of the case, can only follow, and not precede, others. The order of human progress, therefore, may to a certain extent have definite laws assigned to it: while as to its celerity, or even as to its taking place at all, no generalization, extending to the human species generally, can possibly be made; but only some very precarious approximate generalizations, confined to the small portion of mankind in whom there has been anything like consecutive progress within the historical period, and deduced from their special position, or collected from their particular history. Even looking to the *manner* of progress, the order of succession of social states, there is need of great flexibility in our generalizations. The limits of variation in the possible development of social, as of animal life, are a subject of which little is yet understood, and are one of the great problems in social science. It is, at all events, a fact, that different portions of mankind, under the influence of different circumstances, have developed themselves in a more or less different manner and into different forms; and among these determining circumstances, the individual character of their great speculative thinkers or practical organizers may well have been one. Who can tell how profoundly the whole subsequent history of China may have been influenced by the individuality of Confucius? and of Sparta (and hence of Greece and the world) by that of Lycurgus?

Concerning the nature and extent of what a great man

under favourable circumstances can do for mankind, as well
as of what a government can do for a nation, many different
opinions are possible; and every shade of opinion on these
points is consistent with the fullest recognition that there are
invariable laws of historical phenomena. Of course the degree
of influence which has to be assigned to these more special
agencies, makes a great difference in the precision which can
be given to the general laws, and in the confidence with which
predictions can be grounded on them. Whatever depends on
the peculiarities of individuals, combined with the accident of
the positions they hold, is necessarily incapable of being foreseen. Undoubtedly these casual combinations might be eliminated like any others, by taking a sufficiently large cycle:
the peculiarities of a great historical character make their influence felt in history sometimes for several thousand years,
but it is highly probable that they may make no difference at all
at the end of fifty millions. Since, however, we cannot obtain
an average of the vast length of time necessary to exhaust all
the possible combinations of great men and circumstances, as
much of the law of evolution of human affairs as depends upon
this average, is and remains inaccessible to us: and within the
next thousand years, which are of considerably more importance to us than the whole remainder of the fifty millions, the
favourable and unfavourable combinations which will occur
will be to us purely accidental. We cannot foresee the advent
of great men. Those who introduce new speculative thoughts
or great practical conceptions into the world, cannot have
their epoch fixed beforehand. What science can do, is this.
It can trace through past history the general causes which
had brought mankind into that preliminary state, which when
the right sort of great man appeared, rendered them accessible
to his influence. If this state continues, experience renders
it tolerably certain that in a longer or shorter period the great
man will be produced; provided that the general circumstances of the country and people are (which very often they
are not) compatible with his existence; of which point also,
science can in some measure judge. It is in this manner that
the results of progress, except as to the celerity of their pro-

duction, can be, to a certain extent, reduced to regularity and
law. And the belief that they can be so, is equally consistent
with assigning very great, or very little efficacy, to the
influence of exceptional men, or of the acts of governments.
And the same may be said of all other accidents and disturbing
causes.

§ 4. It would nevertheless be a great error to assign
only a trifling importance to the agency of eminent individuals, or of governments. It must not be concluded that
the influence of either is small, because they cannot bestow
what the general circumstances of society, and the course of
its previous history, have not prepared it to receive. Neither
thinkers nor governments effect all that they intend, but in
compensation they often produce important results which they
did not in the least foresee. Great men, and great actions,
are seldom wasted: they send forth a thousand unseen influences, more effective than those which are seen; and though
nine out of every ten things done, with a good purpose, by
those who are in advance of their age, produce no material
effect, the tenth thing produces effects twenty times as great
as any one would have dreamed of predicting from it. Even
the men who for want of sufficiently favourable circumstances
left no impress at all upon their own age, have often been of
the greatest value to posterity. Who could appear to have
lived more entirely in vain, than some of the early heretics?
They were burnt or massacred, their writings extirpated, their
memory anathematized, and their very names and existence
left for seven or eight centuries in the obscurity of musty
manuscripts—their history to be gathered, perhaps, only from
the sentences by which they were condemned. Yet the memory
of these men—men who resisted certain pretensions or certain
dogmas of the Church in the very age in which the unanimous
assent of Christendom was afterwards claimed as having been
given to them, and asserted as the ground of their authority—
broke the chain of tradition, established a series of precedents
for resistance, inspired later Reformers with the courage, and
armed them with the weapons, which they needed when man-

kind were better prepared to follow their impulse. To this example from men, let us add another from governments. The comparatively enlightened rule of which Spain had the benefit during a considerable part of the eighteenth century, did not correct the fundamental defects of the Spanish people; and in consequence, though it did great temporary good, so much of that good perished with it, that it may plausibly be affirmed to have had no permanent effect. The case has been cited as a proof how little governments can do, in opposition to the causes which have determined the general character of the nation. It does show how much there is which they cannot do; but not that they can do nothing. Compare what Spain was at the beginning of that half century of liberal government, with what she had become at its close. That period fairly let in the light of European thought upon the more educated classes; and it never afterwards ceased to go on spreading. Previous to that time the change was in an inverse direction; culture, light, intellectual and even material activity, were becoming extinguished. Was it nothing to arrest this downward and convert it into an upward course? How much that Charles the Third and Aranda could not do, has been the ultimate consequence of what they did! To that half century Spain owes that she has got rid of the Inquisition, that she has got rid of the monks, that she now has parliaments and (save in exceptional intervals) a free press, and the feelings of freedom and citizenship, and is acquiring railroads and all the other constituents of material and economical progress. In the Spain which preceded that era there was not a single element at work, which could have led to these results in any length of time, if the country had continued to be governed as it was by the last princes of the Austrian dynasty, or if the Bourbon rulers had been from the first what, both in Spain and in Naples, they afterwards became.

And if a government can do much, even when it seems to have done little, in causing positive improvement, still greater are the issues dependent on it in the way of warding off evils, both internal and external, which else would stop improvement altogether. A good or a bad counsellor, in a single city at a particular crisis, has affected the whole subsequent fate of the world. It is as certain as any contingent judgment respecting historical events can be, that if there had been no Themistocles there would have been no victory of Salamis; and had there not, where would have been all our civilization? How different again would have been the issue if Epaminondas, or Timoleon, or even Iphicrates, instead of Chares and Lysicles, had commanded at Chæroneia. As is well said in the second of two Essays on the Study of History,* in my judgment the soundest and most philosophical productions which the recent controversies on this subject have called forth; historical science authorizes not absolute, but only conditional predictions. General causes count for much, but individuals also " produce great changes in history, and colour its whole complexion long after their death. . . . No one can doubt that the Roman republic would have subsided into a military despotism if Julius Cæsar had never lived ;" (thus much was rendered practically certain by general causes): "but is it at all clear that in that case Gaul would ever have formed a province of the empire? Might not Varus have lost his three legions on the banks of the Rhone? and might not that river have become the frontier instead of the Rhine? This might well have happened if Cæsar and Crassus had changed provinces; and it is surely impossible to say that in such an event the venue (as lawyers say) of European civilization might not have been changed. The Norman Conquest in the same way was as much the act of a single man, as the writing of a newspaper article; and knowing as we do the history of that man and his family, we can retrospectively predict with all but infallible certainty, that no other person " (no other in that age, I presume, is meant), " could have accomplished the enterprise. If it had not been accomplished, is there any ground to suppose that either our history or our national character would have been what they are ?"

As is most truly remarked by the same writer, the whole stream of Grecian history, as cleared up by Mr. Grote, is one

---

* In the *Cornhill Magazine* for June and July 1861.

series of examples how often events on which the whole destiny of subsequent civilization turned, were dependent on the personal character for good or evil of some one individual. It must be said, however, that Greece furnishes the most extreme example of this nature to be found in history, and is a very exaggerated specimen of the general tendency. It has happened' only that once, and will probably never happen again, that the fortunes of mankind depended upon keeping a certain order of things in existence in a single town, or a country scarcely larger than Yorkshire; capable of being ruined or saved by a hundred causes, of very slight magnitude in comparison with the general tendencies of human affairs. Neither ordinary accidents, nor the characters of individuals, can ever again be so vitally important as they then were. The longer our species lasts, and the more civilized it becomes, the more, as Comte remarks, does the influence of past generations over the present, and of mankind *en masse* over every individual in it, predominate over other forces: and though the course of affairs never ceases to be susceptible of alteration both by accidents and by personal qualities, the increasing preponderance of the collective agency of the species over all minor causes, is constantly bringing the general evolution of the race into something which deviates less from a certain and preappointed track. Historical science, therefore, is always becoming more possible : not solely because it is better studied, but because, in every generation, it becomes better adapted for study.

CHAPTER XII.

OF THE LOGIC OF PRACTICE, OR ART; INCLUDING MORALITY AND POLICY.

§ 1. IN the preceding chapters we have endeavoured to characterize the present state of those among the branches of knowledge called Moral, which are sciences in the only proper sense of the term, that is, inquiries into the course of nature. It is customary, however, to include under the term moral knowledge, and even (though improperly) under that of moral science, an inquiry the results of which do not express themselves in the indicative, but in the imperative mood, or in periphrases equivalent to it; what is called the knowledge of duties; practical ethics, or morality.

Now, the imperative mood is the characteristic of art, as distinguished from science. Whatever speaks in rules, or precepts, not in assertions respecting matters of fact, is art: and ethics, or morality, is properly a portion of the art corresponding to the sciences of human nature and society.*

The Method, therefore, of Ethics, can be no other than that of Art, or Practice, in general: and the portion yet uncompleted, of the task which we proposed to ourselves in the concluding Book, is to characterize the general Method of Art, as distinguished from Science.

§ 2. In all branches of practical business, there are cases in which individuals are bound to conform their practice to a pre-established rule, while there are others in which it is part of their task to find or construct the rule by which they are to

---

* It is almost superfluous to observe, that there is another meaning of the word Art, in which it may be said to denote the poetical department or aspect of things in general, in contradistinction to the scientific. In the text, the word is used in its older, and I hope, not yet obsolete sense.

govern their conduct. The first, for example, is the case of a judge, under a definite written code. The judge is not called upon to determine what course would be intrinsically the most advisable in the particular case in hand, but only within what rule of law it falls; what the legislator has ordained to be done in the kind of case, and must therefore be presumed to have intended in the individual case. The method must here be wholly and exclusively one of ratiocination, or syllogism; and the process is obviously, what in our analysis of the syllogism we showed that all ratiocination is, namely the interpretation of a formula.

In order that our illustration of the opposite case may be taken from the same class of subjects as the former, we will suppose, in contrast with the situation of the judge, the position of a legislator. As the judge has laws for his guidance, so the legislator has rules, and maxims of policy; but it would be a manifest error to suppose that the legislator is bound by these maxims in the same manner as the judge is bound by the laws, and that all he has to do is to argue down from them to the particular case, as the judge does from the laws. The legislator is bound to take into consideration the reasons or grounds of the maxim; the judge has nothing to do with those of the law, except so far as a consideration of them may throw light upon the intention of the law-maker, where his words have left it doubtful. To the judge, the rule, once positively ascertained, is final; but the legislator, or other practitioner, who goes by rules rather than by their reasons, like the old-fashioned German tacticians who were vanquished by Napoleon, or the physician who preferred that his patients should die by rule rather than recover contrary to it, is rightly judged to be a mere pedant, and the slave of his formulas.

Now, the reasons of a maxim of policy, or of any other rule of art, can be no other than the theorems of the corresponding science.

The relation in which rules of art stand to doctrines of science may be thus characterized. The art proposes to itself an end to be attained, defines the end, and hands it over to the science. The science receives it, considers it as a pheno-

menon or effect to be studied, and having investigated its
causes and conditions, sends it back to art with a theorem of
the combination of circumstances by which it could be produced. Art then examines these combinations of circumstances, and according as any of them are or are not in human
power, pronounces the end attainable or not. The only one
of the premises, therefore, which Art supplies, is the original
major premise, which asserts that the attainment of the given
end is desirable. Science then lends to Art the proposition
(obtained by a series of inductions or of deductions) that the
performance of certain actions will attain the end. From
these premises Art concludes that the performance of these
actions is desirable, and finding it also practicable, converts
the theorem into a rule or precept.

§ 3. It deserves particular notice, that the theorem or
speculative truth is not ripe for being turned into a precept,
until the whole, and not a part merely, of the operation which
belongs to science, has been performed. Suppose that we
have completed the scientific process only up to a certain
point; have discovered that a particular cause will produce
the desired effect, but have not ascertained all the negative
conditions which are necessary, that is, all the circumstances
which, if present, would prevent its production. If, in this
imperfect state of the scientific theory, we attempt to frame a
rule of art, we perform that operation prematurely. Whenever any counteracting cause, overlooked by the theorem,
takes place, the rule will be at fault: we shall employ the
means and the end will not follow. No arguing from or about
the rule itself will then help us through the difficulty: there is
nothing for it but to turn back and finish the scientific
process which should have preceded the formation of the rule.
We must re-open the investigation, to inquire into the
remainder of the conditions on which the effect depends; and
only after we have ascertained the whole of these, are we prepared to transform the completed law of the effect into a
precept, in which those circumstances or combinations of cir-

cumstances which the science exhibits as conditions, are prescribed as means.

It is true that, for the sake of convenience, rules must be formed from something less than this ideally perfect theory; in the first place, because the theory can seldom be made ideally perfect; and next, because, if all the counteracting contingencies, whether of frequent or of rare occurrence, were included, the rules would be too cumbrous to be apprehended and remembered by ordinary capacities, on the common occasions of life. The rules of art do not attempt to comprise more conditions than require to be attended to in ordinary cases; and are therefore always imperfect. In the manual arts, where the requisite conditions are not numerous, and where those which the rules do not specify are generally either plain to common observation or speedily learnt from practice, rules may often be safely acted on by persons who know nothing more than the rule. But in the complicated affairs of life, and still more in those of states and societies, rules cannot be relied on, without constantly referring back to the scientific laws on which they are founded. To know what are the practical contingencies which require a modification of the rule, or which are altogether exceptions to it, is to know what combinations of circumstances would interfere with, or entirely counteract, the consequences of those laws: and this can only be learnt by a reference to the theoretic grounds of the rule.

By a wise practitioner, therefore, rules of conduct will only be considered as provisional. Being made for the most numerous cases, or for those of most ordinary occurrence, they point out the manner in which it will be least perilous to act, where time or means do not exist for analysing the actual circumstances of the case, or where we cannot trust our judgment in estimating them. But they do not at all supersede the propriety of going through (when circumstances permit) the scientific process requisite for framing a rule from the data of the particular case before us. At the same time, the common rule may very properly serve as an admonition that a certain mode of action has been found by ourselves and others to be well adapted to the cases of most common occurrence; so that if it be unsuitable to the case in hand, the reason of its being so will be likely to arise from some unusual circumstance.

§ 4. The error is therefore apparent, of those who would deduce the line of conduct proper to particular cases, from supposed universal practical maxims; overlooking the necessity of constantly referring back to the principles of the speculative science, in order to be sure of attaining even the specific end which the rules have in view. How much greater still, then, must the error be, of setting up such unbending principles, not merely as universal rules for attaining a given end, but as rules of conduct generally; without regard to the possibility, not only that some modifying cause may prevent the attainment of the given end by the means which the rule prescribes, but that success itself may conflict with some other end, which may possibly chance to be more desirable.

This is the habitual error of many of the political speculators whom I have characterized as the geometrical school; especially in France, where ratiocination from rules of practice forms the staple commodity of journalism and political oratory; a misapprehension of the functions of Deduction which has brought much discredit, in the estimation of other countries, upon the spirit of generalization so honourably characteristic of the French mind. The common-places of politics, in France, are large and sweeping practical maxims, from which, as ultimate premises, men reason downwards to particular applications, and this they call being logical and consistent. For instance, they are perpetually arguing that such and such a measure ought to be adopted, because it is a consequence of the principle on which the form of government is founded; of the principle of legitimacy, or the principle of the sovereignty of the people. To which it may be answered, that if these be really practical principles, they must rest on speculative grounds; the sovereignty of the people (for example) must be a right foundation for government, because a government thus constituted tends to produce certain beneficial effects. Inasmuch, however, as no government produces all possible beneficial effects, but all are attended with more or fewer inconveniences; and since these cannot usually be combated by means drawn from the very causes which produce them; it would be often a much stronger recommendation of some practical arrangement, that it does not follow from what is called the general principle of the government, than that it does. Under a government of legitimacy, the presumption is far rather in favour of institutions of popular origin; and in a democracy, in favour of arrangements tending to check the impetus of popular will. The line of argumentation so commonly mistaken in France for political philosophy, tends to the practical conclusion that we should exert our utmost efforts to aggravate, instead of alleviating, whatever are the characteristic imperfections of the system of institutions which we prefer, or under which we happen to live.

§ 5. The grounds, then, of every rule of art, are to be found in the theorems of science. An art, or a body of art, consists of the rules, together with as much of the speculative propositions as comprises the justification of those rules. The complete art of any matter, includes a selection of such a portion from the science, as is necessary to show on what conditions the effects, which the art aims at producing, depend. And Art in general, consists of the truths of Science, arranged in the most convenient order for practice, instead of the order which is the most convenient for thought. Science groups and arranges its truths, so as to enable us to take in at one view as much as possible of the general order of the universe. Art, though it must assume the same general laws, follows them only into such of their detailed consequences as have led to the formation of rules of conduct; and brings together from parts of the field of science most remote from one another, the truths relating to the production of the different and heterogeneous conditions necessary to each effect which the exigencies of practical life require to be produced.

Science, therefore, following one cause to its various effects, while art traces one effect to its multiplied and diversified causes and conditions; there is need of a set of intermediate scientific truths, derived from the higher generalities of science, and destined to serve as the generalia or first principles of the various arts. The scientific operation of framing these intermediate principles, M. Comte characterizes as one of those results of philosophy which are reserved for futurity. The only complete example which he points out as actually realized, and which can be held up as a type to be imitated in more important matters, is the general theory of the art of Descriptive Geometry, as conceived by M. Monge. It is not, however, difficult to understand what the nature of these intermediate principles must generally be. After framing the most comprehensive possible conception of the end to be aimed at, that is, of the effect to be produced, and determining in the same comprehensive manner the set of conditions on which that effect depends; there remains to be taken, a general survey of the resources which can be commanded for realizing this set of conditions; and when the result of this survey has been embodied in the fewest and most extensive propositions possible, those propositions will express the general relation between the available means and the end, and will constitute the general scientific theory of the art; from which its practical methods will follow as corollaries.

§ 6. But though the reasonings which connect the end or purpose of every art with its means, belong to the domain of Science, the definition of the end itself belongs exclusively to Art, and forms its peculiar province. Every art has one first principle, or general major premise, not borrowed from science; that which enunciates the object aimed at, and affirms it to be a desirable object. The builder's art assumes that it is desirable to have buildings; architecture (as one of the fine arts), that it is desirable to have them beautiful or imposing. The hygienic and medical arts assume, the one that the preservation of health, the other that the cure of disease, are fitting and desirable ends. These are not propositions of science.

Propositions of science assert a matter of fact: an existence, a coexistence, a succession, or a resemblance. The propositions now spoken of do not assert that anything is, but enjoin or recommend that something should be. They are a class by themselves. A proposition of which the predicate is expressed by the words *ought* or *should be,* is generically different from one which is expressed by *is,* or *will be.* It is true that, in the largest sense of the words, even these propositions assert something as a matter of fact. The fact affirmed in them is, that the conduct recommended excites in the speaker's mind the feeling of approbation. This, however, does not go to the bottom of the matter; for the speaker's approbation is no sufficient reason why other people should approve; nor ought it to be a conclusive reason even with himself. For the purposes of practice, every one must be required to justify his approbation: and for this there is need of general premises, determining what are the proper objects of approbation, and what the proper order of precedence among those objects.

These general premises, together with the principal conclusions which may be deduced from them, form (or rather might form) a body of doctrine, which is properly the Art of Life, in its three departments, Morality, Prudence or Policy, and Æsthetics; the Right, the Expedient, and the Beautiful or Noble, in human conduct and works. To this art, (which, in the main, is unfortunately still to be created,) all other arts are subordinate; since its principles are those which must determine whether the special aim of any particular art is worthy and desirable, and what is its place in the scale of desirable things. Every art is thus a joint result of laws of nature disclosed by science, and of the general principles of what has been called Teleology, or the Doctrine of Ends;* which, borrowing the language of the German metaphysicians, may also be termed, not improperly, the Principles of Practical Reason.

A scientific observer or reasoner, merely as such, is not

---

* The word Teleology is also, but inconveniently and improperly, employed by some writers as a name for the attempt to explain the phenomena of the universe from final causes.

an adviser for practice. His part is only to show that certain
consequences follow from certain causes, and that to obtain
certain ends, certain means are the most effectual. Whether
the ends themselves are such as ought to be pursued, and if
so, in what cases and to how great a length, it is no part of
his business as a cultivator of science to decide, and science
alone will never qualify him for the decision. In purely
physical science, there is not much temptation to assume this
ulterior office; but those who treat of human nature and
society invariably claim it; they always undertake to say,
not merely what is, but what ought to be. To entitle them
to do this, a complete doctrine of Teleology is indispensable.
A scientific theory, however perfect, of the subject matter,
considered merely as part of the order of nature, can in no
degree serve as a substitute. In this respect the various
subordinate arts afford a misleading analogy. In them there
is seldom any visible necessity for justifying the end, since in
general its desirableness is denied by nobody, and it is only
when the question of precedence is to be decided between that
end and some other,' that the general principles of Teleology
have to be called in: but a writer on Morals and Politics
requires those principles at every step. The most elaborate
and well-digested exposition of the laws of succession and
coexistence among mental or social phenomena, and of their
relation to one another as causes and effects, will be of no
avail towards the art of Life or of Society, if the ends to be
aimed at by that art are left to the vague suggestions of the
intellectus sibi permissus, or are taken for granted without
analysis or questioning.

§ 7. There is, then, a Philosophia Prima peculiar to
Art, as there is one which belongs to Science. There are not
only first principles of Knowledge, but first principles of
Conduct. There must be some standard by which to determine the goodness or badness, absolute and comparative, of
ends, or objects of desire. And whatever that standard is,
there can be but one: for if there were several ultimate principles of conduct, the same conduct might be approved by

one of those principles and condemned by another; and there would be needed some more general principle, as umpire between them.

Accordingly, writers on moral philosophy have mostly felt the necessity not only of referring all rules of conduct, and all judgments of praise and blame, to principles, but of referring them to some one principle; some rule, or standard, with which all other rules of conduct were required to be consistent, and from which by ultimate consequence they could all be deduced. Those who have dispensed with the assumption of such an universal standard, have only been enabled to do so by supposing that a moral sense, or instinct, inherent in our constitution, informs us, both what principles of conduct we are bound to observe, and also in what order these should be subordinated to one another.

The theory of the foundations of morality is a subject which it would be out of place, in a work like this, to discuss at large, and which could not to any useful purpose be treated incidentally. I shall content myself therefore with saying, that the doctrine of intuitive moral principles, even if true, would provide only for that portion of the field of conduct which is properly called moral. For the remainder of the practice of life some general principle, or standard, must still be sought; and if that principle be rightly chosen, it will be found, I apprehend, to serve quite as well for the ultimate principle of Morality, as for that of Prudence, Policy, or Taste.

Without attempting in this place to justify my opinion, or even to define the kind of justification which it admits of, I merely declare my conviction, that the general principle to which all rules of practice ought to conform, and the test by which they should be tried, is that of conduciveness to the happiness of mankind, or rather, of all sentient beings: in other words, that the promotion of happiness is the ultimate principle of Teleology.*

---

* For an express discussion and vindication of this principle, see the little volume entitled "Utilitarianism."

I do not mean to assert that the promotion of happiness should be itself the end of all actions, or even of all rules of action. It is the justification, and ought to be the controller, of all ends, but is not itself the sole end. There are many virtuous actions, and even virtuous modes of action (though the cases are, I think, less frequent than is often supposed) by which happiness in the particular instance is sacrificed, more pain being produced than pleasure. But conduct of which this can be truly asserted, admits of justification only because it can be shown that on the whole more happiness will exist in the world, if feelings are cultivated which will make people, in certain cases, regardless of happiness. I fully admit that this is true: that the cultivation of an ideal nobleness of will and conduct, should be to individual human beings an end, to which the specific pursuit either of their own happiness or of that of others (except so far as included in that idea) should, in any case of conflict, give way. But I hold that the very question, what constitutes this elevation of character, is itself to be decided by a reference to happiness as the standard. The character itself should be, to the individual, a paramount end, simply because the existence of this ideal nobleness of character, or of a near approach to it, in any abundance, would go further than all things else towards making human life happy; both in the comparatively humble sense, of pleasure and freedom from pain, and in the higher meaning, of rendering life, not what it now is almost universally, puerile and insignificant—but such as human beings with highly developed faculties can care to have.

§ 8. With these remarks we must close this summary view of the application of the general logic of scientific inquiry to the moral and social departments of science. Notwithstanding the extreme generality of the principles of method which I have laid down, (a generality which, I trust, is not, in this instance, synonymous with vagueness) I have indulged the hope that to some of those on whom the task will devolve of bringing those most important of all sciences into a more satisfactory state, these observations may be useful; both in removing erroneous, and in clearing up the true, conceptions of the means by which, on subjects of so high a degree of complication, truth can be attained. Should this hope be realized, what is probably destined to be the great intellectual achievement of the next two or three generations of European thinkers will have been in some degree forwarded.

THE END.

LONDON:
SAVILL, EDWARDS AND CO., PRINTERS, CHANDOS STREET,
COVENT GARDEN.